For "Marcy",

Inspired and inspiring writer, freshman Roomie, and forever strong...

All my best always,

Margaret ☺

Peripheral Wonders

The Bucknell Studies in Eighteenth-Century Literature and Culture

The Bucknell Studies in Eighteenth-Century Literature and Culture aims to publish challenging, new eighteenth-century scholarship. Of particular interest is critical, historical, and interdisciplinary work that is interestingly and intelligently theorized, and that broadens and refines the conception of the field. At the same time, the series remains open to all theoretical perspectives and different kinds of scholarship. While the focus of the series is the literature, history, arts, and culture (including art, architecture, music, travel, and history of science, medicine, and law) of the long eighteenth century in Britain and Europe, the series is also interested in scholarship that establishes relationships with other geographies, literature, and cultures for the period 1660–1830.

Recent Titles in This Series

Mark Blackwell, ed., *The Secret Life of Things: Animals, Objects, and It-Narratives in Eighteenth-Century England*
Chris Mounsey and Caroline Gonda, eds., *Queer People: Negotiations and Expressions of Homosexuality, 1700–1800*
Susan Manning and Peter France, *Enlightenment and Emancipation*
Evan Gottlieb, *Feeling British: Sympathy and National Identity in Scottish and English Writing, 1707–1832*
Roland Racevskis, *Tragic Passages: Jean Racine's Art of the Threshold*
Lesley H. Walker, *A Mother's Love: Crafting Feminine Virtue in Enlightenment France*
Ron Broglio, *Technologies of the Picturesque: British Art, Poetry and Instruments, 1750–1830*
Will Pritchard, *Outward Appearances: The Female Exterior in Restoration London*
David Duff and Catherine Jones, eds., *Scotland, Ireland, and the Romantic Aesthetic*
Anne Milne, *"Lactilla Tends her Fav'rite Cow": Ecocritical Readings of Animals and Women in Eighteenth-Century British Laboring-Class Women's Poetry*
Gavin Budge, ed., *Romantic Empiricism: Poetics and the Philosophy of Common Sense, 1780–1830*
Shelley King and Yaël Schlick, eds., *Refiguring the Coquette: Essays on Culture and Coquetry*
Margaret R. Ewalt, *Peripheral Wonders: Nature, Knowledge, and Enlightenment in the Eighteenth-Century Orinoco*

http://www.bucknell.edu/universitypress/

Peripheral Wonders

Nature, Knowledge, and Enlightenment in the Eighteenth-Century Orinoco

Margaret R. Ewalt

Lewisburg
Bucknell University Press

Associated University Presses
2010 Eastpark Boulevard
Cranbury, NJ 08512

The paper used in this publication meets the requirements of the American National Standard for Permanence of Paper for Printed Library Materials Z39.48-1984.

Library of Congress Cataloging-in-Publication Data

Ewalt, Margaret R., 1970–
Peripheral wonders : nature, knowledge, and Enlightenment in the eighteenth-century Orinoco / Margaret R. Ewalt.
p. cm. — (The Bucknell studies in eighteenth-century)
Includes bibliographical references and index.
ISBN 978-0-8387-5689-8 (alk. paper)
1. Gumilla, Joseph, d. 1750. Orinoco ilustrado . . . 2. Orinoco River Valley (Venezuela and Colombia) — Intellectual life — 19th century. 3. Enlightenment — Orinoco River (Venezuela and Colombia) 4. Jesuits — Orinoco River Valley (Venezuela and Colombia) I. Title. II. Series.
F2331.G86O7534 2008
987′.03 — dc22

2008013035

PRINTED IN THE UNITED STATES OF AMERICA

For David

Contents

Acknowledgments

THERE ARE SEVERAL PROFESSORS, COLLEAGUES, FRIENDS, AND FAMily members without whom this book would not exist. First and foremost I am greatly indebted to Ruth Hill, whose immense generosity of intellect and unfailing support of and confidence in this project have sustained me over the years. Other professors at the University of Virginia important to the development of this project include David T. Gies, Fernando Operé, Ricardo Padrón, Karen Parshall, and Alison Weber, all of whom gave valuable feedback on earlier portions of this manuscript.

I owe much to the enthusiastic members of the Iberoamerican Society within the American Society for Eighteenth-Century Studies who have encouraged me since I joined the group in 2001, especially Karen Stolley, Ana Hontanilla, Jonathan Carlyon, Catherine Jaffe, Enid Valle, Yvonne Fuentes, Mark Malin, David Slade, Elizabeth Franklin Lewis, Kathleen Fueger, Ana Rueda, and Claudia Gronemann. Members of the Colonial Americas Study Organization have also inspired me: Raquel Chang-Rodríguez, Evelina Guzauskyte, Eyda Merediz, Beatriz de Alba-Koch, Santa Arias, Mariselle Meléndez, Stacey Schlau, Carmen Millán de Benavides, Luis Fernando Restrepo, and Verónica Salles-Reese. I benefited greatly from the 2003 International Seminar on the Eighteenth Century at UCLA, especially the feedback from Philip Stewart, Liam Chambers, Fiona Clark, Doris Garraway, and Mary Helen McMurran. Thank you to the organizers and participants of the 2005 symposium, "Jesuit Accounts of the Colonial Americas," at the William Andrews Clark Memorial Library, especially Clorinda Donato, Marc-André Bernier, and Hans Juergen Luesebrink. Thanks also to Luis Millones.

My colleagues at Wake Forest University have also helped a great deal. Among them, particular thanks to Candelas Gala, Mary Friedman, Anne Hardcastle, Sol Miguel-Prendes, Kendall Tarte, Stan Whitley, and Byron Wells for their mentorship and support. Thanks to Ola Furmanek and Rebekah Morris for help with the translations. Any remaining errors are my own. Thanks to my student assistant, Lily Cottrell, for double-checking the page numbers of so many quotes from Gumilla. Special thanks most of all to my writing group, Renee Gutiérrez and Kathryn

Mayers, from whom I learned so much and without whom I never would have finished this book! Anonymous readers for Bucknell University Press and my editor, Greg Clingham, also made invaluable suggestions that contributed greatly to the quality of the book. However, all opinions herein and any errors that remain are my own.

My appreciation to the directors, curators, librarians, and support staff of the Alderman Library and Special Collections at the University of Virginia, the Z. Smith Reynolds Library at Wake Forest University, and the John Carter Brown Library at Brown University. Special thanks also to Domingo Ledezma, Kenneth Mills, and Elvira Vilches for their encouragement at the John Carter Brown Library. The support of a University of Virginia Society of Fellows Forstmann Fellowship, a Wake Forest Z. Smith Reynolds Foundation Junior Faculty Leave for Research, and several Wake Forest University Archie Research and Professional Development Travel Grants all contributed greatly to advancing this project.

I acknowledge my grandparents, parents, brothers, aunts, uncles, and cousins for instilling in me a love of learning and of Maine. Finally, this book is dedicated to my husband, David, who taught me how to enjoy all the simple pleasures of daily life outside of academia. Thank you for all your love, support, patience, and all the sacrifices you have made for my career.

A portion of chapter 1 appears as "Father Gumilla: Crocodile Hunter? The Function of Wonder in *El Orinoco ilustrado*" in pages 303–333 of *El saber de los jesuitas, historias naturales y el Nuevo Mundo / Jesuit Knowledge, Natural Histories and the New World,* ed. Luis Millones Figueroa and Domingo Ledezma (Madrid: Iberoamericana, 2005), 303–33. Reproduced with the kind permission of the publishers.

Portions of chapter 4 appear as "Frontier Encounters and Pathways to Knowledge in the New Kingdom of Granada" in *Colorado Review of Hispanic Studies* 3, (Fall 2005): 41–55.

Earlier versions of portions of chapter 1 and 4 appear as "Crossing Over: Nations and Naturalists in *El Orinoco ilustrado;* Reading and Writing the Book of Orinoco Secrets," in *Dieciocho: Hispanic Enlightenment* 29, no. 1 (Spring 2006): 1–25.

A portion of the conclusion will appear as "The Legacy of Joseph Gumilla's *Orinoco Enlightened,*" in *Jesuit Accounts of the Colonial Americas: Textualities, Intellectual Disputes, Intercultural Transfers* (Toronto: Toronto University Press, forthcoming).

I would like to thank the editors of the above journals and Toronto University Press for the permissions to publish overlapping content.

My thanks also to the John Carter Brown Library at Brown University for permission to reproduce all the images in this book.

Peripheral Wonders

Introduction: Pathways to Modernity

THIS BOOK FOCUSES ON A JESUIT CONTRIBUTION TO THE HISPANIC Enlightenment from the first half of the eighteenth century, a period in the intellectual history of Spain and Spanish America until recently generally neglected. For the better part of the last two hundred years, most critical examinations of the Hispanic eighteenth century have focused on its later years and have downplayed evidence of transitions to modernity in the first decades. Until very recently, furthermore, studies have maintained a culturally constructed mythology of a teleological, linear progress toward modernity conceived within a paradigm defined by reason, moderation, and control of emotions. This traditionally secular paradigm posits Catholic Spain and her colonies as resistant to or even lacking in Enlightenment. Critics have instead privileged a hegemonic theory of Enlightenment based on the French and English models, thus restricting other European nations' entrances into critical modernity to an Age of Reason narrowly defined within the borders of only a few cultural contexts.[1] Twentieth-century scholarship has been indelibly marked by Marcelino Menéndez y Pelayo, who decreed that an absence of Enlightenment thought in Spain had "contaminated" Spanish America.[2]

The present study resists situating Spain and Spanish America within such traditional concepts as a tardy entrance into Enlightenment or a truncated modernity. Instead, it proposes a model that more accurately reflects the different yet successful transitions of countries that mediated modern philosophies with Christianity. Adding to new criticism on Spain and her South American colonies, it explores an eclectic, Catholic Enlightenment that glorifies God alongside moral reforms and empirical scientific inquiries. By examining the role of preexile Jesuit evangelization and science in the Hispanic transition to modernity, this study reveals an alternative Enlightenment that does not divorce sentiments from reason and does not insist on control of emotions in the accumulation, cataloguing, and dissemination of knowledge.[3]

From the earliest years after the founding of the Society of Jesus in 1540, the order's missionaries wrote accounts that mixed classical and contemporary erudition from the Jesuit curriculum with firsthand knowledge of nature and peoples gained in the field. From China and Japan to the Philippines and Peru, Jesuits like Matteo Ricci and Joseph de Acosta mediated European knowledge with local sources while transitioning from early modern science to the systems of modern natural philosophers. Contrary to the obscurantist Jesuit myth delaying Enlightenment until 1767, not only Aristotelian Scholasticism but also the revolutionary ideas of Descartes, Newton, Locke, Leibniz, Gassendi, and Bacon circulated among Jesuits in Spain and her colonies long before the suppression of the order.[4] In Spanish America, Jesuits enjoyed a privileged role in the production of colonial knowledge. The Society of Jesus endorsed a strategy of "inculturation" that, in addition to dialogue with Europeans and American-born *criollos* and appropriation of local knowledge, allowed its missionaries to participate in a unique space of reciprocity, accommodation, and mediation with Amerindian philosophers.[5]

Despite the breadth of information that Jesuit natural histories can offer, historians continue to privilege Jesuit writings as stimuli for the progression toward independent Spanish American republics rather than examining them as documents revealing transatlantic transitions to modernity mediated in a Catholic context.[6] Spain's colonies are seen as enlightened more by revolutionary ideals from France and the United States than as contributors to the modern critical discourse of a cosmopolitan republic of letters. In the eighteenth century, widely circulated Jesuit periodicals such as the *Mémoires pour l'Histoire des Sciences & Beaux-Arts* (1701–67), a primary journal of the French Enlightenment more commonly referred to as the *Mémoires de Trévoux,* provided a key contribution to this modern critical discourse.[7] Independent of the circulation of ideas in the European press and long before the rise of independence movements, the Jesuits played a fundamental role in a Spanish American Catholic Enlightenment that united both scholarly and evangelical aims and whose core goal was knowing, praising, and loving God better. Central to the late seventeenth-century and early eighteenth-century Jesuit conception of Enlightenment was a concern for human welfare, justice, and happiness that included an enlightened impulse to civilize as well as describe and understand Amerindian cultures.

However, by focusing principally on "political" texts such as the former Jesuit Juan Pablo Viscardo y Guzmán's "Letter to the Spanish Americans" (1792) or nostalgic descriptions of Jesuits' South American homelands such as the exiled criollo Rafael Landívar's *Rusticatio Mexicana* (1782), scholars have postponed the Jesuit Enlightenment to the period of exile after 1767 and emphasized its contributions to Spanish Americans seeking political and cultural autonomy from Spain. This emphasis overlooks important Enlightenment manifestations in preexile works. The forced removal of Jesuit missionaries and intellectuals from

Spanish America did inspire a flurry of scholarly and scientific activity.[8] But critical attention on postexpulsion contributions to science, art, and culture does not mitigate the importance of a rich vein of earlier missionary accounts and official reports to Rome that established the Jesuit tradition of scholarly exchanges. While a legacy of negative Jesuit stereotypes and an insistence on the Enlightenment's secular nature have often limited scholarly focus on Jesuits to questions of "Counter-Enlightenment discourses," this book explores instead the stimulus Jesuits provided for different kinds of Enlightenment by focusing on one text from the early part of the eighteenth century, a period when missionary naturalists were successfully reconciling Catholic dogma, modern scientific ideas, and Amerindian knowledge in mission settlements (reductions). Jesuits were bringing the light of Christian knowledge to unenlightened pagan souls and then highlighting both their evangelical successes and their Jesuit provinces' abundant resources by writing natural histories directed at Christian readers but also appreciated by more secular politicians and economists. The years before their expulsion constitute a transitional period whose study illuminates the Hispanic Enlightenment as an evolving intellectual process instead of a historical event and underscores the Jesuit role in reconciling secular reason and spiritual passion during this age of Enlightenment. More attention to the first half of the eighteenth century will contribute to the creation of new models of Enlightenment that defy the long-standing Eurocentric, secularist notions that continue to marginalize Catholic Spain and Spanish America from eighteenth-century interdisciplinary studies. More inclusive models of the Enlightenment can reconsider the assumption of European centrality by taking into account peripheral areas. The Americas, Asia, and Mediterranean Europe constitute geographical, political, and religious peripheries where rather than resisted, modernity was redefined.

The present book examines the theory of knowledge and the rhetoric of wonder in a key Jesuit contribution to the Catholic Enlightenment: Joseph Francisco Tomás Gumilla's *El Orinoco ilustrado: Historia natural, civil y geographica de este gran río y de sus caudalosas vertientes* (1741–45) [*The Orinoco Enlightened: Natural, civil and geographical history of this great river and its abundant branches*].[9] Gumilla's biography begins during the early years of the Enlightenment. He was born on May 3, 1686, in Cárcer, Valencia, and by his early teens had decided to become a Jesuit missionary to South America. He began his Jesuit training in Seville, was admitted to the Society of Jesus in 1704, and left Spain via Cádiz the day after his nineteenth birthday in 1705. After a brief novitiate period at the Jesuit residence in Tunja, Gumilla arrived in Santa Fe de Bogotá in 1706, and there he studied rhetoric, philosophy, and theology at the Javeriana University in Bogotá.[10] He was ordained in 1714, and then returned to Tunja

Portrait of Jesuit Father Joseph Gumilla. The John Carter Brown Library at Brown University.

for his third probation year (tertianship). In 1715 Joseph Gumilla began to work in the plains and jungles of the Orinoco River region missions of present-day Colombia and Venezuela.

By April 1716, Gumilla had founded his first mission, San Ignacio de Betoyes. He became the Father Superior to all the missions in the Orinoco River region of the New Kingdom of Granada in 1723 and founded four new reductions for five native tribes between 1732 and 1738: San José de los Mapoyes, for Mapoye y Otomac indians; Nuestra Señora de los Ángeles and Santa Teresa de Tabaje, for the Sálivas; and San Ignacio de Var y Paos, for Yaruros y Paos tribes.[11] He continued to work as a missionary until 1738, the year he returned to Europe and spent time as Jesuit procurator at the courts of Rome and Madrid. By 1739 Gumilla had finished writing *El Orinoco ilustrado* and, two years later, it was published in Madrid. Official church correspondence places Gumilla back in New Granada as Father Superior to the missions along the Casanare branch of the Orinoco River in 1745. Research by José del Rey locates his death as having taken place among the first tribe he evangelized, the Betoyes, on July 16, 1750.

Although this Spanish-born Jesuit missionary's natural history centers on a particular viceroyalty (the New Kingdom of Granada), Gumilla's work reveals the importance of Jesuit intercultural discourse for the evolution of enlightened thought in Spain and her South American colonies and the role of Jesuit science and evangelism in the development of modernity.[12] Gumilla employs wonder as a key strategy not only to persuade Amerindians of the value of Christianity and to convince Europeans of the value of the human and natural resources of the Orinoco River region, but also to open pathways to scientific and commercial knowledge. He joins other sixteenth- through eighteenth-century Jesuit natural historians who do not divorce curiosity and wonder from inquiries into nature.[13] To provoke European interest and increase the prestige of South America, natural histories evoke wonder through vivid descriptions. Jesuit natural philosophers also interrogate nature to glorify God as the agent of the existence of all things. Since God is the reason all things exist, Jesuits direct human passions and attention toward his magnificence. Thus, while science in service of the nation characterizes the early modern period, Jesuits at all times seek to glorify God and honor the *Constitutions* of their founder, Ignatius Loyola.

Gumilla aims to enlighten the Orinoco with the Catholic faith and European scientific knowledge, and his particularly "Jesuit way of proceeding" owes much to his Jesuit forefather, Joseph de Acosta.[14] Acosta laid out pathways to increasing both wonder and knowledge by investigating the causes of natural phenomena, and his oft-cited *Natural and Moral History of the Indies* (1590) became a touchstone for South American science. Natural historians since Acosta have mediated modern methods for the investigation of the matter, features, and properties of nature with a Catholic dogma that privileges God as both the active and the final cause of inquiry into material and formal causes. In this way, Jesuits direct human wonder toward God as the root of all happiness, justice, wealth, and reason. Like Acosta, Gumilla illuminates pathways to knowledge that do not separate the theological and the scientific, but *El Orinoco ilustrado* establishes him as a new Acosta, creating a new model for natural history writing.

Acosta had determined that man's natural curiosity was not a sin, but rather a useful inspiration to unlock the mysteries of nature in service of God. Gumilla's discourse advances curiosity into modernity, as it participates in the eighteenth-century Enlightenment trend of the study of nature in the service of humankind. It also enlightens the Orinoco with a blend of reason, imagination, and emotion that anticipates the sublime synthesis of art and science achieved by Alexander von Humboldt in the early nineteenth century. Humboldt's descriptions of South America also depend on wonder and the imagination to advance science and art; his visit inspired such aesthetically vivid travelogues that Humboldt remains the most frequently cited Enlightenment figure for enhancing views of the New World for both Europeans and Americans.[15] The perception remains common that his celebrated visit to South America ushered in its Enlightenment and then its Independence, and that Humboldt's writings in fact marked the "change in thinking" or "new spirit" of the era.[16] However, over sixty years before Humboldt traveled to the Orinoco River region, Gumilla embodied the Prussian's blend of reason and emotion.[17] This book's examination of Gumilla's vivid persuasion through rhetorical, theological, scientific, and commercial uses of wonder reveals Gumilla as a heretofore unrecognized transition between Acosta and Humboldt.

The Enlightenment has traditionally been defined as a period of critical reform brought about by the secularization and rationalization of culture and limited by geography and chronology. Seminal studies by Ernst Cassirer, Paul Hazard, Frank Manuel, and Peter Gay have defined the philosophy of the Enlightenment in terms of a "rebirth of 'paganism' " during an "age of reason" brought on by secular liberalism. Recent approaches have sought to expand the borders of Enlightenment studies. Jessica Riskin challenges the hegemonic "reputation for rationalism" of "the sciences, the Enlightenment, and the modern French intellect."[18] Works by Jonathan Israel, Dorinda Outram, and Margaret Jacob have argued for a broader, more inclusive Enlightenment by reconsidering relationships between science and religion, between public and private spheres, as well as between men and women. Unfortunately, these recent studies take little account of developments in Spain and Spanish America.

Those who do study the Hispanic Enlightenment have generally adopted four approaches. Traditional views such as those expressed by Marcelino Menéndez y Pelayo, José Ortega y Gasset, and Américo Castro effectively deprive Spain—and, by extension, her American colonies—of an Enlightenment by defining the movement as almost exclusively anticlerical and privileging the enlightened philosophies of French philosophes such as Voltaire and Rousseau.

A second approach comprised principally of the pioneering works of Jean Sarrailh and Arthur Preston Whitaker restricts the Hispanic Enlightenment to the later years of the eighteenth century.[19] Sarrailh postpones Spain's Enlightenment until the years of governmental support for scientific expeditions following the 1759 ascension to the throne of Charles III and defines it in terms of the elite's attempts to enlighten the irrational masses through a partial and belated reception of the French encyclopedists. Many other studies associate the Hispanic Enlightenment with the ultimate failures of Bourbon reforms.

Since the early years of Latin America's new republics, a third approach has postponed Enlightenment until after their independence. This liberal strand of Enlightenment theory has supported nationalistic censure of Spanish colonialism in part by negating Enlightenment within Spain's conservative Universal Catholic Monarchy and instead locating it in the early national period of the nineteenth century.[20]

The fourth approach defines Spanish and Spanish American Enlightenment as a Christian movement and therefore "delayed" and "limited."[21] Critics who merely assert obscurantist dogma's defeat of seventeenth-century empiricism and ancient pagan skepticism focus almost exclusively on the Catholic Enlightenment's limitations. While some recent historians have confirmed the need to understand Spain and Spanish America's pathways to modernity within a wider context of successful Catholic Enlightenment, others continue to describe the Hispanic Enlightenment as "insufficient" and its attempts at modernity as failures.[22] Some of these scholars praise the Society of Jesus as an intellectually liberal sect and a pillar of "the modern lettered city" while condemning the Catholic Enlightenment as "mutilated."[23] For them, modernity can still only be achieved by establishing clear limits between the secular and the sacred.

New trends in scholarship, however, have blurred chronological limits as well as underlined the importance of clerics to the Hispanic Enlightenment. Instead of perpetuating the rhetoric of failure, Jorge Cañizares-Esguerra, Ruth Hill, and Jesús Pérez Magallón have provided new perceptions about the Hispanic transition to modernity within the late baroque and early Enlightenment periods. Pérez Magallón theorizes a "pluridiscursive" modernity and reveals a pathway to knowledge at the turn of the eighteenth century that combines reason, observation, and sensibility.[24] Cañizares-Esguerra confronts Eurocentric conceptions of modernity and posits "alternative critical discourses" for Spanish America including a "patriotic epistemology" that is "broadly clerical rather than merely Creole."[25] Hill debunks the myths of Hispanic primitivism by clarifying the middle path taken by Catholic intellectuals during the "emergent and transitional" period of 1680 and 1740, a transatlantic path of mitigated skepticism that mediates the rationalist and imaginative poles and explores the religious and ethnic consciousness central to the new philosophy's reception in the Spains. The value of these recent studies is indisputable, for they respond to the limits of previous criticism and the need to "decentralize Enlightenment studies" and point the way for scholars to examine the Enlightenment's complex and diverse processes and to give voice to the production of knowledge outside of the more hegemonic European Enlightenment patterns.[26]

Joseph Gumilla's *El Orinoco ilustrado* can further our understanding of an Enlightenment that resists previous limitations of borders marked by geography, chronology, and religious beliefs, for it demonstrates both regional particularities and cosmopolitan aspects that transcend national and spiritual borders. Gumilla's natural history reflects a Christian humanism that supports an interaction of Catholic dogma, local traditions, and modern science and reveals much about the nature of transatlantic knowledge during first few decades of the eighteenth century. In addition to putting the Jesuit missions and province of New Granada figuratively as well as graphically on the map, it defends both scientific and evangelical Jesuit Enlightenment in the viceroyalty's previously untamed, peripheral regions and argues for further missionary and political support. In return, the Orinoco River promises a route to great commercial potential. By focusing on the commercial, scientific, and religious discourses in *El Orinoco ilustrado,* this book maps an alternative route to Hispanic modernity enlightened by Jesuit knowledge and reform.

In the prologue to *El Orinoco ilustrado,* Gumilla explains how he will enlighten and illustrate (*ilustrar*) the Orinoco region so that this great river, up to now practically unknown, will be reborn in his book "con el renombre de *ilustrado*" [with the distinguished fame of enlightened].[27] This "enlightenment," Gumilla affirms, will save his Jesuit province from "del caos del olvido" [from the chaos of oblivion] by bringing it "a la luz pública" (31) [into the public light]. Even though definitions of *ilustrar* can still be found that include both the spiritual and intellectual avenues of enlightenment,[28] a basic modern understanding of the verb *ilustrar* (to enlighten, to illustrate) the past participle *ilustrado* (enlightened, or illustrated), and the adjective *ilustre* (illustrious) cannot reveal the complexity of Gumilla's intentions. Covarrubias' *Tesoro de la Lengua Castellana o Española* (1611) definition ("Dar lustre y resplandor a alguna cosa [Put a shine and radiance on something])[29] prefigured key eighteenth-century uses of the verb *ilustrar* that include shining light upon something either physically or spiritually. Gumilla carefully chooses his words as well as what he describes in order to "illustrate," in a sense of the word that *The Oxford English Dictionary* makes clear was shared in Spanish and English during the eighteenth century: "To set in a good light; to display to advantage; to show up. To shed lustre upon; to render illustrious, renowned, or famous; to confer honor or distinction upon."[30]

The definitions published by the Spanish Royal Academy of Languages in their *Diccionario de la lengua castellana* (1726–39) clarify Gumilla's use. The eighteenth century maintained a pious purpose alongside the more general effects of ennobling something, and the light of knowledge could enlighten men both intellectually and spiritually about the divine light of God. According to the Royal Academy of Languages definitions, "illustrations" include both an embellishing linguistic style and a useful purveyor of knowledge, and the verb "to illustrate" enlightens and amplifies: "Dar luz o aclarar alguna cosa, ya sea materialmente, ya en sentido espiritual de doctrina o ciencia; Se usa también por inspirar, o alumbrar interiormente, con luz sobrenatural y divina; Vale también engrandecer, o ennoblecer alguna cosa" [To enlighten or clarify something, be it materially or in the spiritual sense of doctrine or knowledge; also used to inspire or enlighten the interior with supernatural and divine light; it can also mean to magnify or ennoble something].[31] The Royal Academy defined the adjective *ilustre* as "Magnífico, noble, claro, o elevado sobre los demás, notoriamente por naturaleza, o méritos" [Magnificent, noble, clear, or elevated over the rest, notoriously by its nature or merits].[32] Finally, the Latin roots, "*illustrare: in-*, in+*lustrare,* to make bright < *lustrum,* purification" speak to two of Father Gumilla's goals in writing *El Orinoco ilustrado* with his rhetoric of wonder. Not only does he enlighten and evoke wonder in his readers by illustrating and praising the region, but as a missionary he seeks to purify it. Gumilla's intentions are to enlighten (*ilustrar*) his readers and glorify the Orinoco region by setting it in a good light. By "illustrating" the Jesuit province of New Granada both with an actual foldout map and drawings as well as with wonderful descriptions, he intends to make it famous or illustrious. At the same time he intends to defend Jesuit enlightening of Amerindians and emphasize the great potential for future successful evangelization along the river. To this end, he employs diverse rhetorical tools that both express and evoke wonder.

Throughout *El Orinoco ilustrado,* Gumilla engages wonder, alternating between descriptions of Amerindian tribes and mores, fauna, flora, and minerals. The twenty-five chapters of part 1 of his book and the twenty-seven chapters of part 2 illustrate the actual and cultural geography of the Orinoco River region. A brief discussion of the contents and layout of Gumilla's entire text here can help to contextualize the more focused textual analyses that will follow in this present book. The first three chapters of *El Orinoco ilustrado* chart a course from the mouth of the Orinoco to its interior and enumerate geographical details praising the Orinoco's grandeur. If Gumilla drew the cartographic *Mapa de la Provincia y Misiones de la Compañía de IHS del Nuevo Reyno de Granada* [*Map of the Province and Missions of the Society of Jesus in the New Kingdom of Granada*] included from the first edition in a manner that situates the Orinoco River superior to (and separate from) the Amazon River, his textual descriptions situate the Orinoco's abundant branches and riches above those of all known rivers. The Orinoco panegyric grows with each chapter and reaches a high point at the rhetorical accumulation of North and South America's great rivers: the St. Lawrence, the Río de la Plata, and the Amazon. Gumilla's inventory of rivers reaches its climax with the Orinoco's hyperbolic potential:

> Ahora sale a la luz el gran río Orinoco; no quiere quitar su grandeza a los tres nombradísimos ríos; pero pide (y con razón) que se tomen nuevas medidas, que se atienda a su fondo y caudal para entrar a competir con todos cuantos ríos famosos hasta hoy se han descubierto en los dos mundos antiguo y nuevo. (69)
>
> [Now the great river Orinoco comes to light; I do not want to take away from the grandeur of these three much-renowned rivers; but I ask (and with reason) that new measurements be taken, that its depth and abundance be taken note of so that it can enter into competition with all famous rivers that have been discovered in the two worlds old and new up to today.]

The fourth chapter of *El Orinoco ilustrado* concludes this geographical discussion of the region with a treatment of its tropical climate and a first survey of its abundant fruits. After tempting his readers with brief details about economically viable products, Gumilla quickly turns to the even more valuable "product": human souls. In chapters 5 through 18, Gumilla promotes Jesuit evangelical authority. He lays out a racial topography of the region's men and women, and shares his hard-earned knowledge (both cultural and geographical) about Amerindian tribes, including their customs and temperaments. Chapters 19 through 25 (the end of part 1) focus on the prized by-products of evangelization: "imponderable" natural resources to which the Jesuit missionaries open doors. Right in the middle of these chapters' enumeration of economically viable products, chapter 23 underscores the link between successful conversions and commerce. Here Gumilla inserts a discussion of "practical methods" for Jesuits' first contact with unknown regions and tribes.[33] The placement of this reminder about the missionary's role in opening up pagan (gentile) lands for commercial benefit purposely reinforces connections between evangelical and economic progress.

Part 2 relays further ethnographic details that, while indispensable knowledge for missionaries, constitute curious facts for general readers. Chapters 1 through 13 map several cultural, linguistic, and military challenges to the conversion of South American "bárbaros" (277) [barbarians]. The chapters cover idolatry, diversity of language groups, origins of the Amerindians, war customs, arms, and poisons. Yet at the same time Gumilla outlines several difficulties for civilizing the region, he also offers steps to supersede them. For example, with regard to the apostolic hardship of learning new languages in order to cultivate new souls, Gumilla outlines the spiritual benefits of catechizing in a nation's native tongue. Making use of a Jesuit discourse one might classify as evangelical economics, he invites missionaries to imagine eternal compensation by posing rhetorical questions that paint mission settlements as productive soul farms:

> si la salvación eterna de sola una de aquellas almas fuera superabundante recompensa de muchos años de apostólicas tareas; ¿qué será el ver una continua ganancia de almas para la gloria, no sólo de contado, sino también para lo venidero? Porque ¿qué otra cosa es segregar de las selvas, y domesticar aquellos sañudos genios, sino establecer fincas de inestimable valor, que han de ir tributando anuales réditos de párvulos y adultos para el cielo, no por espacio de uno ni de dos siglos, sino hasta la fin de todos los siglos? (298)
>
> [If the eternal salvation of just one of those souls were superabundant compensation for many years of apostolic work; what compensation will come from seeing a continual winning of souls for the glory, not only those cashed in, but also to come? Because what else can the segregating from the jungles and domesticating of those furious dispositions be than establishing ranches of inestimable value, which will yearly be paying out yields of children and adults destined for Heaven, not only for the space of one or two centuries but until the end of all centuries?]

Chapters 14 through 22 of part 2 turn to nonhuman enemies of progress and colonization that lurk in the tropical biodiversity of the Orinoco River region. And just like the challenges posed by uncivilized Amerindian nations, those posed by dreadful snakes, insects, crocodiles, and the like can be surmounted through evangelization, agricultural reform, and increased population through immigration. Chapters 23 and 24 return to native Amerindian customs, treating marriage practices and superstitious beliefs. Part 2 of *El Orinoco ilustrado* concludes with a three-chapter defense of Spain against European accusations regarding the role that excessive Spanish cruelty has played in the massive loss of Amerindian population. By revealing the "causa genuina" (483) [genuine cause] for the loss of Amerindian populations, Gumilla refutes this aspect of what would come to be known as the *Leyenda Negra* [Black Legend]. These chapters (25 through 27) defend Spain's Catholic agenda and the Jesuit's civilizing role in it, positing a Christian resolution to current cultural challenges to population increase. Finally, in the last several pages of *El Orinoco ilustrado* Gumilla includes two Jesuit-specific discourses: "Apóstrofe a los operarios de la Compañía de Jesús" [Apostrophe to the Workmen of the Company of Jesus] and "Carta de navegar en el peligroso mar de los indios gentiles" [Navigational Map to the Dangerous Waters of Gentile Indians]. These appendices offer advice and inspiration for Jesuit novitiates who answer the call to evangelize the Orinoco River and its Meta and Casanare branches.

Gumilla's chorographic description and actual foldout map of the Orinoco posit the river as the gateway to New Granada commerce, with the Jesuits serving as "mercaderes evangélicos" [evangelical merchants] (103) who cultivate both the spiritual and tangible products of this fertile region. *El Orinoco ilustrado* catalogues Orinoco resources in series of chapters that shift between the human and the natural. These chapters as well as the final appendices are arranged in a manner that reinforces connections between the evangelic and the economic. They suggest a pathway to civilization and modernization in which successful viceregal commerce and international commerce depend on Jesuit conversions and Spanish colonization.

Although outside of the Jesuit community Gumilla's text is virtually unknown today, in his own time it was very widely read. Until the late eighteenth century, *El Orinoco ilustrado* was one of the few Spanish-language histories translated into French (1758) besides those of the Dominican writer Bartolomé de Las Casas and the mestizo chronicler El Inca Garcilaso de la Vega. Gumilla supervised the first two editions of *El Orinoco ilustrado* (1741 and 1745). After the Jesuit's death, his natural history continued widespread circulation in Spain. But thanks to the 1758 French translation (based on the 1745 definitive edition, it even translates the map into French), it also circulated widely in other countries of Western Europe.[34] A third "much more correct edition" was published in 1791 in Barcelona, "adorned with" eight late eighteenth-century illustrations showing, according to the title page, "the rites and customs of those Americans."[35] These etchings are nearly identical to those included

Los medicos del Orinoco llamados Piaches [Orinoco doctors called Piaches]. The John Carter Brown Library at Brown University.

Bayle de los Yndios Mapuyes [Dance of the Mapuye Indians]. The John Carter Brown Library at Brown University.

in the much-cited 1781–82 ethnographic work of Filippo Salvadore Gilij, a young Italian Jesuit whom Gumilla brought to the Orinoco from the court in Rome.[36]

The title to the second edition of *El Orinoco ilustrado* (1745) enacts only a slight change: *El Orinoco ilustrado, y defendido* [*The Orinoco Enlightened, and Defended*]. However, an examination of the contents of the first editions reveals many obvious and subtle differences. Gumilla made few subtractions, instead adding everything from lengthy scientific explanations that answered comments about the perceived overcredulousness of his 1741 publication or its pithiness to a few extra words here and there supplementing descriptions or anticipating future criticisms. The expanded introduction responds to actual reception of *El Orinoco ilustrado* in the Spanish court and beyond and lays out Gumilla's strategy for "defending" the contents of his first edition:

> Algunas personas [. . .] así españolas, como extranjeras de la más sobresaliente literatura, y de la más ilustre nobleza, cultivadas en las bellas letras, se han dignado de reconvenirme sobre lo lacónico de algunas noticias [. . .] por lo cual en esta impresión procuraré dar a todos satisfacción sin detrimento de la brevedad que deseo. Y porque no solo he de responder a las dudas de las personas que dificultan con fundamento, sino también a otras [. . .] de donde nacerá la variedad de frases con que me introduciré en las adiciones que prometo [. . .] en las primeras clásulas [*sic*] de cada adición se verá propuesta la duda, y el modo de dudar, y en el contexto se hallará la respuesta pretendida, roborada y autorizado, por lo cual al título del Orinoco ilustrado, añadí: roborado y defendido (39–40)

> [Some people . . . Spaniards as well as foreigners authors of the most outstanding literature, and of the most illustrious nobility, raised on the fine arts, have deigned to remonstrate about the laconic nature of some notices . . . for this reason in this printing I will try to grant to everyone satisfaction without detriment to the brevity that I desire. And since I have to respond not only to the doubts of people who raise objections with reason, but also with others . . . from this will be born the variety of phrases with which I will introduce myself in the additions that I promise . . . in the first clauses of each addition will be seen the question proposed, and the means of questioning, and in the context will be found the answer insisted, reinforced, and defended, for this reason I added to the title of Orinoco enlightened: Reinforced and defended.]

Additional cites of Jesuit and geographic authorities as well as more margin notes and chapter subdivisions (especially the amplification of the chapter on snakes into eight separate parts) constitute the most obvious additions to the 1745 edition. Other notable 1745 additions include three subdivisions to the last chapter in part 1 that underscore its "imponderable" riches discourse and reaffirm Gumilla's beliefs about El Dorado and title changes for chapters in part 2 that emphasize the horrors of the curare poison.[37]

Both editions end with the "Indice de las cosas más notables" [Index of the Most Notable Things], but the 1745 edition adds several items to the index, such as the *Iuca pan* (yucca bread) and three new types of palms. Seven ethnographic notes on the Amerindians are included that correspond to the five subsections added to part 1, chapter 5. The 1741 edition only includes the "Apóstrofe a los operarios de la Compañía de Jesús" [Apostrophe to the Workmen of the Company of Jesus]. The "Carta de navegar en el peligroso mar de los indios gentiles" [Navigational Map to the Dangerous Waters of Gentile Indians] is added to the 1745 edition. The 1791 edition drops the index and the apostrophe but maintains the navigational map and adds a bust portrait of Father Gumilla.

The only nineteenth-century edition of *El Orinoco ilustrado* was published in Barcelona in 1882 as part of of a series entitled La Verdadera Ciencia Española [The True Spanish Science].[38] There have been nine modern editions lauding Gumilla as one of the most valuable workers of the Company of Jesus. Father José Rafael Arboleda introduced two editions (1944 and 1955) based on the 1741 edition of *El Orinoco ilustrado.* Father Constantino Bayle based his edition on the more extensive 1745 publication (*El Orinoco ilustrado y defendido*).[39] Demetrio Ramos reproduces Bayle's 1945 edition (including introduction) and appends his own and another preliminary study by José Nucete-Sardi for the 1963 version published in Caracas by the Venezuelan National Academy of History.[40] Since then some popular editions have appeared based on the 1791 Barcelona publication as well as *Tribus indígenas del Orinoco,* a truncated edition that includes only selected chapters that treat Orinoco tribes. Despite four eighteenth-century editions, the late nineteenth-century edition, and nine editions published since 1944 in Spain, Venezuela, and Colombia, *El Orinoco ilustrado* has garnered scarce critical attention and existing studies have not appreciated its contributions to preexile Jesuit evangelization and science in the Hispanic Enlightenment.[41]

Because Gumilla combines his knowledge as a missionary and scientific traveler to produce a natural, civil, and geographical history that includes ethnographic description and cartography, his text is best understood through an interdisciplinary approach. My study benefits from interdisciplinary methods and examines *El Orinoco ilustrado* as a literary, religious, and scientific document. The first two chapters employ the tools of literary analysis while borrowing from art history, cartography, ethnography, and missionary discourse. Chapters 3 and 4 blend the strategies of literary and cultural studies and are informed by recent trends in the history of science.

Chapter 1 explores Gumilla's rhetorical construction of missionary authority and of *El Orinoco ilustrado* as a textual natural history cabinet. Chapter 2 reveals how Gumilla's maps link Jesuit missionary strategies to Spanish military and economic considerations and plot a course for exploiting the region's evangelical and commercial potential. The Society of Jesus's successful conversions are aided by cross-cultural communication within the Jesuits' role as mediators who depend not only on linguistic abilities or understanding of tribal customs but also, at times, demonstrations of Spanish troop support and European science. Chapter 3 explores the Jesuits' scientific eclecticism and Spanish Baconianism. The Jesuits' diverse influences on the Spanish and Spanish American elite allowed them a privileged role in the transition to Hispanic modernity. Chapter 4 examines four principal pathways to knowledge and modernity Gumilla employs in *El Orinoco ilustrado:* Christianization, demonization, correction, and validation. Throughout his natural history, Gumilla leads readers on pathways from initial wonder toward natural knowledge. The conclusion outlines readings of Gumilla's text since the eighteenth century by both scientists and novelists and offers a wider perspective on the nature of enlightenment.

Chapter 1, "Father Gumilla's Textual Cabinet of Curiosities, Missionary Authority, and Rhetoric of Wonder," reveals how Gumilla takes advantage of Old World fascination with New World curiosities, constructs a textual natural history cabinet, and employs rhetorical strategies that simultaneously appeal to his readers' intellects, senses, and emotions while edifying his missionary integrity (ethos). It explains the crucial relationship between displays of natural history and displays of power, prestige, and economic potential. Stephen Greenblatt has eloquently discussed how, since Columbus's first voyage, Spanish monarchs valued wonder-provoking displays collected from their divinely sanctioned marvelous possessions overseas.[42] Greentblatt's study moves from "medieval wonder as a sign of dispossession to Renaissance wonder as an agent of appropriation: the early discourse of the New World is, among other things, a record of the colonizing of the marvelous."[43] *Marvelous Possessions* does not discuss, however, the missionary accounts produced by the Society of Jesus, many of which, from its inception, constructed images of exotic locales packaged for consumption. Additionally, sixteenth- and seventeenth-century Jesuits constructed some of the earliest cabinets of curiosities. Working within the *curiositas utilis* paradigm, Jesuit natural histories and cabinets demonstrated potential profits and scientific advances. Gumilla displays the Orinoco's "marvelous possessions" in a textual cabinet for eighteenth-century Jesuits and diverse European and American readers, highlighting both the evangelical and commercial values of Spain's recently established viceroyalty of New Granada. To capture both the intellects and hearts of an unusual variety of potential readers, he employs a wide variety of techniques from classical and Christian rhetoric. Gumilla establishes his eyewitness authority both as an historian and as a missionary to reach armchair travelers, merchants, wealthy nobles, and politicians, as well as Jesuit novitiates and provincial leaders. He captures his readers' imaginations with the same weapons of classical and sacred rhetoric that Jesuits employed to convert Amerindian charges, and bases his rhetoric of wonder on a persuasive combination of ethos, pathos, and logos.

Chapter 2, "Colonization, Commerce, and Conversions: Mapping the Economic and Evangelical Values of the Orinoco," challenges the postponement or negation of Catholic journeys to modernity that has displaced them into the peripheries of Enlightenment studies. It examines the dual nature of Spain and Spanish America's intellectual and spiritual enlightenment and the links between economics and evangelism through an examination of the first modern graphic map of the Orinoco region as well as three types of textual, or descriptive, maps in *El Orinoco ilustrado*. Gumilla published his expanded second edition (1745) amid additional defensive needs for the Orinoco River region, not only because of Spain's war with England (1739–48), French attacks at the mouth of the river, Dutch continued support of fierce Carib plundering of missions, and Portuguese slave-raiding, but also because of encroachment on Jesuit territories by the Capuchin and Franciscan orders. Gumilla ties the entire region's defense not only to increased military and economic investment, but more importantly to an increased presence of the Society of Jesus. Gumilla's principal strategy of defense in *El Orinoco ilustrado* is revealed through a series of maps that expose the inextricability of colonization, commerce, and Catholic conversions in Gumilla's overarching plan for the Orinoco River region and that chart a route for understanding Bourbon Spain and her reigns' entrance into modernity.

Clearly, Gumilla mapped plans for colonization and commercial exploitation of New Granada. As he makes clear, however, Jesuit evangelization of and intercultural discourse with Orinoco River region Amerindian nations constitute an essential step in this process. Chapter 3, "The Jesuits and an Eclectic Enlightenment: Mediating Knowledge and Catholicism," examines the reciprocal nature of the route to modernity between and within Spain and Spanish America. It explores Spanish Baconianism, Jesuit eclecticism, and their mediating role within the wider context of the history of science. The Jesuits' influences as educators, missionaries, naturalists, and confessors to the elite allowed them a privileged role in the advancement of modernity. In his role as a mediator of modern scientific developments and Amerindian knowledge with

Catholicism, Gumilla moves beyond the scholastic tradition and effects mutual exchanges of knowledge that enlighten European and Amerindian souls and intellects.

Chapter 3 discusses the fundamental role Jesuits played in a Hispanic Catholic Enlightenment that united both scholarly and theological aims. Jesuits like Gumilla were both devout Catholics and moderns who understood the role of natural history in a manner similar to Francis Bacon. Under Bacon's influence, across Europe aristocratic and academic natural history cabinets shifted from amateur collections of mere curiosities to categories of modern science. This chapter lays out the empirical methods of Gumilla's time and his own Organon for natural history, which consists of a hunt for knowledge that goes beyond describing and cataloguing and extends to analysis and experimentation. Working within a privileged space of knowledge, Gumilla mediates his roles as both Spanish imperial agent and Jesuit vindicator of converted Orinoco souls and Amerindian populations still resisting Christianity. Gumilla's appropriation and adaptation of local cultural, racial, natural, and geographical data results in a text that offers up such a bounty of information on tribes and natural wonders that Jesuit historians introducing modern editions of *El Orinoco ilustrado* still present it as an enticing encyclopedia of legendary secrets. However, Gumilla's textual natural history cabinet contains much more than a catalogue of curiosities. In fact, *El Orinoco ilustrado* provides a textual example of the movement from curious gentlemen's cabinets of curiosities to scholarly cabinets for the pursuit of knowledge. Gumilla's science combines induction in the field (the Baconian "hunt" for knowledge) with deductive reasoning in his reports (doing science versus writing science), emphasizing experience and direct observation alongside a study of ancient and modern natural philosophers. He aims to enlighten the Orinoco with European scientific knowledge, and his particularly Jesuit way of proceeding mediates his South American experiences and observations not only with fellow Jesuits but also European philosophers such as Francis Bacon, Pierre Gassendi, Isaac Newton, and René Descartes, by way of Benito Jerónimo Feijoo and other Spaniards. Gumilla's approach to the Orinoco River region blends the tools learned during his Jesuit formation with the realities of his experience in the "peripheries" of mission culture. Like other Catholic intellectuals during the first half of the eighteenth century, Gumilla resists subscribing to a single philosophical system for natural philosophy. His eclecticism provides an excellent example of an alternative process for "scientific revolution," since religion and science are linked in the pathways to modernity for Spain and her colonies.[44] *El Orinoco ilustrado* participates in a Jesuit eclectic tradition that mediates the European philosophies central to the evolution of Western science with data gleaned both firsthand and via intercultural conversations. Study of its incorporation of local paradigms beyond Eurocentric concepts of science and Enlightenment helps to redefine modernity.

Chapter 4, "¡Oh Monstruo, Oh Bestia! Pathways to Knowledge in *El Orinoco ilustrado,*" examines four rhetorical pathways to knowledge and modernity in *El Orinoco ilustrado* by centering on Gumilla's Christianization of El Dorado, demonization of curare, the correction and validation of Amerindian medical knowledge, and scientific investigations of war drum acoustics, the boa snake's magnetic breath, and crocodile stones and fangs. When narrating anecdotes that shift from describing monstrous creatures and natural phenomena to field experiments and then to scientific explanations, Gumilla's rhetoric first captures the imagination and then leads it along a route from wonder at (*admiratio*) to knowledge of (*scientia*) nature's marvelous secrets. Gumilla catalogues the Orinoco according to Bacon's idea that "all knowledge and wonder (which is the seed of knowledge) is an impression of pleasure in itself."[45] Yet he expands upon this pleasing accumulation of knowledge, mentioning microscopes and atomism alongside experiments to test his newly discovered virtue of crocodile fangs. Chapter 4 transports readers of this book into Gumilla's unique ways of legitimizing seemingly fantastic American natural phenomena. The Jesuit blends scholastic authorities and the recently evolved empiricism and experimentalism of natural philosophers with Amerindian popular wisdom. His discourse on the sound waves of war drums, for example, blends analysis of man's and God's works—that is, a man-made instrument and the natural phenomena of sound waves—to accommodate Amerindian cultural practices, the pious purposes of natural history, and a modern empirical pursuit of knowledge. The core of this chapter takes its cue from Gumilla's significant additions to his pivotal snake chapter for the second edition (1745). Making moral and scientific use of snakes, Gumilla's chapter's central location underscores its central role in Gumilla's epistemology. In particular, his scientific investigation of folklore about the giant boa's magnetic breath pulls readers from wonder toward knowledge and anthropomorphizes the journeys to natural knowledge seen throughout his natural history. These pathways to knowledge reveal how Gumilla's exposure to Orinoco Amerindian folk knowledge influenced the rhetorical strategies he employs to present scientific data in *El Orinoco ilustrado* and form the core of Gumilla's Orinoco Enlightenment.

In addition to summarizing the findings of this study, the conclusion, "The Legacy of Joseph Gumilla's *El Orinoco ilustrado,*" considers other readings of Gumilla's text and suggests avenues for future research. Eighteenth- and nineteenth-century readers enlightened both by Gumilla's scientific content and literary strategies included renowned historians, scientists, artists, patriots, and clergy. Nineteenth- and twentieth-century novelists borrowed Gumilla's vivid jungle and river descriptions and narration of the Orinoco region's popular myths and lore. Their rhetorical debts to Gumilla are clear, and this highlights the importance of studying *El Orinoco ilustrado*. Equally interesting are the ways in which novelists were enlightened by Gumilla's firsthand scientific knowledge. Not surprisingly, eighteenth-century readers of *El Orinoco ilustrado* included exiled Jesuits who praised and quoted Gumilla's work. More striking, perhaps, are Gumilla's influences on non-Jesuits, particularly on some of the most famous scientific travelers to South America. Gumilla was cited in contemporary reports of the eighteenth-century expeditions of Charles Marie de La Condamine, Jorge Juan, and Antonio de Ulloa, as well as in Alexander von Humboldt's much-celebrated *Personal Narrative of a Journey to the Equinoctial Regions of the New Continent.* Both La Condamine and von Humboldt depended on the Spanish Jesuit's contribution to enlightening the Orinoco. Humboldt's understanding of experiment required him to demand more control, precision, and instruments, but the Prussian baron's vision of science shares much with Gumilla's. Yet just as important as their borrowing of scientific facts are the ways in which these voyagers took advantage of literary strategies to construct an appealing discourse of science, travel, and exploration. As the legacy of Gumilla's literary strategies becomes obvious, critics might cease to attribute Humboldt's own rhetoric of wonder merely to his "romantic" nature. Humboldt's status as amateur scientist, anthropologist, and author of fascinating popular science literature made him a pop icon in his day, but his was no nineteenth-century "rediscovering" of an appreciation of wonders. This book contends that we should view Humboldt's style and attitude as a legacy of Gumilla's enthusiastic synthesis of art and science during an Age of Reason traditionally limited to moderation and control of emotions. In short, the model of Enlightenment embodied by Gumilla that combines both reason and the imagination in the advancement of knowledge helped to shape Humboldt—an important consideration, given that Humboldt's writings have affected perceptions of Latin America on both sides of the Atlantic for generations.

Existing analyses of textual relationships between Humboldt and Andrés Bello suggest yet another possible legacy of Gumilla's *El Orinoco ilustrado,* an interesting one within the context of recent studies illuminating the contributions of Jesuit natural and civil histories to Latin American national identity construction. One twentieth-century Jesuit's valorization of *El Orinoco ilustrado* as "the genesis of national consciousness" in Venezuela invites Gumilla's placement alongside studies detailing Jesuit intellectual subjectivity as differentiated from Spanish imperial authority.[46] Gumilla was transformed by the Orinoco, and experiences like his certainly contributed to the process of emerging national consciousness. However, Gumilla embodies the perspective of a Hispanic transatlantic subject. He was a Spaniard but never considered himself a foreigner in the Orinoco, not only because he spent twenty-three years breaking "American bread" with Amerindians, but also because he maintained an international, universal Jesuit identity, a crucial consideration when researching the Jesuit missionaries to Spanish America.

The goal of this book is to help expand definitions of the Enlightenment by exploring a simultaneously spiritual and scientific theory of knowledge from the first half of the eighteenth century in Spain and Spanish America. By examining the role of preexile Jesuit evangelization and science in the different yet successful Hispanic transition to modernity, it challenges traditionally secular prejudices in Enlightenment theory. Gumilla exemplifies the Jesuit mediation of Catholic dogma with Amerindian knowledge and modern scientific discourse. Analysis of *El Orinoco ilustrado* provides a model for an eclectic, Catholic Enlightenment that unites sentiment and reason, allows for emotion within scientific inquiry, and employs the strategy of wonder to accumulate, enumerate, and disseminate knowledge.

1
Father Gumilla's Textual Cabinet of Curiosities, Missionary Authority, and Rhetoric of Wonder

La Historia, no solo es abonado testigo de los tiempos; es, y debe ser también, luz para todas las edades y generaciones. Y al modo que (si falta la luz) en la más curiosa galería, todo aquel archivo de la más apreciable antigüedad, pasa a un caos de confusión, pareciendo ordinarias las piedras más selectas, y borrón tosco la más sutil miniatura, no de otra manera la más curiosa Historia, si le faltare la luz, claridad, distinción y método, será toda confusión, y origen de muchas dudas contra el fin primario de la Historia, que tira a disiparlas.

—Joseph Gumilla, *El Orinoco ilustrado*

[History is not only a reliable witness of the times; it is, and also should be, light for all the ages and generations. And just as within the most curious gallery (if lacking light) all that archive of the most esteemed antiquities will become a chaos of confusion where the most select stones will seem ordinary and the most subtle miniature a preliminary sketch, so the most curious history; if it lacks light, clarity, distinction, and method, will become complete confusion and the origin of many doubts that work against the primary aim of history, which tends to dissipate them.]

Gumilla's Enlightening New World Natural History Cabinet

The Spanish tradition of gathering up the marvels of the New World began with Charles I (Carlos V, 1516–56), who collected natural, artistic, and ethnographic materials not only for their aesthetic or curiosity value, but also as representations of the abundant riches in his far-off kingdoms. Sixteenth-, seventeenth-, and eighteenth-century displays of natural history objects and human artifacts are tied to the power, prestige, and economic potential of imperial possession. One of the central rhetorical features in *El Orinoco ilustrado* (1741–45) is Joseph Gumilla's construction of his natural history text as a cabinet of curiosities.[1]

This Jesuit's wonder-provoking textual cabinet, however, displays more than the far-off "marvelous possessions" of his monarch, Philip V.[2] Gumilla's natural history maintains a religious context for the acquisition of natural knowledge, and its discourse of utility combines the theological and the economic.[3] Seventeenth-century Jesuit New World natural histories downplayed the potential monetary value of nature to emphasize instead the pious purpose behind God's magnificent displays. They evoked human wonder primarily to increase human piety. Natural historians reading God's book of nature alongside the revealed Word of the Bible performed the crucial task of interpreting flora and fauna allegorically to reveal their moral worth. However, while Gumilla values these moral lessons and justifies knowledge of God's works in order to better know and praise the Creator (*ad majorem Dei gloriam:* for the glory of God), his eighteenth-century natural history recuperates the commercial value of the Orinoco first touted in the fifteenth and sixteenth centuries by explorers such as Columbus and Ralegh. Gumilla posits the Orinoco River region as a providential space of natural knowledge and underlines the Jesuit role in reading it in search of divine secrets. To enlighten Europe, he sheds light on his "most curious gallery": *El Orinoco ilustrado* advances knowledge of and removes former doubts about the value of previously underexplored territories. By employing a rhetoric of wonder that persuades through both passion and reason, Gumilla markets the Jesuit province of New Granada as a privileged site of evangelical and commercial potential.

Joseph Francisco Tomás Gumilla has flown under the radar of historians whose definitions of the Enlightenment limit it to secularization and rationalization. However, Gumilla was a Catholic intellectual who, like Benito Jerónimo Feijoo and Ludovico Antonio Muratori, advanced the Enlightenment through theology, moral philosophy, and modern science.[4] As was central to his Jesuit missionary vocation, Gumilla also enlightened Amerindian souls. He narrates these evangelical successes in *El Orinoco ilustrado.* The "light" promised in this chapter's epigraph, quoted from his introduction, however, are directed primarily at European intellects, both Spanish and American-born Spaniards as well as foreigners.[5] The breadth of Gumilla's readers underlines the key role that preexile Jesuit missionaries and naturalists played in the Hispanic Enlightenment. To fully appreciate Spain and Spanish America's entrance into modernity, both the spiritual and the intellectual Enlightenment must be examined within the context of the "Siglo de las luces," or "Ilustración."[6] This book recovers an invaluable, if now virtually unknown, natural and civil chronicle of the New Kingdom of Granada (today, Venezuela and Colombia) that documents both moral and scientific history during a key moment of transition from the late baroque to the Enlightenment.

This chapter analyzes the rhetorical techniques that bring plants, animals, and native inhabitants vividly before the eyes of the diverse readers of his textual cabinet. Gumilla's narrative includes these readers in a thrilling journey along the Orinoco River region as they see, hear, smell, taste, and touch previously unknown flora and fauna, visit Orinoco River region tribes, and experience exotic indigenous customs. His authority as a historian is intertwined with his missionary ethos, or integrity, valuing both personal experience and character in a manner he compares to that of Christ's apostles. Like his Jesuit forefather Joseph de Acosta in the late sixteenth century, Gumilla presents himself as chosen by Christ to both explore the region and bring its inhabitants to the light of God. His natural history exercises Acosta's "excellent theology" by maintaining its pious purpose and directing readers' wonder at God. But it also modernizes Acosta's model for "good [natural] philosophy."[7] As we will see, Gumilla also inserts himself within a two-century-long European dialogue on the wonders of the New World, drawing upon the weapons of ancient rhetoric to appeal to European and Jesuit desires for possession, power, and potential riches and souls.

Rhetoric and Audiences of Spectacular Natural History Displays

In part because of Gumilla's constant recourse to wonder, from its first appearance in the 1740s to the abbreviated edition published by a Venezuelan newspaper in the 1990s, his natural history has been reduced to a series of "captivating vivid descriptions" and "curious revelations about indigenous customs."[8] Perhaps distracted by the spectacular elements of *El Orinoco ilustrado,* very few scholars have interrogated the nature and organization of knowledge in this Jesuit textual natural history cabinet.

From the earliest explorations of American soil, Spanish royalty asserted themselves as collectors. Their wonder-cabinets (forerunners to natural history cabinets) displayed the raw materials of trade and prosperity to highlight "the commercial advantage of magnificence,"[9] and textual descriptions of their latest possessions excited merchants and general readers alike by emphasizing the wonders of the New World. María Paz Aguiló Alonso details how upper-class gentlemen of the court took their cue from the king's zeal for collecting in her explanation of Renaissance ties between aristocratic wonder-cabinets and Spain's *coleccionismo,* which reached new heights of enthusiasm for the exotic objects gathered from the East and West Indies during the reign of Philip IV (1621–65).[10]

Ever since the founding of the Society of Jesus, authors of its official correspondence understood the value of wonder as a rhetorical device. By the seventeenth century, Jesuits had made actual and virtual displays of the spectacle of nature into a standard strategy for working upon the senses, heart, and mind of readers. Missionary reports abounded with worldwide natural wonders. One French missionary to South America even lamented that to have sufficient room to display the diversity of innumerable species, he would have to construct an amphitheater that would dwarf the greatest Roman stadiums.[11] Such comparisons alongside vivid descriptions heighten the grandeur of the New World nature. However, if at first glance *El Orinoco ilustrado* might be categorized as simply following a religious trope detailing spectacular nature in God's *teatrum mundi,* a closer look reveals that Gumilla's goals for putting the Orinoco region on display reached far beyond enhancing Europe's viewing pleasure.[12] Certainly, Gumilla takes advantage of his readers' curiosity about South America and evokes their wonder through rhetorical strategies for vivid description as a means to heighten their appreciation of the Orinoco region. In fact, his cabinet offers up the wealth of El Dorado by detailing tangible mercantile values within the previously mythical Orinoco region. But his observations, reflection, and experiments also lead readers down pathways for possessing the Jesuit province of New Granada. Wonder and possession are united in the Jesuit's sense of political and commercial purposes and quest for scientific knowledge and religious commitment. The diverse targets of his providential discourse are enlightened through firsthand knowledge of the region and the moral lessons to be taken from it, as well as useful details for political strength and economic wealth.

When Gumilla cultivates wonder through his vivid textual descriptions of natural history objects, he evokes the power and prestige afforded wonder-provoking displays in Europe. *El Orinoco ilustrado* illustrates for European readers an Orinoco River region bursting with exotic flora, fauna, and peoples, and it builds upon both religious and courtly traditions dating back more than two centuries. The obsessive transport of exotica from New World regions, or "God's cabinets," to Christian natural historians' cabinets during the age of discovery and conquest reveals how both curiosity and wonder were tied to acquisition and possession.[13] Gumilla creates a textual natural history cabinet that underscores the king's prestige, just as actual physical cabinets had done in the sixteenth and seventeenth centuries for previous Spanish monarchs, as well as for nobles throughout Europe. As William Eamon explains, during the early modern period a revalorization of the political and economic value of curiosity increasingly prevailed over the medieval suspicion of curiosity as a vice:

> [N]ot only was curiosity considered a virtue worthy of a prince, it was an important symbol of his power. Striking visual demonstrations of . . . curiosity could be seen in the cabinets of curiosities (*Wunderkammern*) that Renaissance princes collected and put on display at the courts. . . . [These] were visual manifestations of the prince's power. In them the prince appropriated and reassembled all reality in miniature, symbolically demonstrating his dominion over the world.[14]

However, when offering textual equivalents of such marvelous possessions in *El Orinoco ilustrado* to King Philip V, Gumilla does so in the belief that the ultimate source of Spain's glory is God. Through the pope's authority, God had permitted Spain to discover and settle the New World for the salvation of New World pagans. By amplifying the greatness of his chosen region of New Granada, Gumilla pleads to his majesty, Philip V, to resist neglecting such a worthy region, for this neglect would risk diminishing the glory of His Majesty, God. Gumilla underscores the Crown's importance for continued missionary successes in New Granada and, by association, its control over the New World. By extension, his cabinet also extols the Society of Jesus. Although more immediate concerns for troop support and Spanish colonization overshadow the discourse of Jesuit apologetics in *El Orinoco ilustrado,* Gumilla does indeed cultivate the political favor wealthy nobles and clerics who have the ear of the king.[15]

Thus, Gumilla follows both Jesuit and noble traditions that reach back to the earliest collectors and tie natural history writing to cabinets of curiosity; like them, he finds inspiration in both princely and Jesuit collections of wonder-provoking *naturalia* and human artifacts. From the first years after Ignatius Loyola's Company of Jesus was recognized by Pope Paul III (1540), a constant flow of Jesuit accounts had provided striking images from missionary work in both the East and West Indies. Early on, an ethnographic tradition emerged, with Jesuits describing "exotic" natives both outside and inside the mission settlements, which, despite being civilized, remained "full of wonders and spectacle."[16]

Gumilla's writing was clearly shaped by his exposure not only to the spectacular Orinoco River region, but also to the Jesuit traditions of natural history cabinets and excellent education. By the late sixteenth century the Collegio Romano had become Loyola's vision of the apex of all Jesuit seminaries and had established Rome as the center of Jesuit collection, observation, and investigation of every kind of natural curiosity. In the seventeenth century, scholars like Juan Eusebio Nieremberg (1595–1658), professor of natural history at the Colegio Imperial de Madrid, and the renowned Jesuit polymath Athanasius Kircher (1602–80), founder of the Collegio Romano's first natural history cabinet, continued the Ignatian tradition of excellence in scholarship. Both during his novitiate in Tunja and Santa Fe de Bogotá and later while completing the manuscript of *El Orinoco ilustrado* in Madrid and Rome, Gumilla benefited from intellectual fellowship with Jesuit educational institutions and visits to cabinets of curiosities. For example, he mentions speaking before "erudites" in Europe and viewing what remained of the original collections displayed in the cabinet constructed by Kircher to showcase works of God and works inspired by God.[17] Gumilla's missionary and intellectual pursuits were motivated by collaboration with the Jesuits and their praxis.

Gumilla's choice to construct his narrative as a textual cabinet of curiosities builds upon both the Jesuit tradition of describing and displaying far-off wonders ripe for investigation and consumption and the "courtly virtuosity . . . and the high esteem it attached to wonderment" within the cultured elites' tradition of cabinets of curiosity.[18] Unlike the royals and aristocrats throughout Europe who built private collections that were seen primarily by the elite, however, Gumilla does not restrict access to his cabinet. By responding to these varied traditions, Gumilla appeals to an unusual variety of potential readers, including Jesuit novitiates and provincial leaders, wealthy nobles, politicians in America and Europe, and even the king and the pope. Among the real and imagined readers Gumilla considers are also the armchair travelers who seek the pure entertainment value of natural histories that enumerate the wonders of far-off lands. He specifies commercial benefits of the region's natural products and geography that benefit naturalists, merchants, and general Christian readers on both sides of the Atlantic. And he comments on the success of this general appeal in the introduction to his second edition of *El Orinoco ilustrado:*

> [E]sta Historia ha corrido por todas manos, ha sido examinada [. . .] por muchos ojos . . . este libro, no sólo se dirigía a los científicos y curiosos de la Europa, sino también para los de la América y para aquellos mismos que moran en los países, sitios y Misiones donde recogí las noticias que refiero, y en donde no causan novedad por ser noticias.

> [This history has traveled into everyone's hands, it has been examined . . . by many eyes . . . this book was not only directed at the scientific and the curious of Europe, but also at those of America and even at those very same inhabitants of the regions, sites, and Missions where I gathered the knowledge I refer here, where its novelty causes no surprise.] (Gumilla, *El Orinoco ilustrado,* 39–40)

This last comment reveals the connection between Gumilla's promises to shed light upon strange and new flora, fauna, and people for all who have not traveled to the Orinoco River region, and Gumilla's construction of authority and insistence on the truth value of his claims. In fact, he consciously edifies his missionary ethos while simultaneously appealing to his readers' intellects and to their senses and emotions.

The rhetorical tools Gumilla uses to evoke wonder at his textual cabinet of curiosities address the diversity of his intended audience. To understand Gumilla's success employing persuasive strategies to produce a desired effect in his readers, the first three parts of oratorical discourse must be kept in mind, because his *inventio* (careful selection) of Orinoco River region subject matter and their well-thought-out *dispositio* (division and arrangement) within his textual natural history cabinet are linked to his *elocutio* (style).[19] That is, the arrangement of *El Orinoco ilustrado* reinforces Gumilla's textual cabinet of natural history as a rhetorical construct, and this organization of materials or topics remained just as important to Gumilla as the style of his writing.

The core of Gumilla's rhetoric of wonder consists of both internal and external proofs combined with amplification. For these internal or "artificial," proofs, Gumilla frequently employs logos, or inductive and deductive reasoning. He also maximizes the emotional appeal, or pathos, of his material when he relates his personal experiences. Gumilla often creates a rhetorical induction based on examples and anecdotes, but he also employs the deductive reasoning of the classic rhetorical syllogism. His external proofs consist of testimonies and published texts on geographical, scientific, and religious topics.

Gumilla's constant use of amplification stirs the emotions, thus rooting his rhetoric in sacred persuasion.[20] Christian orators understood that the will is moved more by the senses than the intellect. In the eighteenth century, Friar Luis de Granada's *Los seis libros de la Retórica eclesiástica o la manera de predicar* remained a key source for acquiring both Roman and Greek prescriptions for rhetoric used for Christian purposes. As Luis de Granada explains, amplification had to appeal to the mind and move the will to ire, compassion, sadness, love, hate, hope, fear, wonder, or any other emotion, and therefore persuasion through amplification depends more on narration and enumeration of the "greatness and amplitude of the thing" than on the syllogisms of argumentation.[21] Gumilla clearly understood that appeals to the emotions of readers could be especially persuasive, and almost always more convincing than appeals to reason alone.

As we will see, amplification is a primary rhetorical technique in *El Orinoco ilustrado,* and Gumilla takes full advantage of its emotional intensification.[22] To the delight of his readers, he re-creates the variety and abundance of the Orinoco River region by heaping up its wonders, just as collectors arranged their curiosities or wonders.[23] This arrangement of wonderful objects privileges *incrementum,* a figure for amplification found often in the *epideictic* (demonstrative) genre of both religious and secular oratory. In accordance with incrementum, Gumilla arranges his amplified topics into a climactic chain of wonders that takes the reader, by degrees, with him along the river and into the elaborate natural history cabinet that is *El Orinoco ilustrado.*[24]

Beneath this overarching technique of amplification, Gumilla's rhetorical strategies reveal both the spectacular abundance of nature meant to awaken the armchair travelers' senses alongside the subtle intellectual arguments and command of rhetoric that would attract fellow Jesuits and other well-educated readers. Four of the Jesuit's principal means of doing this include combining a traditional discourse of novelty with an insistence on veracity and his authority as a Jesuit missionary, painting word pictures in the sermonic style taught in Jesuit institutions, linking his natural history discourse to the force and abundance of the River Orinoco, and using the first-person plural to engage readers' hearts and minds and invite them to share in his experiences. Each of these very different strategies is deployed to appeal, in a particular way, to the diverse collection of readers Gumilla addresses in *El Orinoco ilustrado.*

Constructing a Logos-driven Ethos

Gumilla's first rhetorical strategy—insistence on his veracity and his authority as a Jesuit missionary—appeals primarily to the intellects of fellow Jesuits, Christian readers, and foreign scholars who might doubt him. In his prologue Gumilla underlines the importance of a source's character for judging claims that might seem false:

> Es cierto que la notable distancia no sólo desfigura lo verdadero, sino también suele dar visos de verdad a lo que es falso; pero la prudencia dicta que antes de formar juicio decisivo se haga madura reflexión sobre la persona que da la tal noticia. (Gumilla, *El Orinoco ilustrado,* 32–33)
>
> [It is true that notable distances not only disfigure the true but also often give glints of truth to that which is false; however, prudence dictates that before forming a decisive judgment one maturely reflect on the person who imparts the new knowledge.]

Gumilla's long-term residence in the wild and wonderful Orinoco River region establishes him as a key source of information about this still barely known area. He treats things that "a los europeos pareció y realmente es nuevo" (33) [to Europeans seemed and really are new] and provides his readers privileged access to "naciones, animales y plantas, incógnitas casi enteramente hasta nuestros días" (37) [nations, animals and plants, almost entirely unknown up to our days] in one of the last untamed South American frontiers. By consciously exploiting the distance, and therefore strangeness, of the Orinoco region, he creates a pleasing blend of strange yet useful natural history topics. However, since these new Orinoco birds, beasts, and vegetation are as distant (and different) from Europe as the climate of the Orinoco region is distant (and different), Gumilla must address potential concerns about the veracity of this news from far-off lands. To this end, he constructs a rhetorical ethos that underscores the value of personal character and personal experience as a basis of authority while engaging the literary traditions of novelty and the Gospels.

The construction of this rhetorical ethos draws on a number of techniques. First, from the opening lines of *El Orinoco ilustrado,* Gumilla combines the authority of his firsthand observations with the modest and humble character expected of a missionary to create ethical appeal and capture the goodwill of all readers. This technique allows him to underline his authority as a Jesuit and an eyewitness, and thus cite himself throughout *El Orinoco ilustrado* as proof positive: "No pido ni quiero que se me dé más fe ni más autoridad a mi dicho que la que se me debe por testigo ocular, por sacerdote y por religioso, aunque indigno, de la Compañía de Jesús" (181) [I do not ask for nor want that anyone believe or grant any more authority to what I say than that which I deserve as eyewitness, priest, and religious, however unworthy, of the Company of Jesus].[25]

Second, Gumilla bolsters his authority not only with personal internal proofs, but also with external proofs designed to appeal to his readers' reason and convince them that this view of the Orinoco is both objective and truthful. By insisting on the fidelity of what so many authoritative eyes have already seen, this technique simultaneously affirms his own and outside sources' authority. Many of the external proofs he marshals to defend the truth value of his history are based on the findings of other religious men. However, Gumilla does not concentrate on the emotional effect of miraculous occurrences in the way so common to earlier Jesuit accounts. Instead, he focuses on verisimilar or probable demonstrations based on experience and observation and for comparison borrows from other mission accounts based on eyewitness experience, similitude, and support, reinforcing his own experiential claims with esteemed religious external authorities like Bishop Piedrahita of New Granada, the Franciscan father Pedro Simón, and the renowned Jesuit from New Spain, Pérez de Ribas: "De mi sentir son el ilustrísimo Piedrahita y el Reverendo Padre Franciscano Pedro Simón, aunque no se detienen ni dan más prueba que la experiencia, la cual, a mi ver, es la más fuerte. Del mismo parecer es el Padre Andrés Pérez de Roxas [*sic*]." (78). ["The illustrious Piedrahita and the most reverend Franciscan Father Pedro

Simón share my view, even though they do not hold back nor give more proof than their experience, which in my opinion is the strongest proof. Father Andrés Pérez de Roxas {*sic*} also agrees. . . .] Ralph Bauer has articulated "the cultural authority of the Enlightenment historiographer" based on eyewitness experience.[26] Gumilla reinforces this authenticity created by eyewitness experiences in the general historiography of the Enlightenment; but at the same time, he adapts it to a discourse of novelties grounded by his missionary ethos.

Third, Gumilla's rhetorical construction of ethos engages the long literary tradition of a discourse on novelties.[27] Along with affirming humility and citing others' authority, he combines this third technique with eyewitness authority. This discourse commences in the first pages of *El Orinoco ilustrado* when he draws upon traditional topics of invention, as well as an arrangement that heightens expectancy with its incrementum and a delightful style that promises everything new: "ideas nuevas [. . .] nuevas especies [. . .] todo es nuevo (33)" [new ideas . . . new species . . . everything is new]. Gumilla's enumeration of the rare and unusual is directed primarily at armchair travelers and impressionable Jesuit novitiates, and his "Prologue for understanding the work" invites these readers to join him in a journey from the coastline to see all that "son y siempre parecen nuevos" (33) [is and always seems new]. The journey extends from the ocean's "frecuente pesquería de perlas y de nunca vistas margaritas" (33) [frequent fishing of pearls and of never-before-seen pearls] to "los ríos formidables, por el inmenso caudal de sus aguas, por las diversas y jamás vistas especies de peces, por las arenas, ya de plata, ya de oro [. . .]" (33) [rivers formidable for their water's immense flow, for the diverse and never-before-seen species of fish, for the sands, filled with silver or filled with gold . . .]. He provides for his readers a virtual eyewitness experience by leading them away from the river and deep into the jungle and the forests, where they encounter at every step strange animals and remarkable birds and plants so different from those in the Old World:

> Ni causa menor novedad ver hermoseados los bosques y las selvas con árboles de muy diversas hojas, flores, frutos, poblados de fieras y animales de extrañas figuras, y de inauditas propiedades. [. . .] Y aun crece la novedad en cada paso de los que se dan, en las campañas, cuyos naturales frutos y frutas, en la figura, fragancia y suavidad al gusto, se diferencian tanto de los nuestros cuanto aquellos climas distan de éstos. (33)

> [It causes no less novelty to see the forests and jungles beautified with trees of diverse leaves, flowers, fruits, populated by wild beasts and animals with strange shapes and unheard-of properties. . . . And the novelty even increases with each of these steps taken into the countryside, whose native fruits are as

different from our own in shape, scent, and smooth taste as those climates differ from ours.]

As part of his ethos-building strategy, Gumilla allocates authority to the authors referenced in his notes as well as to his readers. He enumerates New World novelties and invites his readers to view them with their own eyes: "Esto sí es cierto, y se puede ver en los autores citados, si hay ojos para ver la verdad" (199–200) [This is definitely true, and it can be seen in the authors cited if there are eyes that can see the truth]. He also shares authority with his readers by encouraging them to believe everything they see, claiming that if had he not seen with his own eyes what he writes about, he himself would not believe it. Thanks to Gumilla's ability to paint the scene as if it were before his readers' eyes, both enjoy full license to believe: "Pues no cito testigos del otro mundo, en éste estoy yo, que refiero lo que he visto; y de no haberlo visto, ni lo creyera, ni lo tomara en boca" (144) [I do not cite witnesses from the other world {i.e., the Old World}; I am in this one {i.e., the New World} and I am reporting what I have seen; and if I had not seen it, I would neither believe it nor repeat it here]. In addition to asserting his eyewitness authority, Gumilla expresses great concern that mistrusting readers might suspect exaggeration, assuming simply because he presents so many new and unusual things "que este mi modo de hablar es hiperbólico" (174) [that this, my way of speaking, is hyperbolic]. Instead, they should have confidence in the virtual eyewitness status Gumilla creates for them. In short, Gumilla does not want the novelties in his textual wonder cabinet to be dismissed as fictional, or as full of the *fábulas* (myths) that had marred earlier accounts. To safeguard his truth claims against accusations of hyperbole, therefore, he asserts his missionary ethos and employs *prolepsis,* a rhetorical figure that prevents or anticipates objections readers might have in order to appeal to the readers' sense of authority:[28]

> Debo entre tanto prevenir a los que miran como fábulas las realidades del Mundo Nuevo, con la noticia cierta de que están muy bien correspondidos por otro gran número de americanos, que, con otra tanta impericia y ceguedad, miden con la misma vara torcida las noticias de la Europa. (32)

> [I must warn those who view New World realities as myths that, numerous though they may be, there are just as many Americans who, with equal inexperience and blindness, measure with the same twisted yardstick reports from Europe.]

In this quote Gumilla at first appears merely to concede that European readers who mistake the wonders of the Orinoco region for myths are not alone: many in America suffer this same "blindness" and "inexperience" when they view things European as wonderful and therefore fictional. In fact, he directs this apparent concession, which actually serves as a subtle criticism, at Europeans. He urges them to suspend their disbelief, lest they appear "blind" or more ignorant than Americans.

Finally, with a technique integral to his missionary ethos, Gumilla borrows from the authority of apostolic witnesses by tying *El Orinoco ilustrado* to the Gospels and positioning the actions of himself and other Jesuits in the New World in relation to the missionary journeys narrated in the Acts of the Apostles. This technique links Gumilla to the master Christian orator of the Bible, St. Paul, and his geographical narration of the triumphant progress of the Good News in the continuation to the Gospel of Luke.[29] Thus, in addition to engaging a traditional discourse of novelty, Gumilla maintains both the humble decorum of the missionary and the importance of Jesuit authority. As he protests in his prologue, his observations should be viewed as mere crumbs that have fallen from the "abundante mesa" (30) [abundant table] of the official chronicler of the Jesuit order in New Granada, José Casani. In fact, this humble understatement of his own natural history's worth next to Casani's "abundant Jesuit table" actually amplifies the value of *El Orinoco ilustrado* by echoing John 6:12 and therefore the actions of Jesus's first disciples, who gathered up the "fragments left over" after the miracle of the five loaves. His tying of his own narrative to the Gospels, his protestations of modesty (a humility topos), his citation of outside authorities, and his discourse on novelties authorized by his status as an eyewitness—all construct an ethos designed to appeal to the intellects of erudite readers and fellow Jesuits, as well as to general Christian readers.

Painting to the Mind's Eye

In contrast to his construction of ethos, Gumilla's second rhetorical strategy for amplification appeals primarily to the emotions of his readers. The technique of rhetorical painting moves his fellow Jesuits and Christian readers to praise God's glory and excites the imaginations of European or American armchair travelers and Spanish politicians by painting vivid word pictures of exotic jungles and gives enumerations of the abundance of nature in the king's valuable Orinoco possessions.[30] Gumilla's narrative illustration of the Orinoco follows the rhetorical prescriptions for vivid description referred to in manuals under various names such as *hypotyposis, ecphrasis, evidentia,* and *enargia.*[31]

Since classical times, the various rhetorical treatises of historians, poets, painters, and preachers have valued "vigorous ocular demonstration" and shared a repertoire of strategies that "set things before the eyes."[32] Great Christian orators evoked passions by preaching to the public with techniques from books of rhetoric and handbooks for Christian painters.[33] As successful preachers and rhetoricians, the Jesuits were particularly skilled at painting images with words.[34] Father Gumilla's rhetoric of wonder is rooted in the prescriptions and conventions of classical and sacred oratory that he acquired as a Jesuit seminarian at the Universidad Javeriana in Bogotá.[35]

A standard trope in sacred oratory to draw attention to a preacher's skills as a rhetorician was to compare a sermon's rhetorical description to paintings. Gumilla not only underlines his rhetorical prowess by referring to his textual descriptions as portraits, but also emphasizes his role as a Christian painter whose eyewitness status serves to further magnify his Jesuit authority. Both Jesuit preachers and painters of the seventeenth and eighteenth centuries celebrated the splendor, grandiosity, and emotional power achieved by following the prescriptions for hypotyposis.[36] From the first lines of his prologue to the last lines of his text, Gumilla likens his natural history writing to the act of painting. When he expresses his hopes to clear up the vision of his readers so that they might better appreciate the vivid colors of the novelties he paints, he underlines the importance of their role as virtual eyewitnesses:

> [Q]uisiera hallar algún colirio para aquellos que apenas ven, por más que abran los ojos; y se me ofrece, que para los tales no hay otro, sino ensancharles la pintura, añadir más vivezas a los colores y dar al pincel toda la valentía factible; de modo, que, vista con claridad la existencia innegable del Nuevo Mundo americano, vean que siendo nuevo aquel todo han de ser también nuevas las partes de que se compone; porque no sólo se llama Mundo Nuevo por su nuevo descubrimiento, sino también porque, comparado con este Mundo antiguo, aquél es del todo nuevo y en todo diverso. (33)

> [I would like to find some eyedrops for those who can barely see no matter how much they open their eyes; I would like nothing else than to enlarge the painting, add more liveliness to the colors, and give the painter all the bravery possible; so that, once the undeniable existence of the New American World is seen with clarity, they will see all that being new, so too must be new all the parts of which it is composed; because it is not called the New World only for its recent discovery, but also because, compared with this Old World, that one is new and different in every way.]

Thus, the tropes and figures that Gumilla marshals to paint with words serve complementary ends, from granting authority to and evoking passions in his readers to amplifying and elevating the status of both painters and portraits.[37]

Gumilla also humbly positions himself as an apprentice, reminding his readers that the highest praise always belongs to God, here evoked as the Master Painter who traced out "the first lines" of the world before perfecting it with colors and light.[38] In the last lines of his introduction, Gumilla starts tracing out his own first lines of an Orinoco portrait: "[A]sí comienzo a desenvolver el lienzo y a tirar en él las primeras líneas, abriendo paso franco al dilatado y ameno campo que para historiar sus grandezas ofrece el soberbio río Orinoco" (40) [And so I start unrolling the canvas and tracing out in it the first figures, opening up a free passage into the expansive and pleasant country whose grandeur the magnificent Orinoco River offers up for painting]. This quote uses the verb *historiar* in its contemporary sense—that is, "to paint or represent a historical or fictional event in portraits, engravings, or tapestries."[39] Gumilla employs *historiar* to evoke the action of painting, and understands *historia* as a noun meaning "portrait." In fact, his subtitle of *El Orinoco ilustrado*—namely, *Historia natural, civil y geographica*—echoes the panegyric function of Renaissance portraits.

At the end of his natural history, Gumilla once again amplifies his representation of the Orinoco region portrait through a *peroratio* (rhetorical closing) with a formulaic anticipation of potential criticism.[40] Here he once again makes use of the modesty topos and excuses the "rough canvas" of the historia through which he has provided a window to examine the wondrous display of God's works before humbly directing his readers to praise him:

> Y antes de retirar la pluma, me debo prometer de la benignidad y discreción del piadoso y prudente lector que disimulará los borrones que de ella se hubieren deslizado en el tosco lienzo de esta Historia, en la cual quisiera haber emulado con los rasgos las pinceladas de Apeles, mezclando con tal viveza los colores en la variedad del contexto, que a un mismo tiempo arrebatasen la vista para la honesta recreación, la atención para el aprovechamiento interior y al ánimo para alabar a Dios, siempre admirable en sus criaturas. (491)
>
> [And before withdrawing the pen, I must hope that the kindness and discretion of my pious and prudent reader will excuse the smudges that from it have slipped into the rough canvas of this portrait, in which I would like to have emulated with its features Apelles' brush strokes, mixing with such liveliness the colors in the variety of its context, brush strokes that at the same time capture the sight for honest recreation, the attention for interior benefit, and the soul to praise God, always wonderful in his creatures.]

Within these deferential final lines, however, Gumilla nonetheless compares himself to Apelles, the highly praised master painter given by Pliny as the epitome of greatness in the fourth century BC. By emulating both divine and human painters, Gumilla amplifies and elevates both divine and human art, both God's creation (his worthy subject) and man's (this portrait of the Orinoco River region).

Father Gumilla remains particularly aware of the connection between images and emotions while crafting his historia (portrait), and he employs Jesuit sermonic strategies primarily to appeal to his readers' sense of sight. Just like in a sermon, he clarifies points by painting before the mind's eye both static individual images and active scenes.[41] As we will see a little later in this chapter, Gumilla purposely stirs his readers' other senses by inviting them not only to view his Orinoco portrait but also to enter into it; he rhetorically sweeps them away into the abundant branches of the Orinoco through his syntax and lexicon. When he pauses to paint still-life portraits with words, however, he interrupts the steady pace that moves them swiftly through this region so vividly described. These still-lifes are essentially of three kinds: overarching vistas of the Jesuit province of New Granada from watchtowers that function as evangelical and political propaganda; the cornucopia of natural and cultivated abundance that underline commercial potential; and showcases of plants, animals, and mineral specimens that would also be highly prized in the cabinets of curiosities of armchair travelers or amateur scientists.

Gumilla's overarching vistas emphasize both the literal and figurative fruits waiting to be harvested by inviting readers up to view rhetorically re-created vast expanses from the high perspective of *atalayas* (watchtowers). For example, in a chapter entitled "Resumen de los genios y usos de las demás naciones que hasta el corriente año de 1740 se han descubierto en el río Orinoco" [Summary of the character and customs of the rest of the nations that up to current year of 1740 have been discovered along the Orinoco River] Gumilla emphasizes recent Jesuit successes with peaceful tribes and then directs his readers' gaze up the river to a region filled with several "gentle nations" ready for and awaiting the Catholic faith, if only there were missionaries to cultivate their souls: "gente dócil y tratable, y que recibe bien la santa doctrina. [. . .] Y siguiendo el río agua arriba, viven a sus márgenes varias capitanías de salivas, la gente aturi, los quirrubas, maipures y abaners. Todas son naciones benignas y prontas a recibir la fe, y solo faltan operarios, que la mies madura está ya" (202) [docile and friendly people receive well the holy doctrine. . . . And continuing up the river, various captaincies of Sálivas, the Aturi peoples, the Quirrubas, Maypoyes, and Abaners live on its banks. They are all kind nations ready to receive the faith, and all that is missing are workmen, for the harvest is already ripe]. The chapter opens with Gumilla and his readers climbing up a giant boulder in the shape of a pyramid:

> [S]e levanta en forma de pirámide uno de los más vistosos obeliscos que ha criado la Naturaleza. Tiene su firme base algo más de media legua de circuito, y estribando sobre sí mismo, se levanta la peña, toda de una pieza, a una altura maravillosa: sólo por dos ángulos se permite paso a su cumbre, y para poder subir sin sobresalto de bajar precipitados, es preciso desnudar los pies de todo calzado; vamos subiendo, que en esta elevada cumbre, llamada Pararuma, que más parece idea del arte, concebida en la más amena fantasía, que roca natural. [. . .] (200)
>
> [Rising up in the shape of a pyramid is one of the most spectacular obelisks nature has created. Its firm base is a little bit more than a half league in circumference, and supported by itself, the huge boulder rises up, all in one piece, to a wonderful height: only at its two corners is there a way to its peak, and in order to be able to climb without suddenly falling down, it is necessary to take off all footwear; let's head up, for on this elevated peak, called Pararuma, which seems more like an artistic idea, conceived in the most pleasing fantasy, than natural rock. . . .]

From this natural watchtower, Gumilla fashions a landscape portrait that displays both the literal fruits of the earth as well as the defensive benefits of viewing the river from such a high point:

> En este terreno tienen los salivas una hermosa huerta, siempre fresca. [. . .] Aquí hay plátanos, piñas y las demás frutos de la tierra; pero lo mejor que tiene para nuestro intento es una fresca y amena arboleda silvestre, que han reservado los salivas para lograr el fresco, así de su sombra como del ambiente que en tal altura jamás falta, y para observar desde aquella eminencia las embarcaciones enemigas que suben río arriba. Tomemos aquí nuestros asientos, y a todo placer, y sin dudar un paso, vamos registrando con la vista terrenos poblados de gentiles y de cristianos nuevos, tantos cuantos no pudiéramos visitar en muchas semanas de camino. Al Oriente y al Sur pondremos las espaldas, porque por estos dos vientos se halla atajada la curiosidad con la fragosa serranía que, acompañando al Orinoco desde su primer origen, corre hasta sepultarse con él en el Océano; pero al Norte y al Poniente no hay altura que estorbe la vista, hasta que fatigada se da por vencida entre el cielo y el inmenso llano, uniéndose al parecer uno y otro, para formar el horizonte, nada menos distante que el que registra en alta mar la vista más lince desde el tope del navío (201–2)
>
> [In this patch the Sáliva tribe has a beautiful orchard, always fresh. . . . Here there are bananas, pineapples, and the rest of the fruits of the earth; but the best it has for our purposes is a fresh and agreeable wild grove, which the Sálivas have set aside to get cool air from both its shade and the environment that is never lacking at such heights, and also to observe from that loftiness the enemy vessels that come upriver. Let's take our seats here, and with complete pleasure and without hesitating one step, let's start scanning with our

eyes terrains populated with pagans and new Christians, so many that we could not visit them all by walking for many weeks. We will put our backs to the east and the south, because by these two winds our curiosity finds itself intercepted by the dense hill country that, accompanying the Orinoco from its first origins, runs until burying itself with it in the ocean; but to the north and the west there are no heights that obstruct our view, until wearied, it gives itself as defeated between the sky and the immense plains, which seem to join together to form the horizon, no less distant than that which the most sharp-eyed vision can record from the masthead of a ship.]

Next Gumilla directs his readers' sight toward even more spectacular cliffs made into a fortress by the hands of Jesuits, soldiers, and Amerindians working together. He describes this newly constructed fort and mission of San Francisco Javier and the "invincible" protection it offers both missionaries and their charges:

> En este mismo lado del Sur, donde estamos, siguiendo agua arriba el Orinoco, hallamos otra peña más singular que esta sobre que estamos [. . .] tiene difícil y única subida, y ha de ser a pie descalzo, por su parte oriental. Desde su cumbre hasta dar en el espacioso plan [. . .] en este balcón o plan que ofrece la disforme peña formaron los misioneros una fuerza con tres baterías, cuarteles y casas para una parcialidad de indios salivas, que se han agregado a dicha fuerza. Esta fue más dirigida de la urgente necesidad que del arte, y fabricada por mano de los mismos Padres misioneros, soldados e indios, contra las continuas invasiones de los bárbaros caribes, año de 1736, con tan feliz éxito, que desde que la vieron ningún armamento de ellos se atrevió a llegar; y aunque llegan, es totalmente invencible. [. . .] Con dicha fuerza hemos resguardado gran parte de las Misiones. . . . Llámase esta fuerza y pueblo de San Francisco Javier, la cual, con la casa fuerte enfrente, cierra totalmente el paso al enemigo. (201–2)

> [On this same southern side, where we are, continuing up the Orinoco, we find another cliff even more outstanding than this one on which we are . . . it has a difficult and unique slope, and it must be done barefoot on the eastern side. From its peak until giving way to its spacious level . . . in this level or balcony that the strangely proportioned cliff offers, the missionaries built a fort with three batteries, barracks, and houses for a parcel of Sáliva Indians who had gathered by the fort. This fort was shaped more by urgent necessity than art and was constructed by hand by the same missionary fathers, soldiers, and Indians against the continued invasions by the barbarous Carib tribe, in 1736, with such success that ever since the Caribs saw the fort not one of their armaments has dared to approach; and even if they did approach, it is totally invincible. . . . Thanks to said fort we have shielded a great part of the missions. . . . This fort and mission settlement are named after San Francisco Javier, and with the strong house in front it completely closes the pass off from the enemy.]

Finally, after finishing his portrait of the successful fruits of Jesuit labor (namely, the fort and mission of San Francisco Javier), Gumilla directs his readers' gaze back out at the expansive region, here represented as a *locus amoenus:* "Ahora volvamos la vista a los dilatados llanos [. . .] que interrumpidos con muchos ríos, vegas y bosques, forman un bello país, siempre ameno y verde" (202) [Now let us return our gaze to the expansive plains . . . that, broken up with many rivers, meadows, and forests, form a beautiful country, always green and pleasant].

From high up on watchtowers wrought from both divine and human art, Gumilla paints pictures bursting with natural beauty and bounty, as well as composes panegyrics of evangelical successes. Despite recent horrors of Carib attacks on Jesuit missions, here Gumilla's overarching vistas downplay past challenges to Orinoco region settlements and instead focus on its potential for both commerce and conversion. By focusing on plentiful harvests of native fruits and souls, Gumilla hopes to be particularly persuasive, given that Jesuits had many appealing and well-established options for their missionary careers. Young novitiates could potentially join successful projects everywhere from India to Paraguay without risking their lives along the Orinoco. As we will see clearly in the next chapter, Gumilla literally and figuratively maps out the Jesuit province to influence fellow missionaries and make subtle appeals to both Jesuit superiors and Spanish politicians who might provide funding for the military and the evangelical support necessary to maintain this dominion.

Gumilla's still-life cornucopia portraits likewise serve political and evangelical functions. Gumilla often praises the fecundity of the Orinoco River region, and toward the end of his natural history he climbs down off his watchtower to dedicate several chapters to its natural abundance of fruits and vegetables. One example of Gumilla's appropriation of the Renaissance commonplace of a cornucopia appears in a section that first praises the wild fruits of the region and then extols the civilizing effects of agriculture and the benefits of the technological advances introduced by Jesuits. While the lush jungle already bursts with flora, planting reforms have served to maximize the Orinoco's natural abundance:

> Cuando siembran el maíz, ya la yuca lleva cuarta de retoño, y entre una y otra mata de yuca siembran una mata de maíz; entre la yuca y el maíz siembran batatas, chacos, calabazas, melones y otras muchas cosas. [. . .] Cogida la primera cosecha de todos los frutos dichos, siembran segunda vez los mismos, y antes de cogerlos van interponiendo retoños de plátano [. . .] el fruto más duradero y más útil de cuantos los indios siembran. (435)

> [When they sow the corn, the yucca is already one-fourth sprouted, and between each shrub of yucca they sow a shrub of corn; between the yucca and

> corn they sow potatoes, *chacos*, gourds, melons, and many other things. . . . Once the first harvest of all the aforementioned fruits is picked, they sow the same ones a second time, and before picking those they go about interposing banana shoots . . . the most durable and useful fruit of all that they Indians sow.]

Gumilla dedicates several pages of hyperbolic praise to the all-providing banana groves and the multifaceted uses of this "árbol de la vida" (437) [tree of life] provided by "la liberal providencia del Criador" (436) [generous providence of the Creator] but made to profit by the wisdom of the Jesuits, who understand that quickly planting banana groves serves as one of the most valuable civilizing forces in mission settlement: "[N]o hay fruta más sana en las Américas ni tan sustancial ni tan sabrosa. [. . .] En fin, los plátanos son el socorro de todo pobre; en la América sirven de pan, de vianda, de bebida, de conserva y de todo, porque quitan a todos el hambre" (436). [There is no healthier nor substantial and delicious fruit in the Americas. . . . In short, bananas are the relief of all the poor; in America bananas serve as bread, viands, drink, preserves, and everything, because they take away hunger from all.] He then concludes with the fragrant, delicious, and prolific American pineapple, another wonder of nature; its crowns are replanted once cut and "sin perder de su verdor prende y resulta otra mata de piñas, y de cada mata se siembran tantas cuantas piñas dio, que son muchas; y así es grande la abundancia de esta rica y saludable fruta" (438–39) [without losing any of their lushness take root and provide another shrub of pineapples, and from each shrub are sowed as many pineapples as it gives, which are many; and so great is the abundance of this rich and healthy fruit].

Along with *atalayas* and the cornucopia of figurative and literal harvests, Gumilla highlights several items recognizable as standard for natural history cabinets. Static descriptions throughout *El Orinoco ilustrado* catalogue curious and unknown natural productions. These word paintings offer glimpses at, for example, the natural armor of the armadillo, the medicinal value of iguana stones, and the black-and-white stripes of the skunk. Gumilla paints his most prized picture based on his firsthand experience with Orinoco caimans and crocodiles, these four-footed "dragons" whose fangs are, in his estimation, more valuable than unicorn horns (427). The following portrait of the crocodile shares elements with natural history texts both ancient and contemporary. However, Gumilla's rhetoric of wonder capitalizes on the spectacular and horrifying in order to thrill both armchair travelers and gentleman naturalists:

> No puede idear la más viva fantasía una *pintura* más propia del demonio que *retratándolo* con todas sus señales. Aquella trompa feroz y berrugosa [*sic*], toda

negra y de duro hueso, con quijadas, que las he medido, de cuatro palmos, y algunas algo más; aquel laberinto de muellas, duplicadas las filas arriba y abajo, y tantas, no sé si diga navajas aceradas, dientes o colmillos; aquellos ojos, resaltados del casco, perspicaces y maliciosos, con tal maña, que sumida toda la corpulenta bestia bajo del agua, saca únicamente la superficie de ellos para registrarlo todo sin ser visto, aquel dragón de cuatro pies horribles, espantoso en tierra y formidable en el agua. [. . .] (419; emphasis mine)

[The most lively imagination cannot imagine a *portrait* of the demon more fitting than *portraying it* with all its signs. That fierce and warty snout, all black and hard bone, with jawbones, which I have measured myself at four palms length, and some even more; that labyrinth of teeth, both top and bottom equally sharp and of such quantity that I don't know whether to say steely-sharp razors, molars, or fangs; those eyes, jutting out shrewdly and maliciously from his skull with such cunning that even when his great beastly body lies under the water, the eyes can break the surface and record everything without ever being seen, that dragon of four horrible feet, frightening on land and dreadful in water. . . .]

Crocodiles had been privileged in New World iconography long before Gumilla's vivid portrait. In Europe stuffed specimens were prized in the most curious cabinets of natural history and even hung from the ceilings of cathedrals. But Gumilla's portrait supersedes the biblical dragon of four feet to bring the Orinoco caiman to life before his readers' eyes. The answer to his opening rhetorical question imagines a caiman's defining moment in the "mirror of nature":

¿Qué definición se podrá hallar que adecuadamente comprenda la fealdad espantosa del caimán? Él es la ferocidad misma y el aborto tosco de la mayor monstrosidad, horror de todo viviente; tan formidable, que si el caimán se mirara en un espejo, huyera temblando de sí mismo. (419)

[What definition can be found that can adequately convey the fearful ugliness of the caiman? He is ferocity itself, a crude freak of the greatest monstrosity, a horror to all living creatures, so fearsome that if the caiman could see himself in a mirror, he would flee trembling from the sight of himself!]

From these first lines of his portrait, Gumilla demonstrates his awareness of wonder's dual propensity for both dread and delight by providing a terrifying yet fascinating spectacle of nature. He will eventually turn away from such strange and novel fauna to invite his readers to examine human artifacts on display in his textual cabinet—"Pero dejemos estas curiosidades de los animales para reír y llorar otras en los indios y en otras gentes" (457) [But let's leave behind these curiosities of the animals to laugh and cry about other curiosities of the Indians and other peoples]—but Gumilla never backs away from his strategy of collecting Orinoco wonders (represented through striking images) for his rhetorically constructed cabinet of curiosities. He fills the chapters of his textual natural history cabinet with curious objects described in a series of word pictures painted with rhetorical techniques for vivid description. From the opening to the closing pages of *El Orinoco ilustrado* Gumilla rhetorically establishes himself as a Christian painter who imitates God's creation within his text and decorates the walls of this narrative/gallery with static word pictures both delightful and horrifying that capture the readers' imagination and direct their wonder at God.[42]

Appeals to All the Senses

However, Gumilla's Orinoco portrait contains much more than static images.[43] While his textual illustrations re-create with words panoramic landscapes and isolate distinct objects of divine and human art, they also include action-filled scenes. Thus, in addition to appealing to the mind's eye, Gumilla also appeals to his readers' other four senses. In his descriptions of his own walks through the Orinoco forests, Gumilla appeals to his readers' sense of smell. For example, these fragrant forests have often overcome his senses: "[A] mí me ha sucedido muchas veces quedarme absorto en medio de aquellos bosques y embargado el movimiento de una tal fragancia y suavidad de olores exquisitos, que no hallo con qué explicarme" (*El Orinoco ilustrado,* 213) [Many times I have found myself amazed in the middle of the forest, overcome by such smooth and exquisite fragrances that I am frozen and unable to find a way to account for it]. In this passage Gumilla first details the fragrances that have left him in a state of ecstasy, suspending his faculties of speech and movement. He then amplifies this anecdote by relating a typical, yet fictional, exchange with the Amerindians who accompany him on such walks: "Preguntaba entonces a los indios compañeros de dónde salía aquel bellísimo olor, y la respuesta era: *¿Odi já, Babi? ¿Quién sabe, Padre?*" (213) [Then I asked my Indian companions where that incredibly beautiful smell was coming from, and their response was: *Odi já, Babi? Who knows, Father?*]. He introduces dialogue with fictitious persons (*prosopopoeia*) within his natural history to narrate more vividly his own wonder experience. When Gumilla puts a rhetorical question in the mouths of his companions ("Who knows, Father?") to confirm his own awestruck description of the beautiful aroma, his use of invented dialogue (*sermocinatio*) intensifies the description of natural wonders and the story itself.[44]

With anecdotes like this, Gumilla magnifies his own and others' wonder. Here the new Christian Amerindians' speech seems to evoke Ecclesiastes 3:14, "God has done this, so that all should stand in awe before him," a subtle reminder that while sometimes humans can explain wonder-provoking natural phenomena, at other times we must direct our wonder to God's incomprehensible majesty. Throughout *El Orinoco ilustrado* Gumilla first searches for and attains knowledge about his physical surroundings but then properly directs man's wonder to God. In this way he shows his piety to church leaders, affirming his respect for Catholic orthodoxy. Still, his re-creations of his new converts' speech function not only to piously remind readers that man cannot always know the causes and therefore should at times simply "stand in awe before" the works of the Creator; they also speak to successful Jesuit evangelization in the Orinoco. Thus, apart from inviting his readers to share in his wonderment, anecdotes like this reinforce Gumilla's missionary authority.

At the same time Gumilla shares wonderful scents with his readers, he also evokes sounds and colors. In fact, Gumilla showcases New World wonders in ways that had been standard since Columbus's diaries. For instance, he evokes the by then long-standing tradition of imitating the Genovese explorer's sense-provoking rhetorical strategies, evoking the "marvelous scents" from "thousands of trees and their fruits," a standard in the earthly paradise trope.[45] Gumilla elevates the Orinoco region by representing it as a tropical paradise; Columbus had portrayed it as possibly the site of the Garden of Eden.[46] He further engages Columbus's strategies by embellishing his natural history descriptions with another key trope of the tropics: wild, loud, and colorful birds.[47] When Gumilla sings praises of the Orinoco's curious birds of paradise sporting the liveliest colors, he clearly evokes Columbus's "thousands of birds so diverse from ours that it is a marvel"[48] decorating the forest: "y hermoseados y aun matizados de aves singularísimas en sí, en la variedad de sus vivísimos colores y en la gallardía de sus rizados plumajes" (33) [embellished and even tinted by the most outstanding birds in the variety of their liveliest of colors and in the gallantry of their ruffled elegant plumage]. The "gallantry" ("gallardía") of these birds "ruffled feathers" also ingratiates them to persons of noble class, and the superlatives "most lively and most outstanding" ("vivísimos" and "singularísimos") that accompany this gracefulness communicate the rareness of the birds. Gumilla echoes Columbus when he writes that these are unlike any birds seen in Europe, and he suggests to his readers that they will only find birds of such colors and shapes in the Orinoco River region:

> papagayos, loros, guacamayos, patos de varias especies, cigüeñas y garzas, grandes y pequeñas, y otras muchas aves, que es para alabar al Criador, así por la multitud, como por lo exquisito de sus plumas, matizadas de vivísimos colores, y principalmente por sus especiales figuras. (449)

> [parrots, macaws, various species of ducks, storks, and herons, large and small, and so many other birds that one should praise their Creator above all for their special shapes but also for both their multitude and the exquisiteness of their feathers, which are blended with vibrant colors.]

Not only the quantity—Gumilla enumerates numerous species here—but also the brilliant feathers and the singular shapes of these birds set them apart. This dual emphasis on quantity and quality underlines the Orinoco as a providentially privileged site, and Gumilla repeats it when praising other Orinoco animals, minerals, and plants, as with his treatment of the palm and banana trees that provide the region with, in his words, limitless manna sent from God.

These echoes of the tradition of marvelous description launched nearly three hundred years before please the well-read intellectual and invite the general reader to discover the Orinoco region for himself or herself, thus appealing to both his readers' thought and feeling. They also provide clear examples of Gumilla's rhetorical stimulation of his readers' senses and bring to light the commanding rhetoric of wonder with which he takes his readers on a virtual journey along the Orinoco. Using a careful combination of tropes and figures, throughout his natural history narrative Gumilla creates a kinesthetic effect of sweeping his readers along the river, in and out of danger, and up over the riverbanks into fragrant forests filled with multicolored birds, all-providing palms, and abundant natural resources. A brief analysis of the Jesuit's literary strategies will demonstrate how his rhetoric of wonder effectively plunges them into a thrilling expedition along the powerful Orinoco River and serves up a feast that seduces their senses.

One core rhetorical strategy for amplification that creates this kinesthetic effect and surrounds his readers with natural abundance is the heaping up of topics in single, lengthy passages. In the following example, the parallelism set up by the phrases and clauses separated by commas and colons (*commata* and *cola*) are indicative of the periodic style that orators from classical to contemporary times have employed to achieve grandeur or awesomeness. Additionally, Gumilla's use of semicolons makes clauses that could easily have been independent instead depend on the reader continuing along the rest of the extended sentence:

> [C]oncatenaré las cosas singulares que observé y noté acerca de las aves, animales, insectos, árboles, resinas, hierbas, hojas y raíces; demarcaré también la situación del Orinoco y de sus vertientes; apuntaré el caudal de sus aguas, la abundante variedad de sus peces, la fertilidad de sus vegas y el modo rústico de cultivarlas; hablaré (con alguna novedad) del temperamento de aquellos climas, de los usos y costumbres de aquellas naciones; daré mi parecer en

> algunas *curiosas y útiles* disertaciones; y, por último, insinuaré de paso algo de lo que fructifica en aquellas almas la luz del Cielo por medio de los operarios, no sólo de la Compañía de Jesús, sino también de otras esclarecidas Religiones; en cuya confirmación referiré no pocos casos singulares, todo el cual conjunto y agregado de noticias dará motivo para que el gran río Orinoco, hasta ahora casi desconocido, renazca en este libro con el renombre de *ilustrado,* no por el lustre que de nuevo adquiere, sino por el caos del olvido de que sale a la luz pública. (31; emphasis mine)
>
> [I will link together the outstanding things that I observed and noted about the birds, animals, insects, trees, resins, herbs, leaves, and roots; I will also demarcate the location of the Orinoco and its branches; I will show the flow/wealth of its waters, the abundant variety of its fish, the fertility of its meadows, and the rustic means of cultivating them; I will speak (with some novelty) about the nature of those climates, and about the usages and customs of those nations; I will give my opinion in some *curious and useful* dissertations; and, finally, I will hint at in passing some of what the light of the heavens has yielded in those souls thanks to the workers, not only from the Company of Jesus, but also from other ennobled orders; for whose confirmation I will recount not a few very singular cases, all of which together and joined with knowledge will provide the reason for which the great River Orinoco, up until now almost completely unknown, will be reborn in this book with the distinguished fame of *enlightened,* not for the luster it again acquires but instead for the chaos of oblivion from which it departs toward the public light.]

The above passage consists of one lengthy, splendid sentence. As long as this quote is, however, it does not contain the entire periodic sentence that Gumilla crafted, a structure meant to foreshadow the abundance of the Orinoco region and its topics of natural history. And Gumilla's division of his separate tasks as historian into an orderly enumeration of topics further extends the passage, making it fuller. For this he employs a series of verbs in the future tense (*concatenaré, demarcaré, apuntaré, hablaré, daré, insinuaré, referiré*) that expand the passage further. Gumilla blends the repetition of first-person future tense endings (*-aré*) with the first-person preterite-tense ending (*-é*) to make use of the rhetorical figure *homoioptoton* (also known as *similiter cadens*), which creates a rhythm that is both beautiful and strong. In this way, Gumilla purposely creates the effect that he and his readers are being swept away by a wave.

Gumilla also establishes a parallel between the river's force and the actions of his pen. His natural history text will rush forth to reflect the river's natural force, and his *pluma* (pen, feather) will paint its abundance, "unas veces ande y otras veces corra" (31) [sometimes walking and sometimes running]. Here again Gumilla elevates his style with careful use of rhetorical figures for amplification, including a crescendo. Erudite readers would have recognized his explicit use of *syncrisis,* or comparison in parallel clauses, use of anaphora with the repetition of *ya* (now, or already) as well as polysyndeton in the repetition of *con . . . para* structures (with . . . in order to) that amplify through the repetition or rephrasing of closely related ideas (*synonymia*) and the qualification of a statement by recalling it (*metanoia*). But they also would not have been able to avoid being swept away by the abundant textual rhythm that flows like unlimited water:

> [P]rocuraré que mi pluma unas *veces ande y otras veces corra* al paso del río Orinoco, cuyas vertientes sigue. [. . .] *Ya* aplica sus caudales a enriquecer y fecundar sus deliciosas vegas; *ya* los explaya en anchurosos lagos; y *ya* con furia los aparta, destrozados del duro choque de incontrastables rocas: variedad natural, que, si hermosea el flujo natural del caudaloso Orinoco, debe dar el ser y la hermosura a la Historia Natural, que el mismo río nos ofrece *con amena variedad, para evitar el fastidio,* y *con novedad, para conciliar la atención.* (31–32; emphasis mine)

> [I will strive to make my pen *sometimes walk and other times run* along the passage of the Orinoco River, whose edges it follows. . . . *Now* its abundant flow enriches and fertilizes its delicious water meadows; *now* it extends this abundance into wide lakes; and *now* it parts them with fury, broken up by its cruel crash into unyielding rocks: natural variety that, just as it beautifies the natural flow of the abundant Orinoco, also provides the being and the beauty to the natural history that this very same river offers us *with pleasant variety to avoid boredom and with novelty to capture our attention.*]

Gumilla rushes his readers forward by nurturing the River Orinoco's natural beauty (its delicious water meadows) and thrilling them with its potential danger (the river's fury and direct impact on unyielding rocks). At the same time, Gumilla draws attention away from this skilled and deliberate rhetoric by reassuring the readers that he is simply reproducing in narrative form a portrait (*historia*) of the beauty, variety, and magnificence that exist naturally.[49] Insisting that the Orinoco region itself gives life, beauty, and abundance to his natural history, Gumilla minimizes the rhetorical skills that his educated readers would have noted. While the textual force (or awesomeness) of his rhetorical evocation of the natural force of the River Orinoco demonstrates erudite rhetorical strategies, it clearly stirs the senses of courtly and popular readers, as well as of young Jesuit novitiates.

The First-Person Plural for Pathos and Logos

Gumilla's final important rhetorical strategy is a pervasive, multifaceted use of first-person-plural verb forms. Some of the pathos-driven examples above, like the commands that direct "our" gaze out toward the plains and the mission settlements, or the one that instructed "us" to leave behind animal curiosities in order to laugh and cry at human ones, generate wonder first through the engaging grammatical "we" and then through the vivid descriptions that follow.[50] Other explicit invitations for "us" to join him while he re-creates his Orinoco journey equally involve the mind as well as the senses. In fact, Gumilla uses the first-person plural to accomplish three main goals. First, as a transition between chapters or topics, it functions to increase his readers' expectancy and delight. Second, it invites his diverse readership to actively experience both the Orinoco region cornucopia and natural-history cabinet curiosities. Finally, it functions, in certain instances, to construct a more specific "we" in order to include and build upon a wider community of Jesuits that consists of fellow experienced missionaries, scholars, and young novitiates.

Gumilla almost always closes chapters in his textual natural history cabinet with the promise to open others full of new wonders to behold and experience. His choice to make these transitions through first-person-plural verb forms invites his readers along. At times these pauses to directly involve his readers provide the space necessary to awaken their senses to God's glory, as when Gumilla guides "us" toward close observation of beneficial roots and herbs that provide natural remedies:

> Ahora será bien que salgamos hacia las sementeras de los indios a ver otros árboles frutales, y de paso observaremos varias hierbas y raíces muy medicinales y provechosas, tanto, que excitan mudamente a que alabemos al sabio y próvido Criador del universo. (439–40)
>
> [Now it would be good if we headed toward the Indians' cultivated fields to see other fruit-bearing trees, and along the way we will observe various herbs and roots that are very medicinal and beneficial, so much so that they silently excite us to praise the wise and generous Creator of the universe.]

At other times when Gumilla pauses his narrative to use "we," he clearly means to create suspense. Sometimes he purposely fosters anticipation of what is to follow, and sometimes he intends for "us" to fully appreciate the novelty of what he has just related. One particularly memorable anecdote that demonstrates both functions of the first-person plural is when Gumilla invites his readers to turn "our" eyes upon the Orinoco lagoons and examine the abundance of fish species, a topic that is certain to hold "our curiosity" a good while:

> [V]olvamos los ojos a esos dilatados placeres del Orinoco, y a esa inmensidad de extendidas lagunas, en que divierte sus aguas cuando crece; y a buen seguro que al ir registrando la multitud, variedad y propiedades de tan innu-

> merables especies de peces, como engendra y mantiene el Orinoco en sus vivares, y al ver y reparar las mañosas industrias con que los indios los engañan y pescan, tenga un buen rato nuestra curiosidad. [. . .] (219–20)
>
> [Let us return our eyes to those vast pleasures of the Orinoco, and to that immensity of extended lagoons in which its waters amuse as they grow; and it is quite certain that as we go about examining the multitude, variety, and properties of species of fish too numerous to count that the Orinoco bears and supports in its waters, that as we see and notice the clever skills with which the Indians trick and catch them, our curiosity will be held for a good while. . . .]

The suspense grows along with our curiosity when Gumilla commands "us" to view from a summit an "amusing" wonder never before seen or perhaps even imagined:

> A bien que la cumbre en que estamos, por su altura, amenidad y buena sombra nos convida a divertir la vista. Y así reparen y verán en aquella ensenada [. . .] *la cosa más curiosa, y el modo más raro* de pescar que puede imaginarse; porque [. . .] muchas especies de pescado mediano *saltan de suyo con tanta abundancia en las canoas,* que a no remar con tanta fuerza, y a no navegar con tanta velocidad, hundiera las canoas la multitud que salta en ellas. [. . .] (221–22; emphasis mine)
>
> [The high, pleasing, and well-shaded summit on which we stand well invites us to amuse our sight. So take a look and you will see in that small cove . . . *the most curious and the most strange way* to fish that you can imagine; because . . . many species of fish *jump by their own accord into the canoes with such abundance* that if the Indians didn't row with such force or sail with such velocity, the multitude of fish that jump into their canoes would sink them. . . .]

This example demonstrates how Gumilla actively involves his readers in experiencing an almost dangerous abundance of fish that jump right into a canoe. Even though his narration of such an amazing ease of fishing might make this anecdote seem fabulous to his readers, Gumilla's use of the first-person plural appeals their senses and intellects. With the last "we" of this anecdote, Gumilla underscores how hard it is to express that which even those who can see and touch for themselves find difficult to believe. He poses a rhetorical question that frees the readers from doubt:

> No es ponderable, ni cabe en la pluma expresar la multitud de peces [. . .] de modo que la abundancia de pescado y tortugas del Orinoco apenas es creíble a los que la ven y tocan con sus manos. ¿Qué diremos de los que esto leyeron? (223)

[It can't be exaggerated, nor can the pen express, the multitude of fish . . . just as the abundance of Orinoco fish and turtles is barely believable to those who actually see it and touch it with their hands. What can we ever say with regard to those who only read about it?]

Gumilla's self-conscious references to the difficulty of his task (that is, of making readers believe this narration by feeling as if they were experiencing it alongside him) serve to underline both the wonder of such Orinoco abundance and his rhetorical skills.[51]

In general, when Gumilla combines first-person-plural commands alongside vivid descriptions of both natural history curiosities and potentially profitable flora and fauna, he invites diverse readers—from fellow missionaries and naturalists to general Christian readers and merchants—to join in an evangelical and economic journey along the Orinoco. Gumilla makes possible wonder-provoking virtual-eyewitness experiences for any reader. Still, when the narrative shifts to the final use of first-person-plural verbs the invitation seems directed more at his Jesuit audience. In this final example the "we" functions as a transition from the spectacle of nature embodied in the bumper crop of Orinoco turtles, their wonderful cultivation, and the "singular oil" harvested from their eggs (a reference to economic potential) to a section on "the most feasible method for a missionary's first encounter with the non-Christians." Gumilla marks time to ease his readers down from yet another spectacular high-viewpoint watchtower and leads them back to the current realities and future potential of the evangelical mission:

Ni a nosotros nos resta ya luz del día, sino para bajar a la Misión de que salimos. Vamos por ese otro lado, que, aunque es más larga, es menos pendiente la bajada. Los Padres misioneros ya nos estarán esperando; allá proseguiremos con nuestros discursos más despacio, y trataremos puntos y materias más curiosas y de mayor importancia. (236)

[We now have only enough daylight to go down to the mission from which we left. Let's go down the other side, where, even though it is longer, the slope is less inclined. The missionary fathers will already be waiting for us; over there we will proceed with our discourse more slowly, and we will treat points and subjects more curious and of greater importance.]

While Gumilla will soon acknowledge the recent crises fueled by Dutch support of Carib slave raiders and the need to increase evangelical and political support of the Orinoco mission project, up to now he has downplayed the need for Orinoco region defense. Still, this transition to a chapter on the "practical methods" for entering into new territory initiates a propaganda strategy whereby Gumilla will appeal to young Jesuits' sense of adventure, evoke an existing Jesuit community in the Orinoco, and promise great success despite past challenges.

To magnify both evangelical and economic possibilities along the Orinoco, Gumilla relies on his missionary integrity and appeals to his readers' intellects, senses, and emotions. Although the Jesuit's techniques for rhetorical amplification could be found in abundance within classic and contemporary guides for historians, poets, painters, and preachers, Gumilla employs them in novel ways. At the same time that his rhetorical construction of authority stimulates readers' minds, his recourse to painting with words stimulates their emotions. By vividly displaying the region's cornucopia, novelties, and the force of its river, Gumilla engages his readers' senses of sight, smell, hearing, taste, and touch. He also uses first-person-plural verb forms to capture their intellects and emotions. In fact, his combination of several rhetorical strategies evokes wonder and appeals to reason, creating an unusually persuasive blend of ethos, logos, and pathos.

Enlightenment studies have traditionally assigned wonder a limited role. However, the incredibly wide appeal of Gumilla's text for both Catholics and secular humanists owes much to his appropriation of wonder, an ancient rhetorical strategy deftly employed to lead his readers on a journey from wonder to knowledge about the evangelical, scientific, and commercial possibilities within the Orinoco River region, thus persuading them of the value and use of its abundant human and natural resources for both spiritual and intellectual goals. I contend that Gumilla developed this rhetoric of wonder in order to reach a wide variety of readers who bring diverse approaches to interpreting his natural history. To showcase New World wonders he consciously constructs a textual natural history cabinet for everyone from armchair travelers, collectors of curiosities, amateur scientists, merchants, wealthy nobles, and politicians to Jesuit novitiates, provincial leaders, and even the pope or the king. *El Orinoco ilustrado* resists culturally constructed expectations that place texts of the Enlightenment within the conventional paradigm of moderation, reason, and emotional control. Its departure from these received limitations provides a model for the transitional period of the first part of the eighteenth century whose study can help us understand how the Hispanic Enlightenment reconciled the Catholic and the secular and united emotions and reason in the accumulation, organization, and dissemination of knowledge.

2

Colonization, Commerce, and Conversions: Mapping the Economic and Evangelical Values of the Orinoco

THE SIXTEENTH AND SEVENTEENTH CENTURIES ARE OFTEN CALLED the Age of Discovery, when powerful European monarchs first claimed and mapped their overseas territories. Yet the eighteenth and nineteenth centuries included another key period of discovery that produced many detailed maps. This Age of Enlightenment is known for its classification of knowledge and can be divided into two phases. First, in the eighteenth century, Spanish, Portuguese, French, Dutch, and British kings dispatched scientific voyages to expand their jurisdictions, to exploit natural resources, and to maintain control over these territories.[1] During the second phase, which reached into the nineteenth-century nation-building period, scientific expeditions in South America thoroughly illustrated territories with maps, drawings, and textual descriptions highlighting animal, mineral, vegetable, and human resources. Studies of the Enlightenment have privileged topics from the second phase, such as science in service of the foundation of new republics or the institutionalization of scientific branches like "applied economic botany," and have tended to divorce scientific and commercial interests from spiritual motivations.[2] Consequently, reports from voyages that combine both economic and evangelical missions are not typically studied as Enlightenment texts. The dual nature of Spain and Spanish America's intellectual and spiritual enlightenment has caused scholars to displace them to the peripheries of Enlightenment studies. Attempts to construct a secular Hispanic modernity instead locate its origins in writings produced by prototypical Northern European intellectuals carrying out expeditions that separated commerce from conversions, such as those by Alexander von Humboldt.[3]

To appreciate the Hispanic Enlightenment, however, it is necessary to revisit missionary discourses that, because of their links to evangelical enterprises, have been ignored in favor of secular scientific travelogues and categorized as unenlightened. Joseph Gumilla's eighteenth-century expeditions into the Orinoco River region were primarily evangelical. Yet his *El Orinoco ilustrado y defendido: Historia natural, civil y geographica* reveals inextricable links among conversions, colonization, and commerce, and evidences an evangelical, intellectual, and scientific Enlightenment that bears further scholarly attention. Gumilla underscores the Society of Jesus's role in maintaining a long-standing Spanish imperial agenda that during the Age of Enlightenment continued to combine religious and economic goals, tying evangelical progress to commercial potential.[4] To illustrate this, Gumilla employs maps, both cartographic and descriptive. Taken as a whole, *El Orinoco ilustrado* can be read as one expansive textual map that enlightens his readers, charting not only the region's human and natural resources, but also the Jesuits' and Spain's previous successes and future possibilities in New Granada. To supplement this textual map, *El Orinoco ilustrado* includes a foldout graphic map. In the history of cartography, Gumilla receives credit for drawing the first modern graphic map of the mythical Orinoco River region. This map emphasizes the real potential of a river still enshrouded in fables such as that of El Dorado, and positions the river as a bona fide gateway to colonization, commerce, and conversions. An examination of this actual map alongside *El Orinoco ilustrado,* however, reveals various vulnerabilities. Gumilla publishes his natural, civil, and geographic history as a platform not only to outline past achievements, but also to address present challenges. In it, he proposes solutions aimed at ensuring future political, economic, and religious victories.

This chapter analyzes Gumilla's graphic map and three aspects of his textual map that reveal the inextricable union of commerce, colonization, and conversions in his overarching plan for the Orinoco River region. After a brief historical overview of contemporary challenges to the defense of New Granada, we will consider Gumilla's foldout map. Next we treat his "treasure map" to commercial riches, including indices to the useful plants, medicinal roots, notable animals, minerals, and peoples; his "road map to the future" responding to the challenges to colonization and free commerce with a military defense of and immigration plans to New Granada; and finally his "navigational map" for Jesuit evangelization, an appendix to the second edition that condenses into lists the practical advice for missionaries spread throughout *El Orinoco ilustrado.* Together, these maps expose the links between commerce and Catholicism during the Hispanic Enlightenment in the first half of the eighteenth century. By creating a model of knowing that ties commercial potential to evangelical progress, Gumilla's maps chart a route for understanding the entrance into modernity of Bourbon Spain and her possessions.

The Orinoco Enlightened and Defended

Gumilla wrote, published, and revised *El Orinoco ilustrado* during a period of deepening crisis, when defense needs for New Granada became paramount. His maps identify vulnerabilities from its extensive coastline to the interior Orinoco River region. Several major themes emerge from his geographical and evangelical cartography, including the need for defense against British and French pirates and smugglers, and from Portuguese and Dutch slave raiders and traders. There is also a rhetorical defense of missionaries against accusations of contraband and of the Jesuit province against encroachment by competing Catholic religious orders. Gumilla ties the entire region's defense not only to increased military and economic investment, but also more importantly to an increased presence of the Society of Jesus.

The threats to New Granada's coastline came chiefly from English pirates. In 1739 political and commercial vulnerabilities were intensified when Spain and England broke the Treaties of Utrecht. Irked by the continued seizure of British trading vessels, King George declared war on Spain, and within five weeks his naval commander in chief, Edward Vernon, had captured Puerto Bello (Portobelo) on the northwest of the Darien Peninsula, forcing King Philip V to declare war on England. At the end of 1740 the British attacked from the northeast, entered the Orinoco River as far as its confluence with the Caroní River, and burned Santo Tomé de Guayana, the "city of the holiest sacrament" where Gumilla had served as chaplain after his second major expedition on the Orinoco in 1731. Just a few months before the first edition of Gumilla's natural history appeared in 1741, the commercial hub of Cartagena de Indias had barely resisted Vernon's assault. This "War of Jenkins' Ear" did little to halt foreign contraband (British, Dutch, or French), and Gumilla demonstrates obvious concern for these issues in his writings, proposing that economic battles might be won less in the waters of the Caribbean Sea than in the evangelization, colonization, and population of New Granada from its wide-open coasts to the interior, thus securing and defending an Orinoco River commercial trade route.[5]

Interior attacks by Dutch, Carib, and Portuguese slave raiders coming up the Orinoco River posed an additional threat. Gumilla laments the solid alliance between Dutch Suriname and Guiana and the heathen Carib nation. Additionally, he explains that the greatest commercial competition comes from the Dutch and British colonial Jews of the coast and Caribbean islands, who are motivated by purely economic goals and therefore sin with their military aid to the Carib cannibals. In contrast, a Catholic combination of commerce with conversions redeems Spain spiritually and financially. Gumilla underscores the role of the Society of

Jesus for protecting Spain's interests. As described in chapter 1 of this book, in the face of Carib destruction of several Orinoco missions, Spanish soldiers and Jesuits had joined forces with their Amerindian charges to construct the fortified mission San Francisco Javier in 1736, which at the time of the publication of *El Orinoco ilustrado* continued to successfully defend the missions upriver from this site.

Jesuit mission projects also required vindication from competing religious orders. Gumilla addressed this third important aspect of Orinoco defense in separate letters as well as within the pages of *El Orinoco ilustrado.* In 1740 the Society of Jesus claimed another key strategic point (Cabruta) as a defensive site along the Orinoco River. The 1731 restoration of Jesuit missions had sparked a polemic between the Jesuits, Capuchins, and Franciscans, and this 1740 claim on Cabruta intensified strife.[6] One the goals assigned to Gumilla as Jesuit procurator to Spain included defending the society's eighty-year Orinoco legacy in the face of a judicial conflict with these orders that continued even after Gumilla's first edition appeared in Madrid. Although the Council of the Indies ruled in favor of the Capuchin order in 1742 and the courts reaffirmed this in 1744, the viceroy in Bogotá and the governor of Cumaná supported the Jesuits, who held on to the Cabruta reduction until their expulsion in 1767.

Additionally, Jesuits in New Granada were the targets of numerous allegations of contraband activities. Lance Grahn notes in *The Political Economy of Smuggling: Regional Informal Economies in Early Bourbon New Granada* that "illegal trade occurred more along the Caribbean coast of Colombia in the early eighteenth century than in any other part of the Americas."[7] And as a representative of the British South Sea Company affirmed, the Jesuits were " 'the greatest traders in America.' "[8] In *El Orinoco ilustrado,* Gumilla defends his order by responding to accusations of commercial fraud directed against the Jesuit brother Bernardo Rotella, who was active in the legal battle between Capuchins and Jesuits. To curry the Bourbon monarch's favor, the Capuchins pointed fingers at Jesuit excess and subversion. As Grahn makes clear, this was an era of increasing tensions among viceregal political leaders frustrated by the inability to enforce secular laws that prohibited clerical trafficking and "ecclesiastical smugglers."[9] Thus, governmental fingers pointed not only at Jesuits, but also at the Franciscans and Capuchins who, alongside the Society of Jesus, "invoked both corporate privilege and spiritual authority to shield their smuggling activities" in New Granada.[10]

As this brief overview indicates, Gumilla published his expanded second edition (1745) amid additional defensive needs for the Orinoco River region because of Spain's war with England (1739–48), French attacks at the mouth of the river, Dutch continued support of fierce Carib mission-plundering, Portuguese slave-raiding, and encroachment on Jesuit territories by the Capuchin and Franciscan orders. He makes his text's "defensive" purpose explicit by changing the title from its first publication to include the words "and defended." This second edition, *El Orinoco ilustrado y defendido,* further promotes its own merits as a source of military defense plans for the Orinoco River region while also defending the Jesuits as key agents in the solutions to current challenges facing New Granada. By framing the Society of Jesus's project of civilizing the Orinoco frontier as service to God and to the pope and the Jesuits themselves as the religious Order most capable at serving the king and "al bien público" (213) [the public good], Gumilla crafts an apologetics that ties the Jesuits' successful "paz evangélica" (324) [evangelical peace] to Spain's economic success. This recipe for success prescribes a series of three essential steps: Jesuit conversions clear the road to Catholic colonization, which in turn defends New Granada, which in turn opens it to infinite commercial productivity. As Gumilla notes, the steps are difficult, but Jesuit missionaries are happy to commence them: "[C]uestan las paces muchos pasos [. . .] que dan con mucho gusto, porque por Isaías saben que son preciosos los pies de los que evangelizan la paz [. . .] no solo eterna para las almas, sí también la temporal, porque con el bautismo se unen entre sí las naciones más enemigas" (324) [Peace takes many hard steps . . . that missionaries give with much pleasure, because from Isaiah they know that precious are the feet of those who preach peace . . . not only eternal peace for souls, but also temporal, because with baptism the most hostile nations unite among themselves]. With cartographic and textual maps Gumilla defends both Spain's interests and the Society of Jesus's role in maintaining them. His overarching plan in *El Orinoco ilustrado* defends the commercial potential of New Granada first through Jesuit pacification and then through Spanish population, and he reveals his principal strategy of defense through a series of maps that link conversions, colonization, and commerce.

Gumilla's Graphic Map of the Jesuit Province of New Granada

The most literal and also the most studied map in *El Orinoco ilustrado* is the foldout *Mapa de la Provincia y Missiones de la Compañía de IHS del Nuevo Reyno de Granada* [*Map of the Province and Missions of the Society of Jesus in the New Kingdom of Granada*] that has been included in every edition since 1741. This map has been studied in several articles by twentieth-century historians and Jesuit scholars who esteem it as the first modern cartography of the Orinoco River region.[11] Drawn by Gumilla with copper plates en-

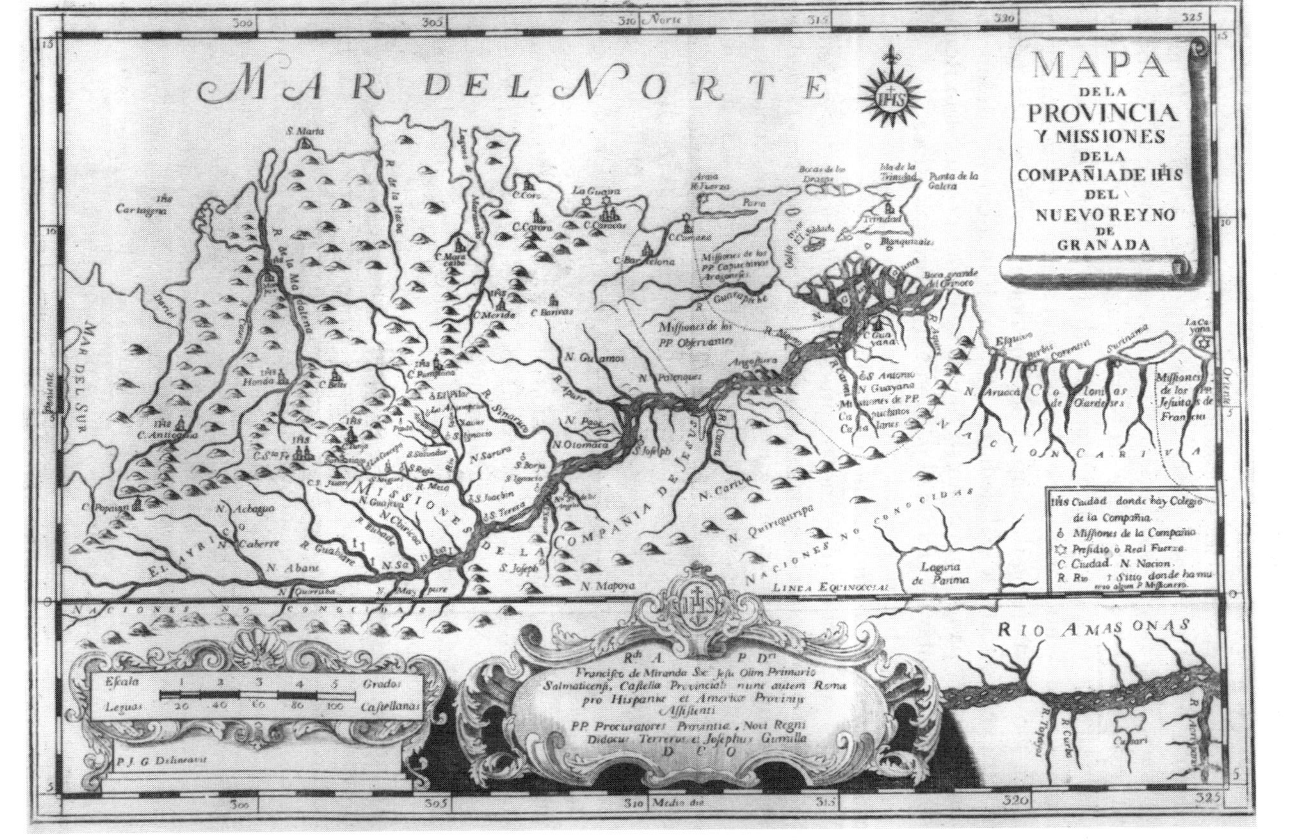

Mapa de la Provincia y Missiones de la Compañía de IHS del Nuevo Reyno de Granada [*Map of the Province and Missions of the Society of Jesus in the New Kingdom of Granada*]. The John Carter Brown Library at Brown University.

graved in Madrid by Pablo Minguet, the foldout *Mapa de la Provincia* measures 39½ by 28½ centimeters and covers a latitude from 15° north to 5° south and a longitude from 295° west to 325° east.[12] It centers on the Orinoco River, emphasizes several mountainous regions, and contains a series of symbols including crosses, circles with crosses emerging from the top, stars, and the letters *C, N,* and *R* (abbreviations for "city," "nation," and "river," respectively). "Mar del Norte" (labeling the Caribbean Sea) runs in large letters across the top, along with one of ten "IHS" letterings (repeating the symbol of the Jesuit order) and a representation of an unfolding scroll containing the map's title. The bottom includes from left to right the map's scale, a seal dedicating the map to a current Jesuit administrator of Spanish and Spanish American Jesuit provinces in Rome, and the representation of a small portion of the Amazon River.

During the eighteenth century, Gumilla's graphic *Mapa de la Provincia* received as much attention as the natural history it accompanied. Though it included errors of latitude and longitude, it was immediately valued as the first serious contribution in transforming an enormous, untamed region from mythical status to reality.[13] Not only Jesuit missionaries and Creoles, but also European explorers and cartographers paid careful attention to Gumilla's detailed map of New Granada. Many read, studied, praised, and sometimes corrected it. Local writers such as the Jesuit Filippo Salvatore Gilij, who drew his own *Carta Corográfica* of the Orinoco region and New Granada-born naturalists such as Francisco José de Caldas worked with it. European scientists such as Charles Marie de La Condamine, Jorge Juan and Antonio de Ulloa, José Celestino Mutis, Pehr Loefling, José de Iturriaga, Aimé Bonpland, and Alexander von Humboldt—that is, all of those who charted the most famous voyages to and within this region during the age of Enlightenment—also depended on Gumilla's map.[14]

Gumilla's *Mapa de la Provincia* thus served a crucial role in depicting the realities of the Orinoco River, which passed through geography still fabulous in many European and American imaginations. He presents the Orinoco as a valuable route to colonial, economic, and evangelical success. Yet a quick glance reveals the perpetuation of two remarkable geographical errors that highlight Jesuit success as well as the river's commercial superiority. First, Gumilla denies any connection between the Orinoco and Amazon rivers. A second geographical error reinforces the evangelical and commercial appeal of the Orinoco River region over the Amazonian jungles by inserting a chain of mountains as the border between the two. Both errors mold the readers' perceptions of the region, as do the graphic symbols on the map.

Gumilla's map repeats a misconception still accepted by most geographers early in the eighteenth century: the separation of the Orinoco and Amazon rivers. His map focuses entirely on the Orinoco and its branches and relegates the "Rio Amasonas" to the bottom right sector, where its disconnection from the Orinoco River is reinforced by the map's scale and an elegant seal of the Society of Jesus, which identifies its cartographer in Latin ("P.J.G. Delineavit" and "Iosephus Gumilla") and obscures the Amazon by covering up all but a small portion of the more-celebrated river. The fact that Gumilla would make this geographical error in both his map and in its accompanying textual description is not surprising, because at the time it was not at all clear to Europeans or Americans that the two rivers were indeed connected. The Jesuit published his map amid geographical debate in Europe about the size and shape of the earth, a debate intensified by South American exploration. For example, during La Condamine's quest to measure the circumference of the earth near the equator, the French explorer suggested the possibility of a connecting channel. The Casiquiare Canal turned out not only to connect these two great South American rivers but also to be the world's only known natural canal. Although La Condamine neglected to draw this channel between the Orinoco and the Amazon on his 1743–44 map of the Amazon River, he specifically corrected Gumilla in a philosophical transaction to the Royal Academy of Sciences in Paris that was later excerpted for the French translator's preface to the 1758 Paris edition of *El Orinoco ilustrado*.[15] La Condamine's hypothesis was famously proven during Humboldt and Bonpland's 1799 expedition down the Casiquiare and into the Negro River, an Amazon River tributary. Humboldt boasts in writings about the Orinoco cataracts that his journey had "entirely refuted" doubts raised during years of geographical debate about the rivers' separation.[16] Yet we should recall that Humboldt not only read Gumilla's natural history and used his map, but also that he cited Gumilla's Jesuit protégée, Gilij, whose *Saggio di storia Americana* had explained that Gumilla was aware of the Jesuit, Manuel Román's 1744 daring navigation through this same canal nearly sixty years before Humboldt, and that before Gumilla died he had intended to revise this geographical error for a third edition of *El Orinoco ilustrado*.[17] Gilij's testimony suggests that we cannot be sure of Gumilla's intentions. But it is possible that he perpetuated this geographical error to highlight the Orinoco River's superiority by graphically marginalizing the Amazon.

The second major error on Gumilla's *Mapa de la Provincia* further underscores the separation of the Orinoco and Amazon rivers. This fabulous geography, a chain of mountains from the Andes to the sea, creates a fictitious border between a civilized region labeled "Jesuit missions" ("missiones de la Compañía de Jesús") and the untamed, "barbarous" unknown nations ("Naciones no conocidas"). This invented mountain chain might have been inspired by Gumilla's desire to separate Christian reductions from unenlightened Amerindian nations or to elevate the status of Orinoco region missions over the Jesuit provinces farther south. More likely, however, Gumilla was simply following an existing cartographic and chorographic tradition that included his Jesuit forefather Samuel Fritz's Amazon River map as well as earlier efforts by more famous cartographers who also imagined mountains here, such as Blaeu, Laet, and de Bry.[18] Intentional or not, Gumilla's geographical errors function to frame the Jesuit province of New Granada as superior to the Amazon, both literally and symbolically.

A third notable characteristic of Gumilla's *Mapa de la Provincia y Missiones de la Compañía de IHS del Nuevo Reyno de Granada* is its actual symbols, which mark a chain of Jesuit missions, churches, and colleges and defend his order's preeminent role in unlocking routes along the Orinoco River and her tributaries, especially the Meta and Casanare branches. Above the visible portion of the Amazon River and to the right of his society's seal, Gumilla places a key that explains how to read his map in a manner that praises the Jesuits' evangelical sacrifices, agenda, and potential. Tiny crosses mark the sites of missionaries martyred in service to Catholic Spain. Stars along the coast reveal existing sites of royal military forts. The cities keyed by the "IHS" label mark sites of Jesuit knowledge from the coast (Cartagena) to the interior (Monpox, Antioquia, Honda, Santa Fé, Tunja, and Merida) and tie the past progress of Spanish imperial expansion to the Society of Jesus. Circles with crosses signal Jesuit mission sites that promise future potential cities and mark the steady pace of the Society of Jesus up the Orinoco from San Antonio to San Ignacio, San Javier, and Nuestra Señora de los Angeles. Meantime, a visual defense against Capuchin and Franciscan efforts confines these competing orders to small areas near the coast. Their missions are marked without crosses or any other symbols; and instead they are given only textually as "Missiones de los PP Capuchinos Aragoneses," "Missiones de los PP Capuchinos Catalanes," and "Missiones de los PP Observantes." From the early modern through the postmodern eras, cartographic studies have read graphic maps like texts that reveal both literal and figurative meanings. No map, ancient or modern, is truly an objective representation of space, and the *Mapa de la Provincia* is no exception. Nor does Gumilla attempt to portray his map as an objective report of a geographical expedition. His inclusion of Jesuit symbols and exclusion of emblems representing other orders make his intentions clear. By graphically inscribing successful reductions and cities with Jesuit colleges, Gumilla figuratively situates the Society of Jesus as central to Spain's progress in civilizing these final frontiers. This panegyric of Jesuit presence and imperial progress symbolically ties together the interests of the Crown and the Jesuit clerics and emphasizes the Society of Jesus's role in maintaining control over South America.

A TREASURE MAP TO THE "IMPONDERABLE RICHES" OF NEW GRANADA

Just as Gumilla's foldout map graphically links conversions, colonization, and commerce, so do the textual contents and arrangement of *El Orinoco ilustrado.* Even the most superficial examination Gumilla's text reveals a careful rhetorical *dispositio* that maps a journey full of danger and potential through the Orinoco River region, while proposing Jesuit missionaries as the key ingredient for realizing its evangelical, colonial, and economic success. *El Orinoco ilustrado* promotes the evangelization and domestication of Amerindian nations and depicts both Jesuit and Spanish authorities as essential to securing the defense and commercial development of this providentially favored region. Some of the most arresting aspects of how its first editions are arranged include eye-catching notations throughout the margins and two indices placed at the end. Together they create a "treasure map" that points to financially profitable products and particularly curious contents and leads readers into a persuasive narrative about the necessary connections between evangelism, immigration, and economics.[19] The marginalia reveal these connections with phrases such as "Riquezas imponderables" (*El Orinoco ilustrado,* 372) [Imponderable riches][20]; "Medios oportunos para la paz y aumento de las Missiones" (103) [Timely means for the peace and increase of the missions]; "Modo con que los Holandeses fomentan estas guerras" (85) [Means with which the Dutch foment these wars]; "Solo el Santo Evangelio destierra las guerras entre los indios" (84) [Only the Holy Gospel banishes the wars between the Indians]; "Si entablan buenas sementeras danse ya por assegurados" (359) [If they set up good sown fields they can be already taken as secured]; "Caña para azucar" (364) [Cane for sugar]; "Tobacco [. . .] Café [. . .] Abundancia de Añil" (365) [Tobacco . . . Coffee . . . Abundance of Indigo]; "Solo faltan pobladores" (373) [Only settlers are lacking]; and "Madera para un grande Astillero" (12) [Wood for a great Shipyard]. Margin signposts like these call attention to conversions, colonization, and commerce as the three essential steps for success and complement the alphabetical indices.

Gumilla's first index, the fourteen-page "Índice de las cosas más notables que se contienen en este libro" [Index of the most noteworthy things that this book contains] catalogues a veritable cornucopia of valuable natural treasures, including many that have already yielded economic bonanzas elsewhere, like indigo, cocoa, tobacco, sugarcane, sassafras, and cotton.[21] Precious spices also abound in the Orinoco River region.[22] Gumilla often directs his readers' textual treasure hunt back to his graphic map.[23] For example, the cornucopia index entries for cocoa and coffee both lead to chapter 24, whose contents in turn refer to the foldout map. First, "Cacao silvestre, haylo en aquellas vegas" (n.p.) [Wild cocoa, there is some in the fertile plains] (1745) sparks readers' interest by pointing to pages within chapter 24 where a miraculously fertile land is proclaimed: "milagro del terreno, que sin cultivo alguno prorrumpe en bosques. [. . .] ¡Oh, y qué país, si se lograra su fertilidad!" (1963, 247–48) [miraculous land that without any cultivation at all bursts into forest. . . . Oh, and what country, if its fertility is achieved!]. Next, Gumilla stirs their imagination by describing this product and the fertile land in which it grows wild, this time punctuating his vivid description with a rhetorical question: "[S]i aquel fecundo terreno así produce el cacao de suyo, ¿qué arboledas y qué cosechas diera al favor del cultivo y del riego?" (247) [If that fecund land produces cocoa this way by itself, what groves and what crops will it give the favor of cultivation and irrigation?]. Finally, an interaction between his textual geography and his graphic map underscores cocoa's commercial potential if domesticated. The Jesuit invites readers to favorably compare the plantation sites proposed in unpopulated valleys to the already highly productive cocoa bean plantations near Caracas, "donde se da el mejor cacao" (247) [where the best cocoa is yielded]. Gumilla himself scouted out this superior location on "la banda Sur del Orinoco, [donde] hallé en éstos más campo, mejor migajón en la tierra, más fácil y más abundante el riego para inmensos plantajes de cacao" (247) [the southern strip of the Orinoco, where I found more land, better crumbs in the land, easier and more abundant irrigation for immense plantings of cocoa]. The cornucopia index entry for coffee provides another example of this interaction between textual and graphic maps as well as Gumilla's firsthand experience with potential bumper crops. "Café, prueba bien en el río Orinoco" (n.p.) [Coffee, tests well along the Orinoco River] points to a section on coffee beans, which throughout the eighteenth century became a stimulant of increasing economic value that eclipsed chocolate. Gumilla himself cultivated coffee beans while testing the agricultural potential of the land surrounding one of the Jesuit missions located on his *Mapa de la Provincia:* "El café, fruto tan apreciable, yo mismo hice la prueba: lo sembré, y creció de modo que se vio ser aquella tierra muy a propósito para dar copiosas cosechas de este fruto" (249) [Coffee, such an valuable plant, I myself did the test: I planted it, and it grew in such a way that that land was seen as very appropriate for giving copious crops of this fruit]. It is interesting to note that today in a region that still dominates international markets for coffee beans (Venezuela and Colombia), Gumilla's mission settlement crop is celebrated as the first coffee planted there. Whether textual or graphic, Gumilla's maps to commercial success praise the Jesuits' primary role in paving the way for colonization and defense, for domesticating peoples and plants, and as we will see, for mediating Amerindian knowledge about potentially profitable species.

Gumilla's second index, a three-page "Índice de raíces, frutas, yerbas, aceites, resinas y otras cosas medicinales, que se han descubierto en el Río Orinoco y sus vertientes" [Index of the roots, fruits, herbs, oils, resins, and other medicinal things that have been discovered along the Orinoco River and its slopes] maps regional pharmaceutical products that are ripe for commercial exploitation and underscores the Society of Jesus's essential role for Spanish control of these precious remedies.[24] This pharmacopoeia index sends readers sixteen times to part 1, chapter 20, "Resinas y aromas que traen cuando vuelven los indios de los bosques y de las selvas, frutas y raíces medicinales" [Resins and aromas that the Indians bring when they return from the forests and from the jungles, medicinal fruits and roots], whose contents alternate between products known to Europeans and products known only to Amerindians and Jesuits. Here the reader finds natural products celebrated since antiquity, like the famous Arabian dragon-blood tree, cited from Herodotus and Pliny through New World pharmacopoeia such as Oviedo's or Monardes' sixteenth-century botanical reports to the Spanish Crown. The index captures readers' attention with "Sangre de Drago, hay con mucha abundancia" (1745) [Dragon Blood, found there with much abundance]. Once within chapter 20, Gumilla relocates its Old World prestige to the jungles of the Orinoco: "[E]l árbol llamado drago se halla por aquellas selvas con abundancia; el jugo que destila [. . .] se llama sangre de drago, tan apreciable y medicinal, como todos ya saben" (1963, 217) [The tree called dragon is found in those jungles with abundance; the juice distilled . . . is called dragon blood, so precious and medicinal, as everyone already knows]. In addition to these remedies already well-appreciated by readers, hiding within New Granada are pharmaceutical treasures that have remained unknown to even the most erudite Europeans. As Gumilla explains:

> [A] la verdad es muy poco lo que [. . .] se ha descubierto, en comparación del gran tesoro que yace escondido por falta de personas inteligentes. [. . .] Para mí es indubitable que hay entre aquellas vastas arboledas resinas, aromas, flores, hojas y raíces de grande aprecio y que serán muy útiles a la botánica, cuando el tiempo las descubra. Ahora apuntaré lo poco que se ha descubierto, que creo muy útil al bien público. (212–13)
>
> [The truth is that very little . . . has been discovered, in comparison with the great treasure that lies hidden for lack of intelligent people. . . . In my opin-

ion it is undeniable that there are among those vast groves resins, aromas, flowers, leafs, and roots of great value, and they will be very useful to botany when in time they are discovered. Now I will point out the little that has already been discovered, which I believe very useful to the common good.] (1963 ed)

Despite his protestations of how little treasure has yet been discovered, Gumilla piles one secret atop another in a narrative celebrating Jesuit intervention for a "public good" that connects colonists and Christianized Amerindians. One of the recently discovered panaceas that Gumilla reveals possesses healing properties already proving useful to Europeans. Its entry in the pharmacopoeia index simply states, "Currucay, es resina muy util" (1745) [Currucay, a very useful resin]. However, chapter 20 posits the *cabima* or *curucay* oil as a remedy whose high economic value became known only after its appropriation from Amerindians. The Jesuits have christened it *aceite de María,* or Virgin's oil.[25] At the same time Gumilla reveals secrets already mediated by Jesuits, he ties the region's still unknown but seemingly limitless potential to the Society of Jesus. He underscores how little has been discovered beyond the borders of civilization in order to project a bright future of new and useful botanical discoveries once the lack of "personas inteligentes" (213) [intelligent people]—Jesuit missionaries followed by Spanish colonists—has been remedied.[26]

The pharmacopoeia index in particular highlights Jesuit access to Amerindian medical knowledge. Gumilla extends the links between colonization and evangelization by enumerating raw materials that Jesuits deem worthy of cultivating. One such "secret worthy of inquiry" is announced in the index as "Mara, resina muy singular" [Mara, very extraordinary resin] and investigated within the text as "la resina rara. [. . .] Secreto es el de la mara digno de inquirirse" (216) [the peculiar resin. . . . The mara is a secret worthy of investigating]. Jesuits also seek control of medicinal crops traditionally cultivated by Amerindians. For instance, tobacco already enjoyed a local and global market, and Gumilla refers to the potential for great Orinoco harvests: "No se hallará en las Provincias de Tierra Firme terreno ni temperamento más al propósito para copiosas y apreciables cosechas de tabaco, como está ya visto y comprobado en el que siembran y cogen aquellos Indios para su gasto" (249) [In the Tierra Firme provinces will not be found land nor temperament more appropriate for copious and valuable crops of tobacco, as is already seen and confirmed in that which those Indians plant and gather for their consumption]. In fact, several pharmacopoeia index entries lead readers to textual narrations that emphasize the civilizing benefits of agriculture.[27] Whether pointing to the harvest of wild botanicals or the exciting possibilities for cultivated crops, Gumilla's revelation of secret roots, resins, herbs, and more widely recognized medicinal plants like tobacco depends on Jesuit appropriation of Amerindian natural remedies and the geographic area to grow them.[28] Without Jesuit mediation of local medical knowledge disclosed through natural history reports, many treasures would still be covered up. By converting Orinoco tribes and Christianizing their natural remedies, Gumilla says, Jesuits gain access to products of already proven economic weight, to panaceas previously unidentified by Europeans, and to fertile territories.

If *El Orinoco ilustrado* provides a Jesuit treasure map to commercial exploitation, at the same time it also illustrates present realities that demand not only military defense, but also defense against commercial competition by the Dutch, French, and British, as well as against slave raids by the Portuguese. In addition to reporting on past successes and a current demonstration of his Jesuit Province's bounty, *El Orinoco ilustrado* makes a deliberative argument for political and economic action.[29] Besides warning King Philip V not to ignore increasing Portuguese threats, Gumilla cautions readers that the riches of New Granada will not conveniently wait for Spanish colonization. There are already many intelligent people from foreign nations competing for these natural treasures, and Gumilla's maps create a sense of urgency in part by suggesting that the Dutch, French, and British populations on the coast are looking toward the interior.

Dutch colonists, in particular, pose threats beyond arming the Carib tribes attacking Catholic settlements along the Orinoco. They are Spain's most fierce economic competitors, and in an "importante digresión" (249) [important digression], Gumilla refutes foreign newspaper charges of Spanish "lethargy" in the face of vigorous Dutch trade: "[D]e modo que así este cuantioso renglón de las especias, como otros muy considerables, que desprecia nuestra monarquía, no es por vía de letargo" (248) [So thus this substantial line of spices, like others very considerable that our monarch rejects, is not rejected by way of lethargy]. Despite his loyalty to King Philip V, the Jesuit wonders why Spain allows "los holandeses" [the Dutch] to control New Granada's precious line of goods, including spices such as cinnamon, especially when there are very nonlethargic men in Spain and America ready, willing, and able to set plans in motion to reap the commercial benefits of his most recently established viceroyalty: "[N]o faltan ministros, muchos y muy despiertos, y argos vigilantes, que comprenden lo más oculto de los caminos y rumbos más intrincados de la economía y del comercio" (248) [Ministers are not lacking; there are many very smart and watchful argonauts who understand the most secret and most intricate routes and courses of economy and commerce]. They, along with Gumilla, understand how Spain errs. For example, not only do the Spains lag behind in agricultural technology but they also allow foreign nations to monopolize sugarcane harvests. Gumilla calls for productive cocoa bean plantations and lucrative fields of cane, and he suggests that Spain take action planting uninhabited territories and using technological improvements already employed by foreigners:

> Fuera de esto, de la caña dulce, que casi todas aquellas naciones siembran para golosina y entrenimiento [*sic*] de sus hijos, del tamaño de ella y del intenso dulce de su jugo se infiere con evidencia que todos aquellos inmensos y despoblados territorios dieran no menos útil con el azúcar que con el grano de cacao; y más cuando la pendiente de los ríos diera a poca costa copiosos caños de agua para el movimiento de los ingenios y máquinas con que en otros países se beneficia la caña. (249)
>
> [Apart from this, consider the sugarcane, which almost all those nations plant for their children's sweets and entertainment. From the size of it and the intense sweetness of its juice it is clear that all those immense and unpopulated territories would be no less useful with sugar than with cocoa beans; and even more useful when the slope of the rivers produces inexpensive, copious channels of water for sugarcane plantation and the machine operations by which in other countries the cane benefits.]

Jesuit evangelism and productive Amerindian labor combined with advances in agricultural technology and Spanish colonization, Gumilla suggests, will benefit a plantation economy that will allow Spaniards and Americans to control crops that can thrive in the interior as well as effectively "clip the wings" of Dutch commerce by controlling the coasts, "para que se perdiera su ala derecha el elevado vuelo que ha tomado el comercio de Holanda" (249) [so that its right wing loses the lofty flight that the commerce of Holland has taken].[30]

Jesuit conversions of Orinoco region tribes constitute the key ingredient in promoting and defending Spanish commercial and colonial success, and Gumilla's arrangement of *El Orinoco ilustrado* confirms this. The Jesuit arranges his chapters and their contents in a manner that repeatedly reveals the importance of the Society of Jesus. Gumilla even reinforces ties between commercial exploitation of products and the civilizing effects of evangelization within the apparently rigid organization of the treasure map's alphabetical indices. For example, in the cornucopia index under *F* appear these entries: "Frutas silvestres del Orinoco, y sus vertientes" [Wild fruits of the Orinoco and its banks], "Frutas cultivadas de aquella tierra" [Cultivated fruits of the land], "Frutos varios" [Various fruits], and "Frutos espirituales, que se logran de los Indios" [Spiritual fruits that are achieved from the Indians] (1745). Here Gumilla slightly rearranges true alphabetical order to emphasize an interplay of literal and figurative wild fruits that underscores the civilizing benefits of agriculture and directly ties conversions to colonization. The wild and varied "fruits" proven better cultivated through domestication include both the natural and human resources of the Orinoco River region. For example, within chapter 24, "Fertilidad y frutos preciosos que ofrece el terreno del río Orinoco y el de sus vertientes" [Fertility and precious fruits that the land of the Orinoco River and its banks offer], Gumilla evokes goods just waiting to be harvested at the same time he reminds readers to consider the fates of the human souls awaiting religious enlightenment. This rhetorical double entendre (crop harvest and soul harvest) fashions a plea tying commerce to colonization and evangelism. In chapter 24, Gumilla employs a rhetorical pause to interrupt the enumeration of harvests, inserting a comparison between unfortunate wild plants that have been ignored instead of domesticated and uncivilized tribes awaiting domestication by Jesuit missionaries: "la misma desdicha de aquellos hombres, que por mas que lo merezcan, no hallan quien les dé la mano" (1963, 250) [the same misfortune as of those men, who as much as they deserve it, still do not find anyone to give them a hand]. He enhances this rhetorical pause with *negatio,* a classical strategy for argumentation that draws attention to a point by denying it is actually making this point: "No me detengo en apuntar cuánta utilidad diera solo el renglón de esta cosecha, en la suposición de que se poblara aquel inmenso territorio" (250) [I won't dwell on pointing out how useful just the goods of this harvest would be, in the assumption that that immense territory will be populated]. By stating that he will not pause to point out useful by-products of colonization, this *negatio* reminds readers of its necessity following evangelization in order to harvest the most goods.

Gumilla stops the flow of cash crops in chapter 24 to stress the connection between immigration and conversions, pairing discussion of the many precious fruits offered in the fertile terrain along the Orinoco with reminders of Spain's duty to convert the many precious souls to Catholicism.[31] In addition to enumerating these valuable natural resources, Gumilla seeks to convince the Spanish government of inextricable connections between conversions, colonization, commerce, and defense. Two margin notes—"Apóstrofe a la piedad del rey N. Señor"[32] [Apostrophe to the piety of the King our Lord] and "dichas poblaciones fueran de util a su Majestad, y a la conversion de los Gentiles" [said populations would be useful to His Majesty, and to the conversion of the pagans]—signal key moments in an argument that posits conversions and colonization as the remedy for the current challenges to the commercial exploitation of New Granada and the challenges to its defense. Gumilla concludes the argument with a periodic sentence replete with an accumulation of four past-subjunctive verbs followed by a culminating three-word statement of absolute certainty:

> de modo que sin daño de las naciones ya domésticas, y con mucho útil de éstas, y grande esperanza de domesticar otras muchas, se pudieran fundar

> muchas y grandes colonias, con evidente beneficio del comercio de España, y grandes ventajas de la real Corona, fuera de su principal y máxima utilidad que se siguiera (como apunté) en la conversión de nuevas naciones, la cual precisamente se facilitara mucho a la sombra y abrigo de las poblaciones de Españoles. Esto es así. (251)
>
> [[W]ithout harm to the nations already domesticated, and with much usefulness to them and with great hopes to domesticate many others, many and great colonies could be founded, with obvious benefit to Spain's commerce as well as great advantages to the Crown, beyond the principal and maximum utility that would follow (as I already pointed out) the conversion of new nations, which would particularly facilitate the protection and support of the populations of Spaniards. This is certain.]

If Gumilla arranges chapter 24 to emphasize the inextricable ties between conversions, colonization, and commerce, then the *dispositio* of the final six chapters in part 1 further achieves this goal. Chapters 20, 21, 22, 24, and 25 each enumerate economically viable products as well as tie economic to evangelical success. However, inserted directly in the middle of part 1's chain of goods is a chapter on practical missionary methods that clearly designates the Jesuits as essential to any successful treasure hunt. The placement of chapter 23, "Método el más practicable para la primera entrada de un misionero en aquellas tierras de gentiles, de que trato, y en otras semejantes" [Most practical methods for the first entry of a missionary into those pagan lands that I deal with, and in other similar ones], interrupts the flow of products, and suggests the necessity of conversions for productive commerce. Gumilla's rhetorical arrangement reminds readers yet again of the inextricability of colonization, commerce, and Jesuit conversions. Additionally, by situating the very practical enumeration of methods for successful evangelization of untamed regions immediately preceding the details in chapter 24 about the most coveted spices, dyes, sugarcane, cocoa, and coffee beans, Gumilla ties successful evangelization to successful exploitation of untapped resources in New Granada. Here once more we must pause to appreciate Gumilla's masterful rhetoric, as he employs strategies for *inventio* (topic selection), *elocutio* (style), and especially *dispositio* (arrangement) to argue that Jesuit evangelization remains at the core of acquiring and controlling access to the region's treasures.

The Jesuits hold the key to Gumilla's treasure map, which privileges the fertile region's natural abundance and offers remedies for both corporal and economic health. This treasure map consists of marginalia and financially profitable cornucopia and pharmacopoeia indices that lead readers into a persuasive narrative about the necessary connections between economics and evangelism. *El Orinoco ilustrado* reveals the Jesuits' role in unlocking the "riquezas imponderables de el Nuevo Reyno" [imponderable riches of the New Kingdom], "los tesoros de el Nuevo Reyno de Granada" [the treasures of the New Kingdom of Granada], and "el gran tesoro que se sacàra [*sic*]" [the great treasure that will be taken out], all alphabetized in the indices.[33] The cornucopia index emphasizes the bounty of products already proven to carry economic weight and stresses the ties between agriculture and colonization. The pharmacopoeia, in contrast, reveals particular secrets of nature hidden in the Orinoco region and hints at untold others yet to be "discovered" in the wilds while stressing the Jesuits' role as mediator of Amerindian medical knowledge. New Granada's fiscal and physical remedies are consistently tied to Jesuit mission culture, and defending successful treasure hunts for these "imponderable riches" depends on both Amerindian conversions and Spanish colonization. Thus, if at first glance the treasure map can be read as an economic hook directed at merchants or politicians, the *dispositio* of the indices, the chapters, and the arrangement of their individual contents all reveal a carefully constructed argument that repeatedly ties the Jesuits' evangelical agenda to the colonization, defense, and commercial benefits of New Granada.

The Road Map to a Prosperous Future

Like his graphic and treasure maps, Gumilla's third important map—his "road map to the future"—provides spatial references. However, this road map differs from the two previous maps in that it also sets out a chronological plan for peace and prosperity in New Granada. This temporal map employs deliberative calls for action and includes plans for a defensive fort at the mouth of the Orinoco, as well as a three-step chronological road map to secure the region's future. Much of *El Orinoco ilustrado* consists of wonder-provoking panegyrical descriptions. However, Gumilla alternates this demonstrative rhetoric typical to natural histories with deliberative strategies that move readers to take further steps for possessing and protecting the abundant resources on display in his text. This propaganda seeks military troops to protect the interests of the Crown and the Society of Jesus, Spanish immigrants to increase the viceroyalty's population, and more Jesuit missionaries to further evangelism efforts. Gumilla's defense plan and road map chart an optimistic future that expands colonization, commerce, and conversions and will control the geographical spaces and treasures he maps both textually and graphically.

To emphasize the urgency for deliberative action, Gumilla narrates extremely vivid scenes of violence. British aggression in the War of Jenkins' Ear as well as French and Portuguese attacks were retarding economic progress in New Granada.[34] *El Orinoco ilustrado* focuses primarily on remedying the deadly assaults by the Carib nation that were plaguing Christians from the coast to the interior. The Carib nation received support, Gumilla says, from Protestant heretics, some of whom have even "gone native" and joined them in slave raids up the Orinoco: "desde el año de 1731 hasta acá, los herejes, ya holandeses, ya de otras naciones, se embijan, esto es, se pintan al uso caribe" (328) [ever since the year of 1731 until now, the heretics, be they Dutch, or be they from other nations, *se embijan*—that is to say, they paint themselves in the style of the Carib Indians]. In particular, Dutch-backed attacks on mission settlements upriver were challenging further development of the region.[35] Gumilla fills an entire chapter entitled "Daños gravísimos que causan a las Misiones las armadas de los indios caribes que suben de la costa del mar" [Grave damages that the forces of the Carib Indians cause to the missions when they come up from the seacoast] with shocking war stories about the ferocious Carib tribe, stories intended to move readers to fund necessary troop support. In one particularly graphic extended sequence, Gumilla laments the fate of a Franciscan friar who first takes a bullet in the leg, then a whack on the mouth, then a near-fatal blow to the head; finally, he suffers a slow death hanging naked from a tree while his attackers work hard to light a fire under him. After this carnage, the Caribs jump into their boats "contentos con el botín y gran número de esclavos" (333) [happy with the booty and great number of slaves] and head downriver to sack a Capuchin mission. Their goal is twofold: to acquire Amerindians to sell as slaves and to eradicate any progress of Christianity. And one of their most barbarous tactics is to destroy all the civilized foodstuffs: "[T]alaron los caribes las sementeras, arrancaron los frutos y quemaron las trojes; golpe el más fatal con que pensó el enemigo desterrar las Misiones de todo el Orinoco" (331) [The Caribs cut down the sown fields, rip out the fruits, and burn the granaries; with this fatal blow the enemy thought to banish the missions from all the Orinoco]. The "diabolical cleverness" of Carib efforts to combat successful Jesuit reductions is left to the imagination of Gumilla's already horrified readers: "Omito aquí [. . .] otros repetidos asaltos hechos por los caribes que nuevas industrias y sagacidad diabólica contra las Misiones de la Compañía, fomentados con la esperanza . . . de porfiar y proseguir su guerra hasta quitar la vida a todos los Padres misioneros y destruir todos sus pueblos" (334) ["I omit here . . . other repeated assaults carried out by the Caribs with renewed industry and diabolical cleverness against the missions of the Company of Jesus, promoted with the hope . . . of prosecuting their war until taking the life of all the missionary fathers and destroying all of their reductions.][36] After setting such violent scenes of "recent pain," Gumilla lays out his defense plan.[37]

The first part of his plan involves the construction of a stronghold on Limón Island, in a channel near Santo Tomé de Guayana.[38] For *El Orinoco ilustrado,* Gumilla expands his 1739 letter to King Philip V, "Informe [. . .]

sobre impedir a los indios Caribes, y a los olandeses las hostilidades, que experimentan las colonias del gran Río Orinoco [. . .]" [Report . . . on how to impede the hostilities of the Carib Indians, and the Dutch, that the colonies of the great Orinoco River are experiencing . . .] but he eliminates the graphic map detailing the river's mouth and the proposed site in the channel.[39] In order not to offend the Jesuits who had different ideas about the location of this fort, Gumilla had included with the 1739 letter a royal engineer's rendering of his own sketched map. This defense plan was solid, but the government chose not to fortify Limón Island, an error in judgment made abundantly clear when first the British and then the French sacked Santo Tomé de Guayana in 1741. In 1747 another engineer, Gaspar de Lara, would propose Gumilla's plans anew.[40] Thus, any doubts as to the value of this Jesuit's garrison plans for defending the mouth and interior of the Orinoco River were put to rest.

In addition to this actual defense plan, Gumilla also establishes a road map that lays out three consecutive steps that will remedy the horrible violence affecting the region: "lo primero, si se poblara; lo segundo, si se labrasen sus minas; y lo tercero, si se desarraigase el comercio con los extranjeros" (262) [the first, if it is populated; the second, if the mines are worked; and the third, if trade with foreigners is eradicated]. If the indices and marginalia of his treasure map reveal botanical and agricultural bonanzas, this three-step road map charts the route for peace, as well as for extracting enormous wealth from the mines in a region he calls "un Dorado de tesoro inagotable, cual realmente es todo el Nuevo Reino de Granada y Tierra Firme, tan lleno de fecundas minas de oro, plata y esmeraldas [. . .]!" (253) [a Dorado of inexhaustible treasure, which is actually all of the New Kingdom of Granada and its Tierra Firme provinces, so full of fecund mines of gold, silver, and emeralds . . .!"].

The first and most urgent step for exploiting this very real Dorado is to populate New Granada. Gumilla accompanies panegyrics of the empty plains awaiting colonization along the Orinoco River with calls for population.[41] Gumilla had already proposed a population plan to the king in a 1739 letter. In *El Orinoco ilustrado* he repeats it at the start of a subsection entitled "Infiérese el gran tesoro que se sacara si se poblase bien el tal reino" [The great treasure that would be taken out if that kingdom is well-populated is inferred] in chapter 25 of part 1:

> Estas noticias [. . .] evidencian el inmenso tesoro que el Nuevo Reino tiene patente en sus minas abiertas y desiertas [. . .] indican lo mucho que aquellos países retienen oculto y cuán imponderables riquezas darán, si Su Majestad se digna de repartir en aquellos terrenos tantas familias como en Cataluña, Galicia y Canarias están en la última pobreza, por no tener tierras propias en que emplear su trabajo. (263)

[This news . . . evidences the immense treasure that the New Kingdom has in its open and deserted mines . . . it indicates how much those lands have hidden and how imponderable the riches they will yield, if His Majesty deigns to distribute into those lands as many families as are in extreme poverty in Cataluña, Galicia, and the Canary Islands, because of not having their own lands in which to invest their labor.]

Gumilla's concrete plan for immigration of poor Spanish families will serve the king and his loyal subjects, and allow the viceroyalty of New Granada to compete with the well-established viceroyalties of New Spain and Peru.[42]

Gumilla addresses the second step of his road map through another plea to Philip V to populate New Granada for the "public good." If the lack of mining labor were remedied in this new viceroyalty, which Gumilla describes as replete with "so many Potosies," then the New Kingdom of Granada might rival Peru and New Spain. To argue this, the Jesuit rhetorically contrasts the present state of New Granada, whose natural excess is being neglected, with its unlimited potential for future commerce if the mines are worked:

Digo ingenuamente lo que hay, y lo mucho que hubiera, si aquellas riquísimas tierras estuvieran tan pobladas como la Nueva España y el Perú [. . .] todas las riquezas deseables sobran; sólo faltan pobladores que las saquen de los ricos minerales. Ojalá la Majestad de nuestro Católico Monarca vuelva sus piadosos y apacibles ojos hacia aquel pobre reino, sólo pobre por falta de habitadores, y opulentamente rico por sobra de abundantes minas [. . .] en útil del bien común. (254–55)

[I state ingenuously what there is, and how much there can be, if those rich lands were as populated as New Spain and Peru . . . there are more than enough of all the desirable riches; only settlers are lacking to mine the rich minerals. If only the Majesty of our Catholic King would turn his pious and gentle eyes toward that poor kingdom, only poor because of lack of inhabitants, and opulently rich because there are more than enough abundant mines . . . useful to the public good.]

Here Gumilla pays tribute to his king's other viceroyalties in order to then beseech support for New Granada. The best indicator of wealth for any viceroyalty is its commerce, Gumilla argues, and the king need only grant this one similar attention and population to achieve an even more robust economy than those in New Spain and Peru.[43]

The third and final step in Gumilla's road map to a prosperous future protects Spain's economic interests by getting control of foreign smuggling and contraband, which presented an equally important challenge to healthy commerce in New Granada. The Jesuit laments that Spain's restrictive laws, combined with lack of control of its South American coasts, have allowed foreign merchants to dominate, taking advantage of three-fourths of all the commerce in New Granada alone (260). With these plaints Gumilla anticipates the debates and laws of the future reign of Charles III (1759–88) and Charles IV (1788–1808), when calls for unrestricted trade between Spain and her colonies would finally be answered. Gumilla quotes from a propagandistic text translated and published in 1728 by a Spanish Jesuit in Mexico City that argued for free trade and increased manufacturing to halt illegal Dutch, French, and British trade. Addressing "el dicho factor inglés" [the said British factor] (260), especially England's use of Jamaica as a clearinghouse for smuggling, this political and economic pamphlet quotes a supposed confession from a member of Parliament in Britain, which Gumilla reproduces in his text:

> El más considerable ramo de nuestro comercio en la América es el contrabando que nosotros hacemos en los dominios del rey de España. Nosotros enviamos a Jamaica los géneros propios que se consumen en las colonias españolas, y nuestras embarcaciones los llevan furtivamente a los parajes donde tenemos nuestros corresponsales. Nosotros les vendemos allá por plata de contado o a trueque de preciosos géneros, como la tinta fina y la grana, que nos producen muchas y gruesas ganancias [. . .] de suerte que entra más en Inglaterra por la vía de este contrabando, que por Cádiz u otra parte de los dominios de España etc. (259)

> [The most considerable branch of our trade in America is the trade in contraband that we do in the dominions of the king of Spain. We send to Jamaica the typical merchandise that is consumed in the Spanish colonies, and our vessels furtively bring them to the spots where we have our correspondents. We sell to them there for silver cash down or in exchange for precious goods, like fine dye or cochineal, which produce for us many and gross earnings . . . in such a way that more enters England by way of this contraband, than through Cádiz or another part of the dominions of Spain, etc.][44]

Spain's current war with England, he writes, has only exacerbated problems related to this "British factor." And the Dutch, Gumilla warns, have their own Jamaica in the island of Curaçao. These "tricky merchants" support a wealthy colony through ostentatious displays of illegal trade: "[L]os convoyes de marchantes holandeses, que llenan su puerto; la multitud de balandras con que trafican, todo son señales de que no saca Curazao menos millones de la Tierra Firma que Jamaica" (259) [The convoys of Dutch tricksters that fill its port and the multitude of sloops with which they traffic are all signs that Curaçao does not produce less millions from dry land than Jamaica]. According to Gumilla, Dutch financial and physical encroachment on Spain's territories comes naturally to them. After all, they usurp even their homeland from the sea by constructing dikes: "Nadie ignora que el genio mercante de los holandeses es todo su modo de subsistir, pues hasta el suelo de la patria que pisan se lo han usurpado al mar" (259) [Nobody is unaware that the tricky nature of the Dutch is their whole way of surviving, since even the ground of their homeland that they walk on they have usurped from the sea]. The third step of Gumilla's deliberative road map seeks to eradicate foreign contraband as a pernicious threat to economic progress in Spanish New Granada.

At the same time Gumilla calls for tighter control of trade with foreigners, he denies allegations of the Jesuits' own illegal commerce. Gumilla's defense of one Jesuit brother's questionable trade during the current state of crisis caused by Carib attacks attempts to exonerate all Orinoco region Jesuits, including himself: "En este gravísimo aprieto salió el Padre Bernardo Rotella lejos del Orinoco a comprar provisiones [. . .] sin reparar en costos ni en trabajos, a fin de que el hambre fuese menor y no ahuyentase los indios catecúmenos" (331) [In this serious tight spot Father Bernardo Rotella went out far from the Orinoco to buy provisions . . . without considering the cost or hardships, so that the hunger would be lessened and the newly converted Indians would not be scared away]. To be considered contraband, Gumilla argues, Jesuit trade should be contrary to the public good.[45] This understanding of contraband allowed Gumilla to simultaneously decry the trade with foreigners causing Spain's economic woes while at the same time defend the Jesuits against accusations of illegal trade. Due to dire straits and in order to meet the basic needs of Christians in the Orinoco River region, missionaries have been forced to negotiate with foreigners. All three orders were suffering great need. Gumilla expresses outrage at Capuchin slander against Rotella:

> [F]ue levantar el grito contra él tan alto, que se oyó en Caracas, en Santa Fe de Bogotá y mucho más adelante, achacándole que iba con muy diferentes intentos. De modo que se vió su crédito oscurecido y gravemente denigrado, hasta que ejecutoriada jurídicamente en Santa Fe de Bogotá con declaraciones de testigos oculares la inocencia de dicho padre, se le dio competente satisfacción para restaurar su crédito y estimación debida. (331)

> [The screams against him were raised so loud that they were heard in Caracas, in Santa Fe de Bogotá, and much further away, imputing that he went with very different intentions. In this way he saw his reputation sullied and seriously damaged, until a legal judgment in his favor in Santa Fe de Bogotá accompanied by eyewitness testimony of the innocence of said father enabled him to restore his good name.]

This smear campaign by a Catholic order also engaging in questionable commerce did seem to affect opinions in Spain, where the Council of the Indies sided with the Capuchin order. However, the viceroy in Bogotá and the governor of Cumaná maintained their esteem for the Society of Jesus.

Gumilla's three-step road map to a prosperous future fleshes out specific issues within his thesis for a pathway to success in New Granada that combines conversions, colonization, and commerce. The road map expresses the aspects revealed by the treasure map and graphic map, but is much more deliberative. This temporal map functions as propaganda for increasing the number of soldiers, colonists, and missionaries. Yet again, the Jesuits provide the key to this goal. Without Jesuit evangelization all three steps in Gumilla's road map will fail. Converting the Amerindians remains at the core of his master plan, as this opens the coast and interior to Spanish colonies populated by immigrants who will provide additional labor to extract New Granada's many treasures as well as help defend them. Gumilla ties the route to exploit this viceroyalty commercially with greater control of both foreign and domestic trade. His road map and plans for defense are chronological maps that complement the spatial graphic and textual maps in *El Orinoco ilustrado*. If his calls for its commercial and evangelical efforts and also a military defense are answered, the Orinoco River will serve as New Granada's gateway to commerce, colonization, and conversions.

The Navigational Map for Jesuit Success

The final map we will consider differs from the road map, the treasure map, and the graphic map in that it consists of a practical sermon specifically directed at new Jesuit recruits. The nautical allegory of this sermon he calls "Carta de navegar en el peligroso mar de los indios gentiles" [Navigational map to the dangerous waters of gentile Indians] appeals to their desire for heroic missionary adventures while its commands chart the route to successful Jesuit conversion and colonization. This "navigational map" moves beyond the descriptive or deliberative to the didactic, and its overtly religious discourse and intended audience somewhat distances it from royal state interests in a manner not seen in the three previous maps.

In its original 1741 edition, *El Orinoco ilustrado* concludes with a different propagandistic sermon clearly meant to move fellow Jesuits to take action and fulfill the current needs of the Orinoco River region project. In response to a continued shortage of missionaries, for the 1745 edition Gumilla appends his navigational map to the original "Apostrofe a los operarios de la Compañía de Jesús que Dios se sirviere destinar para la conversión de los gentiles" [Apostrophe to the workers of the Company of Jesus that God has supplied to send for the conversion of the pagans].[46] In this second appendix Gumilla condenses the advice for missionaries spread throughout *El Orinoco ilustrado* into a series of commands for successful Jesuit evangelization, which the book's chapters have already established as the key to New Granada's prosperous future. The recommendations in this navigational map "darán más luz al operario deseoso de acertar" (505) [will enlighten further the worker eager to get it right] and are based on the experience of Gumilla and other Orinoco region missionaries: "[D]oy este corto alivio a los nuevos misioneros de indios, con el seguro de que algunos Padres de las Misiones de Orinoco, que trasladaron, al entrar en ellas, esta carta, vieron después en la práctica que son muy importantes sus avisos" (505) [I give this small comfort to the new missionaries of Indians that some Fathers of the Orinoco missions, who upon entering them put this letter down in writing, later saw in practice that their warnings are very important]. This didactic navigational map takes the form of a three-point sermon. The first point in Gumilla's sermon preaches the necessity of strong character, faith, and humility for any missionary "para navegar en un golfo peligroso" (505) [to navigate in a dangerous gulf] both literal and figurative. The Jesuit's second point emphasizes the need to be prepared with information about the cultures and dangers of the region to which the missionary has been called. The third point enumerates the skills required for success; then Gumilla concludes with several stirring reflections meant to fortify young Jesuits' resolve in the face of "inevitable" challenges and doubts. Gumilla mixes two principal rhetorical strategies in his sermon: allegory and commands. He opens and closes with a pleasing allegorical *captatio* that gratifies Jesuits versed in sacred oratory as well as eager for eighteenth-century seafaring adventure. This erudite nautical allegory frames a practical series of maxims, warnings, and meditations.

When Gumilla published the first editions of his text, he was actively recruiting Jesuit novices in Madrid and Rome. As procurator representing his Jesuit province in the European court, Gumilla worked to convince them that they should resist the facile draw of missionary service elsewhere and instead answer the call to New Granada.[47] At this time, the Society of Jesus suffered competition not only from the Capuchin and Franciscan orders inching up the Orinoco River but also the allure of their own enormously successful mission program in Paraguay as well as more established Jesuit reductions in the Quito province and along the Amazon River. His appendix to the 1745 edition acknowledges "las circunstancias que hacen más difíciles las Misiones del Orinoco" (495) [the circumstances that make the Orinoco missions more difficult]. His navigational map responds to this current crisis of the Jesuit missions. Well aware of the particular challenges fresh recruits would face upon arriving in the Orinoco River region, Gumilla composes a sermon plotting a proven route to success among its tribes.

The first principal strategy of this three-point sermon is the allegory of a ship that must navigate troubled waters. Gumilla chooses this thrilling allegory to craft a *captatio* that engages young missionaries' thirst for adventure and challenges them to prove their bravery. His ecphrastic explication of the allegory is found in points 1 and 2 of the sermon. The first point is subtitled "Del misionero, su vocación y aparejo" (505) [On the missionary, his vocation and nautical rigging], and it divides the ship into parts, enumerating all that is needed to face any "borrascas" (505) [squalls] and to find "la Estrella matutina" (507) [the morning star] in the face of any challenge. This list helps new missionaries perform a thorough inspection to be sure their ship is sound before embarking on dangerous evangelical journeys: "[L]o primero y más importante es mirar y registrar con cuidado la nave, poniéndola en estado competente" (505) [The first and most important one is to look at and inspect with care the vessel, putting it in a capable state]. On Gumilla's allegorical ship, the compass is faith, the anchor is firm trust in God, the precious cargo is love and charity for others, the keel and ballast are profound humility before God, and the sails are only kept aloft through "oración fervorosa" (507) [fervent prayer] and fortifying "santos pensamientos" (507) [holy thoughts]. In point 2 of the sermon, subtitled "Causas principales de disturbios" (506) [Principal causes of troubles], Gumilla explains three main causes of "las tormentas y contratiempos [. . .] frecuentes en el golfo inconstante de las naciones gentiles" (507) [the frequent setbacks and storms in the ever-changing bay of the pagan nations]: fragile ships, Satan's tricks, and Amerindians' deceit, for the Amerindians are "primerosa en el arte, así de maliciar como de engañar" (508) [as greatly skilled in the art of wickedness as of trickery]. This entire section maintains navigational terms: the setbacks are "escollos" (505) [reef rocks] that require superior "maniobra" (505) [navigational ability]; if a Jesuit's vessel is not sound, then he had better not take this journey: "[S]i no se halla firme, fuerte y apta para toda la navegación, que es de por vida, hasta dar fondo en el feliz puerto de la eternidad, mejor será que no salga, porque son fuertes y frecuentes los riesgos" (506) [If it is not found firm, strong, and capable for all the navigation, which will be lifelong, until running aground in the happy port of eternity, it will be better that it not depart, because the risks are severe and frequent]. Gumilla's metaphorical ship, waters, and safe port represent both the routes to a missionary's success and a pathway to his salvation, a key point he recaps in the very last words of his sermon, when he returns to the allegory of his opening *captatio:* "y no se puede dudar que todos aquellos a cuya salvación cooperó le servirán de abogados eficaces en todos sus aprietos, y en especial en la hora de su muerte, término de esta breve navegación y puerto seguro en que de la misericordia de Dios esperamos gozar tranquilidad dichosa y descanso eterno. Amén" (519) [And one cannot doubt that all those souls for whose salvation one co-operated will serve him as effective advocates in all his tight spots, and especially in the hours of his death, end point of this brief navigation and safe harbor in which by God's mercy we hope to enjoy joyful peace and eternal rest. Amen]. Well-educated and adventurous missionary recruits would respond to this nautical allegory and its appeal to their sense of religious mission.

The second rhetorical strategy of the sermon that we will analyze is its explicit commands. In contrast to the allegorical frame that appeals to Jesuit erudition, in the longest section of the sermon, point 3, Gumilla abandons his higher, figurative register to dispense direct and very practical advice. The first series of commands are seven "máximas prácticas" [practical maxims] in paragraph form (each one opening with "the first," "the second," etc.) that dictate seven steps to stave off an impending Amerindian rebellion in their reduction. For example, Gumilla commands young missionaries to stay strong and show only love to avoid scaring off their charges. They must never try to reason with an angry mob, but instead should appeal to the women's natural piety and then use these women to help convince tribal leaders that there is no cause for revolt. Most importantly, Gumilla mirrors the content of his commands with their form. His maxims do not simply tell Jesuits how to reason with Amerindian charges. They mimic the simpler style he is prescribing, thus dramatizing how to deal with them in a direct and practical manner. When Gumilla commences point 3 of his sermon, he makes a deliberate shift to didactic language from the allegorical *captatio.* The advice in his maxims mirrors the evangelizing style Jesuits must adopt with Amerindians. For example, Gumilla warns against calling upon the erudition gained by their Jesuit education. Missionaries must tailor their rhetoric to the particular task:

> La séptima máxima, y de mucha importancia, es que en estos lances no haga hincapié en alegar razones fuertes y de peso para convencer a aquellas gentes; busque razones caseras, insista en ellas, y, según ellos usan, repítaselas muchas veces. [. . .] Estas razones perciben y les hacen fuerza. (511)
>
> [The seventh maxim, and of great importance, is that in these incidents do not require putting forth strong and weighty reasons to convince those peoples; seek out homespun reasons, insist on them, and, as they do, repeat them many times. . . . These reasons they receive and are given authority.]

To maintain peace in the mission, Jesuits should not deal with their charges on a figurative or allegorical level, but instead repeat simple and familiar arguments. The practical rhetoric in the series of maxims serves as a model for persuading Amerindians. Gumilla's final example testifies to his own experience reasoning with Betoye tribal leaders in 1719 as proof that these seven steps can save a troubled mission. The ultimate function of these commands is to instruct new arrivals on the best means for protecting their successful conversion of Orinoco tribes.

In the core of the third point of Gumilla's sermon, a series of practical warnings subtly transition from conversion-focused instructions to guidelines for colonization. These eighteen "avisos prácticos" [practical warnings] are formal commands laid out with Roman numerals. For example, his first warning stresses the importance of cultural understanding for the successful conversion of particular tribes: "[T]ire a conocer bien el genio de la nación que cultiva y según él tenga meditados medios proporcionados" (512) [Make an effort to get to know well the nature of the nation you are cultivating and according to it think out the means in proportion]. But by the fifth Gumilla is emphasizing their colonization by mandating the languages for doctrinal instruction: "[L]a doctrina enséñala por la mañana en su lengua natural y a la tarde en castellano, porque en lo primero se sirve a Dios y en lo segundo al rey nuestro señor, que ordena se establezca en las Misiones la lengua española" (513) [Teach doctrine in the morning in their native language and in the afternoon in Spanish, because with the first God is served and with the second the King our Lord, who orders that the Spanish language be established in the missions]. The next warnings focus on converting Amerindians to European mores with the ultimate goal of self-governence: "[C]onviene desde los principios irlos imponiendo en el gobierno políticos y señalar alcaldes" (516) [It is advisable that from the beginnings they get used to political government and appointing mayors]. These newly civilized colonies can best foster productive agricultural labor by designating the first harvest for charity: "para las viudas pobres, para los huérfanos y para los enfermos; y sucede que viendo los indios cuán bien se emplean aquellos frutos, renuevan con gusto la sementera en adelante" (514) [for the poor widows, for the orphans and for the sick; and it happens that the Indians seeing how well those fruits are employed, renew with pleasure the seeding from then on]. Commands 11 through 15 enumerate further civilizing steps toward European colonization: census taking, hospitals, blacksmithing, spinning shops, music lessons, and rules for ordering society. This second series of commands in Gumilla's sermon voices the second essential ingredient in his master recipe for success (conversion, colonization, and commerce) by shifting the focus from successful evangelical tactics to instructions on how to civilize and colonize the converted Amerindians. By reforming their customs, Jesuits prepare them for peaceful coexistence with European colonies.

The final series of commands in Gumilla's navigational map appear under the subtitle "Reflexiones que animan y fortalecen el ánimo del misionero de indios" (517) [Reflections that encourage and fortify the spirit of the missionary of Indians]. Ten final paragraphs set off with Roman numerals conclude step 3 of his sermon. Like the maxims and warnings, these meditations shift registers. Gumilla alternates the addressees in this series of commands between "I," "you," and "we," thus appealing to the wider Jesuit community. For example, Jesuits should repeat the first meditation to themselves—"¡Cuánto debo yo apreciarlas!" (517) [How much I should appreciate them!]—as a constant reminder that Amerindian souls, if sometimes hidden within "conchas toscas" (517) [rude shells], are precious pearls for which Jesus died. Next, through second-person-singular familiar command forms, Gumilla adopts an intimate register with three mandates directed at "tú" [you]. For these, he shifts from an authority figure dispensing practical advice in formal commands to a father figure reassuring young Jesuits that their mission as an instrument of God will save their own souls by saving the "toscos" (517) [rough] ones in their charge. Gumilla returns to his nautical allegory to stress how the dangerous journey of any Jesuit called to the salvation of Amerindians leads to God's "puerto seguro" (519) [safe harbor]. To appeal to the Jesuits' sense of community and purpose, Gumilla extends his meditation on Amerindian souls to the first-person plural: "[M]erecen toda nuestra estimación; y el mirar por ellas es hacer nuestro mayor negocio" (517) [They deserve all our respect; and looking after them is to make our best profit]. The last four meditations shift back to the third-person singular, employing a series of formal commands to dispense more generalized strategies to sail on with humility and generosity of spirit and stay the course of conversion and colonization. Although recalling specific Jesuit triumphs (such as those by Father José Ancheta in Brazil and the heroic martyr Luis de San Victores in the Philippines named by Gumilla in these final reflections) can strengthen a missionary's resolve when one feels alone and uncertain, ultimately "no hallará otra defensa ni otras armas que las del recurso a Dios en la frecuente oración y meditación de algunas de estas reflexiones" (518) [he will not find another defense nor other weapons than those of the appeal to God in frequent prayer and meditation on some of these reflections]. Even though outward Jesuit success abounds—"en todas las provincias de Indias hallará muchos y admirables ejemplares" (519) [in all the provinces of the Indies he will find many and wonderful models]—to ward off the inevitable doubts and fears of this vocation, Jesuits must look inward to prayer and meditation.

In contrast to the actual map, the treasure map, and the road map, this final navigational map in *El Orinoco ilustrado* incorporates allegory and commands to respond to the lack of Jesuit missionaries to New Granada and the lack of skills of those who have answered the call to evangelize along the Orinoco River. If at first glance Gumilla's navigational map appears to focus only on the first of the three essential ingredients for New Granada's successful future (namely, conversion), a closer examination reveals implications for the other two as well (namely, colonization and commerce). Just as the commercially valuable treasure map leads readers to a text that inextricably ties the Jesuits' evangelical agenda to the colonization, defense, and commercial benefits of New Granada, so too does Gumilla's navigational map provide instruction for both conversion and colonization. While his treasure map reveals the region's obvious commercial benefits and his road map to the future overtly blends religious and political discourse, Gumilla's navigational map implicitly links colonization and conversion with commerce. His choice to conclude *El Orinoco ilustrado* with a three-point sermon outlining clear instructions for Jesuit missionaries reinforces their key role for the viceroyalty's success.

Taken together, the four maps we have examined chart enlightenment in New Granada as a three-step process: convert the Amerindians, colonize the coast and interior, and commercially exploit the viceroyalty. *El Orinoco ilustrado* responds to current challenges to Spanish and Jesuit interests, and Gumilla centers both his graphic map and his textual maps on defending the Orinoco River as the entrée to colonization, commerce, and conversions. Throughout *El Orinoco ilustrado,* the Society of Jesus remains at the core of Gumilla's argumentation. Still, this Jesuit targets a diverse audience by defending his order, touting the region's commercial and spiritual treasures, and proposing military and political tactics. In the first modern cartographic map of the region, Gumilla symbolically represents past Jesuit progress and future possibilities for the scientific, spiritual, and economic enlightenment of New Granada. This foldout visual aid complements the entire text of *El Orinoco ilustrado,* which itself can be read as a descriptive map that reveals the Orinoco River as the gateway to colonial, evangelical, and commercial success. While on one level *El Orinoco ilustrado* represents a straightforward plan for colonization in order to best exploit commercial potential, this agenda is, in fact, rooted in evangelization and exemplifies the dual nature of the Hispanic Enlightenment, which is both spiritual and intellectual. Behind Gumilla's plans for immigration, commerce, and defense—unquestionably Enlightenment issues for many countries—lies Spain's Catholic agenda. A

Catholic Orinoco is the goal, and Gumilla defends not only the region but also the Jesuits' role in its protection and development.

The Enlightenment cannot be reduced to a secularization of society and its corresponding economic growth and reform. Studying texts previously not regarded as enlightened can broaden traditional conceptualizations of the Enlightenment. The next two chapters expand on this inextricable link between commerce and evangelism that helps us understand the Hispanic Enlightenment by interrogating the Jesuits' role in the formation of a modern scientific identity for Spain and her colonies. They continue our examination of the dual nature of the spiritual and intellectual Enlightenment in Spain and her possessions by revealing pathways to knowledge that alter restrictive definitions of modernity and refute stereotypes about tardy or nonexistent Enlightenments in Catholic countries.

3
The Jesuits and an Eclectic Enlightenment: Mediating Knowledge and Catholicism

Introduction

In the last quarter century, historians of science have discredited the myth that natural philosophy developments (the "new sciences") were unilaterally transmitted from centers of knowledge in northern Europe to southern European nations and peripheral colonies.[1] By eschewing the teleological sequence of heroic discoverers who maintained, from the sixteenth through the seventeenth centuries, a steadfast progression away from purely biblical explanations toward purely empirical laws of nature, historians such as Peter Dear, Bruno Latour, and Simon Schaffer are altering the traditional narrative of Western Europe's "scientific revolution" and, by extension, opening the door to broader interpretations of the Enlightenment.[2] Studies by Lorraine Daston, Dorinda Outram, and Roy Porter have affirmed for northern Europe the continued importance of amateur naturalists in the face of increased professionalization of the sciences during the long eighteenth century.[3] They also affirm the continued moral and symbolic value of nature despite lasting perceptions about the supposed development of a passionless rationality free from religious dogma during the Enlightenment. Current trends in the social history of science are replacing the fallacy of Eurocentric linear transitions to a modernity rooted in secular shifts in epistemology[4] with "the *spatial* and the *local* . . . a more contingent, more cultural image of what the sciences are, and how they have been constructed."[5] However, as several historians of science have recently reaffirmed, Spain and Spanish America continue to be marginalized or excluded from these productive reevaluations that are expanding the master narrative of the construction of modern science.[6]

As new scientific methodologies circulated, eighteenth-century Catholic natural philosophers mediated innovations within Christian paradigms instead of divorcing religion from science. Their preservation of nonsecular pathways to knowledge has caused historians to exclude the Iberian world from Enlightenment-era progress and reform. In the narratives of modernity that have their origins in the eighteenth century, explains Walter Mignolo, starting with the Catholic Counter-Reformation "Spain and the Latin Countries . . . began to slip out of the march of progress," falling victim to a myth of northern Protestant progress in contrast to southern Catholic "backwardness" that continues today.[7] In fact, many current scholars' assumptions echo the negative accounts of Spanish scientific and social progress first published over two hundred years ago in encyclopedias. As Masson de Morvillier famously concluded in his 1782 encyclopedia entry on Spain for the *Encyclopédie méthodique:* "Que doit l'Europe à l'Espagne?" The answer to this rhetorical question was "Very little."[8] Disregard for Spain and Spanish America persists today; they have been largely omitted from rigid conceptions of the scientific revolution and displaced into the peripheries of definitions of the Enlightenment. As Juan Pimental reminds us: "Western historiography has always had its centers and peripheries. . . . It is not by chance that a triple perspective (scientific, colonial, and historiographic) from the north of Europe (and by extension, from the Anglo-Saxon world), has tended to consider the Iberian world as episodic and marginal, without a doubt peripheral, and even lacking in interest."[9]

Currently, postmodern and postcolonial scholars are challenging the myths of center and periphery. Persistent stereotypes about the consolidation of "new sciences" in the north but not the south (both in Europe and the Americas) underscore the need for further attention to the multicultural realities at play in so-called peripheries in order to provide alternative narratives that expand cultural perspectives and debate the centrality of northern Europe as site of the production of "modern" knowledge.[10] As Ralph Bauer notes, by examining "the cultural developments on either side of the ocean within the context of transatlantic imperial systems" scholars can reveal how "modernity is the product of the complex and inextricable *connectedness* of various places and histories, of the way in which these places *acted upon each other.*"[11]

In the effort to overcome stereotypes about Spanish and Spanish American resistance to modernity and ignorance of modern science, one of the principal stumbling blocks is religion. Traditional narratives of the early modern world's transition from the Renaissance to the Enlightenment cite the Inquisition and increased Catholic conservatism after the Council of Trent as preventing scientific advances in the Iberian world. Even though historians of science have studied the accommodations and reconciliations of Spain's eclectic philosophy (*filosofía ecléctica*), or eclecticism (*eclecticismo*), the complexities of its blend of Catholic faith and modern science are only beginning to be fully appreciated.[12] Further study of Jesuit contributions to scientific progress and their influence on

Catholic intellectuals can help us understand Spain and Spanish America's uniquely eclectic Enlightenment. Far too often, the Jesuits have been "assigned the role of dry and uncreative opponents to progress" during the scientific revolution and the Enlightenment by historians willing "to credit the Protestants with the rise of modern science" while deliberately ignoring Catholic contributions.[13] However, multinational Jesuits greatly increased the reciprocal flow of science, both transatlantic and within Europe. Eclecticism took root from the first years of the Society of Jesus, thanks to its mediation of knowledge gathered in mission settlements in Asia and America with Catholicism and European natural philosophy.[14] Early on, the Jesuits extended their educational mission for younger members of the society to the general public. Their *Ratio Studiorum* (1599), implemented in colleges and universities on both sides of the Atlantic, was conceived as a flexible curriculum that could be adapted and revised; in practice it would deviate from Aristotelianism when deemed necessary *ad majorem Dei gloriam* (for the glory of God).[15] The Jesuits' diverse influences as educators, missionaries, naturalists, and confessors to the elite allowed them a privileged role in the transition to Hispanic modernity.

Recent studies of the Society of Jesus's preexile global networks and of the scholarly work of Jesuit university professors and amateur naturalists offer new insights into a transatlantic Catholic Hispanic Enlightenment. For example, in a 2005 article in *ISIS,* Stephen J. Harris emphasizes the continued ties between science and Jesuit evangelism in Spain and Spanish America, "despite the modernist perception of the Enlightenment as a fundamentally secular, and secularizing, movement."[16] As he concludes, "Catholic education at home, edifying reports on the wonders of the natural world from abroad, and conversion of nonbelievers throughout the world to the Catholic faith underscores the intricacies of the science-and-religion complex forged by the Jesuits of the period."[17] Diverse scholars are taking a fresh look at the Society of Jesus and resisting a long-standing dichotomy in scholarship that either vituperates Jesuits or excessively praises their virtues. More objective studies are superseding both habitual condemnation as well as Jesuit apologetic discourse, and instead are employing a cross-disciplinary analysis of their constant scientific activity alongside theological and philosophical teaching. This welcome shift in scientific historiography has naturally generated increased attention to Iberian colonial science. Prominent journals, chapters in anthologies, and a new volume on eighteenth-century science in the *Cambridge History of Science* series are contributing to a wider account of Europe and the Americas.[18] This trend in the history of science is less technical and more social and has encouraged scholars to examine scientific networks and reciprocal exchanges of knowledge in both the intercontinental and transoceanic realms. Increased attention to Jesuit mediation of non-European ways of knowing and the corresponding role of "peripheral" types of knowledge in the evolution of the Enlightenment in Europe will further expand our understanding of the development of modernity beyond traditionally Eurocentric paradigms.

In the transition to modernity for Spain and her colonies, the production of knowledge served both intellectual and religious functions. This chapter argues that long before exile the Jesuits played a central role in the formation of a modern scientific identity for Spain and Spanish America, and that they did so in a way that contributed to a different form of the Enlightenment. Starting in the sixteenth century, Catholic intellectuals mediated pious purposes for natural history and natural philosophy with evolving empirical methods. The Society of Jesus was essential to this mediation and to Spain's Catholic philosophy of knowledge, which, alongside the intellectual and economic utility of knowledge, maintained the Christian moral value of investigating nature during first half of the eighteenth century. The pages that follow will take a closer look at how the Jesuits greatly influenced Hispanic eclectic methodology, as revealed in scientific treatises published by Catholic intellectuals during this formative period in the eighteenth century. Additionally, in service to its pious acknowledgment of the limits of human understanding, Spain adopted its own version of Francis Bacon's skepticism.[19] The Benedictine friar Benito Jerónimo Feijoo helped remove any anti-Christian undertones to the label of skeptic and fostered the spread of Spanish Baconianism.[20] Notably, both Feijoo and Bacon were eclectics indebted to the Jesuits.[21] Although much work remains before we can properly appreciate the Hispanic Enlightenment's debts to the Society of Jesus, the second part of this chapter as well as the examples in chapter 4 offer Joseph Gumilla's contributions to this eclectic Enlightenment as an important lens for viewing Jesuit mediation of Catholic dogma with European scientific discourse and Amerindian knowledge. By examining Gumilla's eclecticism and roles as a mediator in his natural and civil history, we can appreciate how *El Orinoco ilustrado* presents an eclectic Jesuit theory of knowledge that further rejects assumptions about centrality and periphery, contradicts canonical Enlightenment patterns, and helps us redefine the concept of modernity.[22]

The Jesuit Eclectic Tradition

As Amos Funkenstein reminds us in *Theology and the Scientific Imagination from the Middle Ages to the Seventeenth Century*, the Jesuits' symbiotic accommodation between religion and science plus symbolic interpretations of nature through which man can perceive God's presence have roots in the church fathers' Christianization of pagan philosophers. Medieval Thomism, Renaissance Aristotelianism, and Augustinian Neoplatonism all subordinate natural philosophy to theology.[23] Within the hermeneutic of early modern Christian humanism, knowledge is received either directly from the divine through scripture or by way of reading God's "book of nature." Authoritative readers such as Aquinas and Augustine mediate the revelation of *scientia,* the knowledge of nature, in order to advance *Scientia,* the pious comprehension of God's omnipotence over his creation. In general, sixteenth-century Christians forced data gleaned by their firsthand experiences with nature within known paradigms to avoid heresy, whereas during the seventeenth century natural philosophers began to allow the possibility of challenging received authorities when necessary to explain phenomena that did not fit within known philosophies of nature.[24] By the early eighteenth century, epistemology was reduced to two general means of knowing: the systematic and the experimental. Post-Cartesian systems modernized ancient systems of knowledge while also maintaining Aristotelian syllogisms and deductive reasoning as the path to knowledge. Via mathematical logic, particular hypotheses were deduced from general cases or "universal essences." In contrast to this systematic analysis (which had as its starting point existing axioms), Enlightenment-era empiricism (experimental methodology) offered novel ways of establishing "truths" about the world. Inductive reasoning formed the core of this new scientific method, which began with a hypothesis based on firsthand observations. Ideally, general conclusions were not formed before direct experience with a particular case or multiple experiments tested a hypothesis. The experimental pathway led to laws, principles, or even mathematical formulas. The new empiricism depended on induction and advanced from particulars to generalizations about nature and natural phenomena instead of deducing particulars from established principles.

Far from being a Jesuit invention, philosophical eclecticism can be traced back to the origins of Christianity.[25] The first Christians mediated pagan knowledge with Christianity. Later, throughout the early modern period, Catholic missionaries maintained this early tradition by reconciling their charges' "pagan knowledge" with European natural philosophies. From the sixteenth century until their disbanding, Jesuits mediated Asian and Amerindian knowledge with European natural philosophies. Jesuit eclectics also reconciled divine and human authorities and mediated emerging systematic and experimental methodologies. The main difference between ancient eclecticism and modern eclecticism is that the latter mediated the new philosophies of the seventeenth and eighteenth centuries. In Spain, Jesuit scholars served as model participants in a *via media* approach of the eclectic philosophy that would be adopted by Catholic intellectuals.[26] By the first half of the eighteenth century the Society of Jesus was at its evangelical and scientific peak, and Jesuit eclecticism was at the center of a Catholic Hispanic Enlightenment that continued to mediate between religious dogma and emerging experimental methods, and between post-Cartesian rationalism and Baconian skepticism. And Spanish intellectuals were very aware of its importance for the advancement of knowledge: "Jesuit eclecticism and the cautious but progressive manner whereby members of the Order approached modern science were well suited to the Spanish environment. . . . Thus, Spaniards who favored innovation in these areas, even those who were not members of the Society, embraced the Jesuits for their effort to introduce the new science in Spain."[27]

From its earliest years, the Hispanic Enlightenment fostered a flourishing culture of knowledge accumulation and production for Spain and her colonies. One example was the Spanish Royal Academy of Language (an academic society founded in 1714), which brought together Jesuits, Crown officials, noblemen, university scholastics, and independent scholars. Among the topics discussed under the rubric of "new philosophy" were efforts to reconcile competing scientific methodologies and modern systematic and experimental developments with received knowledge from the ancients and church fathers. In the open philosophical climate of this Catholic *tertulia,* challenges to "ancients" such as Aristotle were not automatically branded heretical. Instead, "moderns" such as Bacon, Descartes, Gassendi, Malebranche, and Newton were debated, mediated, and integrated into natural histories and theological dissertations.[28]

The Jesuits fostered a pious eclecticism throughout the Spanish territories. And at the same time, their epistemology clearly benefited from overseas evangelism. Jesuit missionaries adhered to a strategy of "inculturation" that allowed their New World mission sites to become unique spaces of reciprocity, accommodation, appropriation, and mediation. In fact, mission culture proved particularly important to Spanish science. By studying previously unknown minerals, plants, animals, and men, the Society of Jesus not only opened Spain to copious new information, but also to innovative ideas. Jesuits were constantly negotiating new data with new philosophies, not only those emerging in Europe, but also Amerindian approaches to knowledge. At the same time the Jesuits adapted empirical data gathered in the missions to their Christian Neoplatonic tradition and European hierarchy of knowledge, they also had to reconcile these Old World philosophies with startling New World realities. Although Jesuit authors of natural histories and professors of natural philosophy consistently wrote and taught under the rubric of Western theories of knowledge, an insistence on knowledge accumulation alongside evangelization fostered a syncretic Jesuit epistemology that mediated between local *scientia* and strict academic training. In the second part of this chapter, we will consider Gumilla's syncretic eclecticism. But first, we will look at the state of Spanish eclecticism in the first half of the eighteenth century, when both Jesuits and Jesuit-educated intellectuals were mediating Spain's centuries-old Christian heuristic tradition with the "new philosophies," in contrast to some "moderns" who adopted them in a secular manner that displaced God. Spanish and Spanish American philosophers were well aware of the epistemic tensions between divine and human authority and between deduction and induction while establishing an enlightened path for discerning the laws behind natural phenomena. However, instead of breaking away from medieval and Renaissance Christian paradigms, these Catholic intellectuals consciously mediated rational, analytic systems and experimental quests for general principles for nature with their belief in God as the Creator of all nature. Jesuit educators and missionaries were role models for this eclectic, Catholic theory of knowledge.

The Rhetoric of Enlightened Eclecticism

Instead of secularizing knowledge, Spain's eighteenth-century understanding of philosophical eclecticism encouraged a productive overlap of Catholicism and natural science. As role models for eclecticism in Spain and Spanish America, the Jesuits stimulated eclectic methods that maintained inextricable links between religion and science. Their theory of knowledge proved flexible, mediating the scholastic tradition of deductive reasoning with the new discoveries afforded by firsthand experience and inductive reasoning and justifying new scientific methodologies to better know and glorify God. Catholic intellectuals echoed this justification in order to incorporate the "new tools" of modern science into Spain's long-standing Aristotelian tradition. A brief examination of contemporary natural philosophical treatises reveals the legacy of a Jesuit discourse that closely ties natural philosophy to theology. Under a rubric in which all philosophical pursuits of "truth" find value in supporting the Catholic faith but the modern experimental path is suggested as the most valuable, these writings also share a few basic rhetorical strategies that defend methodological innovations in the face of Catholic and scholastic traditions. The first strategy is to root modern philosophies in these ancient traditions. The second entails challenging the human authority of the church fathers with regard to earthly truth when the study of nature introduces new knowledge that conflicts with patristic teachings.

The Jesuit theory of knowledge (*ad majorem Dei gloriam*) clearly tied natural philosophy with theology. During the first half of the eighteenth century, the Hispanic Enlightenment maintained these close ties, which proved central to the eclectic approach of Catholic intellectuals, Jesuits, and other religious scholars such as Feijoo. Among the Catholic eclectics who did not belong to religious orders, the most prominent was the Valencian Andrés Piquer y Arrufut (1711–72), who held a key position in the courts of Philip V and Ferdinand VI. Spanish historians of science dedicate entire chapters to this royal physician and author of several scientific treatises published between 1745 and 1768.[29] They hail Piquer and his mentor, Gregorio Mayans y Siscar, as well as Tomás Vicente Tosca and Juan de Cabriada, as the leaders of Valencia's flourishing Catholic culture of philosophical eclecticism.[30] Piquer's natural philosophical writings particularly embody Spain's eclectic philosophy. They consistently echo the Jesuit maxim that justifies quests for knowledge by framing any human examination of nature as a means to glorify God, as when they say, "[L]evanta el entendimiento del hombre à considerar los inmensos atributos de Dios"[31] [Raise man's intellect to consider the immense attributes of God"]. Piquer clearly reflects the contemporary belief that the primary purpose of natural philosophy (or, *física:* "physics" in the early modern sense of the word) was to better know the divine. For example, in his *Física moderna, racional y experimental* [*Modern, Rational, and Experimental Physics*] (1745) Piquer voices a stance that could just have easily been a lecture topic in a Jesuit college: "Debe, pues, ser el fin principal de este estudio el conocimiento de aquel Sér infinito"[32] [The principal purpose of this study should be, then, knowledge of that infinite Being]. Additionally, by arguing not only that natural philosophy and Catholicism reinforce each other but that modern philosophical inquiries could serve as useful tools to improve Christian persuasion, Piquer voices the evangelical value of science: "[L]a Philosophia [. . .] sirva para mejor persuasion de las verdades Christianas, y para mayor claridad, è inteligencia de ellas"[33] [Philosophy . . . serves the better persuasion of the Christian truths, and the greater clarity, and intelligence of them], a stance reaffirmed by Jesuit missionaries around the globe.

We can also see the link between natural philosophy and theology in Spanish ecelctics' stated preference for experimentation and incorporation of inductive reasoning in order to better appreciate God. While the overarching position behind eclecticism clearly encouraged mediation of various philosophical approaches, in practice Catholic intellectuals of the first half of the eighteenth century privileged experimental strategies.

That is, the pathway to knowledge of Spanish eclecticism was generally antisystematic. In the opening pages of his *Física moderna, racional y experimental* Piquer affirms that, for Spanish natural philosophers interested in reaching the "truth," the experimental path proves much more valuable than the systematic one.[34] In his widely read textbook, *Discurso sobre la aplicación de la Philosophia a los assuntos de religion para la juventud española* [*Discourse directed at Spanish youths on the application of Philosophy to religious topics*] (1757), the Valencian doctor brings together the thoughts of many eclectic philosophers greatly indebted to the Jesuits. Here again, the experimental methodology is privileged and praised for opening the door to many "marvelous operations of nature" that had been hidden from the ancients and even the fathers of the Catholic Church.[35] The Catholic eclectics' turn to experimentalism and to the process of induction culminates with a paradigm shift away from Scholasticism's abstract systematic thinking and Aristotelian deduction.

Not only did the Jesuits' flexible incorporation of new experimental tools for increasing the glory of God serve as a model for eclectics, but their writings modeled effective strategies for introducing modern methodologies. The first rhetorical strategy that eclectic natural philosophers borrowed from Jesuits continues to be mistaken for a conservative resistance to innovation. Jesuit natural philosophers such as Noël Regnault (1683–1762) famously claimed ancient origins for the "new physics." Yet Regnault's influential treatise, *L'origine ancienne de la physique nouvelle* [*The ancient origin of the new physics*] (1734) should not be interpreted to mean that with eclectic philosophers "there is nothing new, nor any rupture" with the ancients.[36] Affirmations like these from today's historians help to preserve the stereotype that Catholic conservatism prevented scientific renovation in Spain and her colonies. Some modern historians of science continue to mention that Spanish eclectics merely completed and clarified ancient thought and practice, introducing new knowledge only when it did not conflict with received authorities.[37] No doubt, some concepts in modern philosophy had roots in the ancients.[38] But when Catholic intellectuals throughout the first half of the eighteenth century underscored ancient roots for modern philosophy, they were repeating a Jesuit rhetorical strategy used to appease contemporary fears of heresy by pointing out pious, ancient roots for modern philosophies. For example, when the Jesuit chair of mathematics in Madrid's Imperial College wrote in his review of a 1738 treatise on "modern experimental physics" that "la Philosophia moderna (que yo no sé por qué se llama assi, siendo tan antigua; mas propio le era el nombre de resucitada)"[39] [modern philosophy (and I do not know why it is called that, it being so ancient; more appropriate for it was the name of "resuscitated")], he was not truly claiming that there was nothing new in modern philosophy. This was actually a rhetorical justification for methodological renovation.

Eclectics did not evoke the ancients to prevent incorporation of the moderns, a fact made even clearer by considering a second, simultaneous rhetorical strategy for justifying their mediation of modern methodologies: questioning the church fathers' human authority. In order to depart from the Aristotelian philosophy upheld by St. Thomas Aquinas yet remain good Catholics, Jesuits and eclectics established clear boundaries between the earthly truths and divine truths. Again, Piquer's *Discurso sobre la aplicación de la Philosophia a los assuntos de religion para la juventud española* offers an excellent example. Here, the study of nature affords understanding of "las cosas humanas, y terrestres, como la naturaleza de los Elementos, la fabrica y composicion de los cuerpos, y otras cosas semejantes" [the human, and earthly things, like the physical constitution of the elements, the production and composition of the substances, and other similar things],[40] which differs greatly from the expression of church doctrine, or, "las cosas Divinas, y verdades de la Religion Cristiana" [the divine things, and truths of the Christian religion]. This rhetorical separation allows eclectics to argue that the church fathers' opinions about nature do not hold the same authority as their expression of "Christian truths." At the same time their religious writings are indisputable for representing the divine, their scientific writings represent human opinions and are thus open to dispute just like any other natural philosopher's.[41] For this reason, experimental philosophers in Spain and Spanish America could openly challenge these patristic authorities when their experience and new knowledge contradict them. As Piquer reaffirmed: "[N]adie debe extrañar, que los modernos se aparten algunas veces de Santo Thomas, de S. Buenaventura, y a otro SS. Doctores, quando sus opiniones no convienen con la experiencia, ó se hallan otras mas propias para explicar la naturaleza"[42] [Nobody should be surprised that the moderns sometimes move away from Saint Thomas, Saint Bonaventure, and other holy doctors when their opinions do not agree with experience, or other more appropriate theses are found to explain nature]. Piquer's *Lógica moderna* (1747) provides an excellent example of the articulation of this Jesuit approach to knowledge that blended reason, authority, and experience. For natural philosophical inquiries, received authorities held less weight than observation-based and experimental reasoning.

Jesuits and eclectic natural philosophers privileged their personal experience and reason over outside authorities. During the first half of the eighteenth century they further distanced themselves from scholastic traditions by rhetorically questioning the human authority of Aquinas. For example, Piquer suggests that Aquinas is even more human than other church fathers (*Santos Padres*) by claiming that Aquinas was only a "church doctor" (*Doctor de la Iglesia*) and not a real church father (*Santo Padre*) and that St. Bernard was the last church father.[43] Treating Aquinas as a "doctor" rather than a "father" of the church was an extension of the rhetorical challenge to church fathers. This rejection of Aristotelianism allowed a methodological preference for the "moderns" over the "ancients."[44] Piquer's textbook, *Discurso sobre la aplicación de la Philosophia,* openly rejects Aristotelian philosophy as insufficient:

> La Philosophia Aristotelica no es suficiente [. . .] porque como en su Physica no se descubre la naturaleza por el camino de la experiencia, ni se averigua la fuerza, y resistencia de los entes corporeos, ni sus movimientos generales, y proprios, y mucho menos las leyes con que el Criador los ha enlazado entre sì para componer la gran fábrica del Universo, antes bien toda ella consiste en formalidades, y abstracciones, que conducen poco para el examen de estas cosas, por esso es insuficiente para la averiguacion de ellas.[45]

> [Aristotelian philosophy is not sufficient . . . because since in his *Physics* nature is not revealed by the experimental path, nor is the force, and resistance of corpuscular bodies, nor their general and own movements, investigated, and much less the laws with which the Creator linked them together to compose the great fabric of the universe; quite the opposite, it all consists of formalities, and abstractions, that lead almost nowhere for the study of these things. For this reason it is insufficient for the investigation of them.]

Piquer's express purpose in writing his earlier treatise, *Física moderna, racional y experimental* had been to address the deficiency of Aristotle's system by providing the first comprehensive treatment in Spanish of "modern experimental physics."[46] Other contemporary treatises by Spanish eclectics championed "modern physics" over Aristotelian philosophy. For example, in his *Physica moderna, experimental, systematica donde se contiene lo mas curioso, y util de quanto se ha descubierto en la Naturaleza* [*Modern, experimental, and systematic Physics, containing the most curious and useful of all that has been discovered in Nature*] (1738), Antonio María Herrero framed his support of the moderns' approaches to studying nature as more useful to theology than the philosophy of the ancients: "[L]a Philosophia moderna es mas util, que la antigua, para el estudio de la Theologia, para el conocimiento de la naturaleza, y para la comodidad de la humana vida"[47] [M]odern philosophy is more useful than ancient philosophy for the study of theology, for knowledge of nature, and for the convenience of human life]. A related strategy for justifying a turn away from Aquinas and Aristotelianism was to note that even the church fathers judged Aristotle harshly. In his *Discurso sobre la aplicación de la Philosophia,* Piquer writes:

> [E]s una valiente impugnacion de la materia, forma, y privacion, en el modo que los propone Aristoteles, como principios del cuerpo natural. Yo quisiera la leyessen con cuidado los Philosophos demasiadamente afectos à Aristoteles, para que viessen, que los Padres que defendieron la Religion Christiana con tanto espirituo de verdad, y de sabiduria, impugnaron à este Philosopho. [. . .][48]

> [It is a powerful refutation of the material, form, and deprivation, in the manner in which Aristotle proposes them, as principles of the natural body. I would like the philosophers too fond of Aristotle to read it carefully, so that they can see that the fathers who defended the Christian religion with so much spirit of truth, and of wisdom, refuted this philosopher. . . .]

As noted above, eclectics believed that to better understand God's laws of nature, the modern experimental path supersedes systematic abstractions, whether ancient or modern.[49]

By rhetorically rooting eighteenth-century eclecticism in ancient eclecticism, yet also challenging received authorities' human wisdom, Catholic intellectuals justified renovations in the advancement of natural knowledge. Piquer again serves us well as the quintessential eclectic asserting his freedom to pick and choose from whatever approach best serves the particular situation: "Yo sigo la Filosofia Eclectica, esto es, aquel modo de filosofar, que no se empeña en defender sistema alguno, sino que toma de todos lo que parece mas conforme á la verdad"[50] [I follow the eclectic philosophy, that is, that manner of philosophizing that does not insist on defending any system, but instead takes from all that which seems most in accordance with the truth"]. Piquer's approach also models the reverence for church fathers other than Aquinas who practiced earlier forms of philosophical eclecticism. For many Catholics, St. Augustine offered the original eclectic model:

> San Agustin, despues de su conversion, no fuè addicto à ninguna Philosophia [. . .] hizo lo mismo que hacen los que professan la Philosophia Eclectica; pues sin atarse à Systèma ninguno, de todos toman lo verdadero, y estiman en mas aquel Philosopho, cuyos dictamenes les parece se acercan mas à la verdad.[51]

> [Saint Augustine, after his conversion, was not a follower of any philosophy . . . he did the same thing that those who profess the eclectic philosophy do; that is, without tying themselves to any system, from all they take that which is true, and they respect even more that philosopher whose opinions seem to them to come closer to the truth.]

Again, claiming ancient roots for modern methodologies (for example, positing Augustine and others as model eclectics) did not restrict modern eclectics to the philosophical schools of the ancients. Rather, it was a rhetorical justification for their eighteenth-century advances that opened the study of nature to modern European ideas and practices that also departed from Thomism. By embracing the experimental method as a pious tool for increasing natural knowledge, and rhetorically underscoring ancient foundations for modern philosophies while also engaging the church fathers' traditional authority for divine and earthly teachings, Spanish Catholics justified their eclectic approach to scientific methodology and evoked the simultaneously spiritual and scientific discourse of Jesuits and Jesuit-educated intellectuals.

Feijoo and Spanish Baconianism

The rhetorical strategies above demonstrate a few ways that Catholic intellectuals distanced themselves from Scholasticism in order to embrace experimental over systematic methodology. When Spaniards turned away from Aristotelianism and toward renovated empiricism, they privileged Francis Bacon over modern systematic philosophers. For example, Piquer rejected Descartes' systematic deduction in his *Física moderna, racional y experimental,* contrasting it with Baconian empiricism: "[A]penas puedo conformarme con una pequeña parte de su Filosofia. [. . .] Lo contrario aconsejaba Bacon el Verulamio, repitiendo muchas veces, que no debe en la naturaleza buscarse lo que piensa el entendimiento, sino lo que enseña la experiencia. [. . .]"[52] [I can barely resign myself to a small part of Descartes' philosophy . . . Francis Bacon, Lord Verulam, advised the opposite, repeating many times, that one should not search for what the mind thinks, but instead what experience teaches]. Another famous Valencian professor of medicine who affirmed reason, experience, and experiments over received authorities prefaced his natural philosophical treatise by noting "ecclesiastical approbations declaring Bacon's principles entirely safe with respect to 'our holy faith and good customs.' "[53] Neither Piquer nor his fellow Valencian, Juan de Cabriada, concerned themselves with Bacon's Protestantism, his appearance on indices of prohibited books, and his reputed separation of theology and natural knowledge. In fact, eclectics on both sides of the Atlantic seemed unconcerned with Bacon's provenance. They simply chose the aspects of Bacon's philosophy that fit their Catholic goals and values. Jonathan Israel recently explored this trend and concluded that "empiricism and British ideas were indeed the lever which shattered the scholastic stranglehold on Iberian culture and shaped the Iberian and Ibero-American Enlightenment."[54] Other scholars have explored Catholics' "cultural reduction" of Baconian philosophy.[55] Spanish eclectics reduced the British empiricist's natural philosophy to a mantra in favor of observation, experience, and experiments over "imaginative systems" like Aristotle's or Descartes'. As we will see, they revered Bacon as the model of "good taste" in inquiries in natural philosophy. The Spanish conception of Baconianism was influenced by the Jesuits, but it was the Benedictine father Feijoo who provided the best-known mediation of Jesuit sources and Baconianism.

Following the same course as most eighteenth-century Jesuit intellectuals when challenging Aristotelianism, Catholic eclectics favorably evoked Bacon's "new" inductive tool. Along with modern systems, Baconistas openly challenged Aristotle's deductive organon, or tool for natural philosophical inquiries. Instead they emphasized firsthand experience and experiments and crafted their own version of the Baconian induction prescribed by Bacon's *New Organon.* As noted above, Piquer emphasizes experimental over systematic methodology in his writings. To praise European academic societies' "useful investigations" (especially those of the Baconian Royal Society of London) in his *Discurso sobre la aplicación de la Philosophia,* Piquer quotes a leading Catholic philosopher, Ludovico Antonio Muratori. As early as 1708, Muratori had framed Bacon's inductive path within an enlightened decorum of "good taste," suggesting that Bacon's works "have been and will always be a *Seminario* on excellent laws to govern *buon gusto.*"[56] Piquer echoes the renowned Italian's vituperative claim that the "imaginative, systematic" approach to natural philosophy was merely building "castles in the air."[57] Like Piquer, Muratori denied that systematic reasoning could lead to any one great "truth," and instead he prescribed close observation of "particular truths" in nature. As Piquer explained, Bacon's seventeenth-century reform of natural history and natural philosophy had revealed this "pathway of experience" as the "true means" for advancing knowledge: "[E]l verdadero modo de adelantarla era por el camino de la experiencia"[58] [the true means for advancing it was by the pathway of experience]. This stance exemplifies Catholic eclectics' rejection of abstract rationality and preference for firsthand experience, close examination of particular cases, and induction to reach general conclusions.

Eighteenth-century Catholic eclectics' particular conception of Bacon proved fundamentally important to the Hispanic Enlightenment. The Benedictine friar Benito Jerónimo Feijoo, thanks to his monumental and widely read *Teatro crítico universal* (1726–40) and extensive *Cartas eruditas y curiosas* (1742–60), has received the most credit for transforming natural philosophy in Spain and Spanish America.[59] As Piquer himself put it: "El P.M. Feyjoó ha probado la importancia de la Física para la Teología Moral"[60] [Father Feijoo has proved the importance of physics for moral theology]. Feijoo was instrumental in crafting a specifically Catholic definition of Bacon's pathway to knowledge that challenged both ancient and modern systematic philosophies. One important contribution of Feijoo was to tie the Baconian method to skepticism and further remove negative undertones against religion by reducing skeptic goals to a "watchful" objectivity and suspension of judgment during philosophical inquiry. This pious skepticism acknowledged the limits of human understanding while prescribing experimentalism as the best means to know all that God allows so that he may be glorified. Countless intellectuals on both sides of the Atlantic read Feijoo's opinions about Bacon's path offering the only sure route to philosophical truth: "[E]l plan de Bacon, el único que puede dar algún útil, y seguro conocimiento de la Naturaleza"[61] [Bacon's plan, the only one that can provide some useful and reliable knowledge of nature]. This inductive path for reasoning contrasted greatly with systematic methods that deduced particulars from generalizations or "universal essences."

Feijoo was dubbed the new Spanish Bacon ("nuevo Verulamio español") for his mediation of Bacon's methodology with Catholicism. This must have pleased him greatly, given that he had dubbed Bacon the heroic "discoverer" of "experimental physics."[62] First in his *Teatro crítico* and then in his *Cartas eruditas y curiosas,* Feijoo maintains a hyperbolic praise of Bacon while expressing his "disenchantment" with both ancient and modern systems.[63] By Feijoo's account, Bacon's method helped Spaniards realize that all the "systems" laid out erroneous routes to understanding nature. Thanks to Bacon's writings (as mediated by Feijoo's Spanish writings), they could now find the correct path to knowledge: "Advirtió el primero el Canciller Bacon, que eran descaminados los rumbos de todos los Sistemas; y en varias Obras suyas mostró a los Filósofos la senda por donde debían caminar"[64] [Chancellor Bacon first warned that all the systems' courses were misleading; and in various of his works he showed to philosophers the path where they should walk]. Like Piquer and Muratori, Feijoo underscored the role of academic societies in fostering Baconian experimental methods, with the royal societies of London, Paris, and Bologna as models for Spain to follow.[65]

Feijoo, certainly today's most widely known eclectic philosopher of the Hispanic Enlightenment, helped to situate Bacon's inductive method as going hand in hand with religion.[66] However, Feijoo was greatly indebted to the Jesuits. Ramón Ceñal has studied Jesuit sources for Feijoo's knowledge of, among other "moderns," Bacon, Gassendi, Malebranche, and Leibniz.[67] The Benedictine never hid his respect for the Jesuits, whom he widely praised as "wise and scholarly."[68] Instead, Feijoo freely borrowed passages from the Jesuits' key eighteenth-century periodical, the *Mémoires pour l'Histoire des Sciences & Beaux-Arts* (1701–67), commonly referred to as the *Mémoires de Trévoux* for its site of publication. Although it was published in French, this official publication of the Society of Jesus included notices and analyses of philosophical advances from all of

Europe. In his philosophical writings, Feijoo credits the Jesuits for greatly impacting Spanish science.

A Jesuit Eclectic Methodology

On both sides of the Atlantic, the Jesuits' accommodating attitude in the face of varied philosophical systems stimulated an eclectic Enlightenment in Spain and Spanish America. Gumilla was trained in a flexible approach that allowed for respectful modification of classical and patristic authorities when new knowledge of nature and methodologies promised increased piety and knowledge of God. *El Orinoco ilustrado* culminates years of firsthand observation, analysis, and experimentation in the field, and is of course greatly influenced by Gumilla's university education as an eclectic Jesuit.

When Gumilla's Orinoco observations and experiments contradict what he has been taught, he respectfully modifies the authorities, departing from the scholastic "libros sagrados y canónicos"[69] [sacred and canonical books] and writings of previous missionaries yet all the while asserting his respect for the Jesuit parameters of *ad majorem Dei gloriam.* When challenging previous or contemporary authorities, he maintains a tone of veneration. For example, at the close of a discussion affirming new insights gleaned from his personal experiences and observations of the Orinoco River region's climate and weather, Gumilla writes:

> No obstante todo lo dicho, cedo alegre, y voluntariamente mi parecer a los doctísimos autores citados [. . .] pero también les he de merecer el favor de que no nieguen los experimentos expresados, y más cuando sin profesarla, los hice con toda la refleja que pude, a favor de su noble y apreciable ciencia. (*El Orinoco ilustrado,* 79–80)
>
> [Despite everything said, I happily and voluntarily yield on these matters to the most expert authors cited . . . but I also deserve from them the courtesy that they do not negate the stated experiments, especially since, without professing it, I did them with full consideration of their noble and appreciable knowledge.]

Gumilla's simultaneous reverence for and respectful departure from Catholic authorities typified the technique employed by Jesuit eclectics when their experiences or experiments challenged prior theories. His display of erudition and accumulated firsthand knowledge about the region's natural phenomena and resources in *El Orinoco ilustrado* participates in a strong tradition of Jesuit writings that reveal an eclectic Jesuit theory of knowledge. Analysis of *El Orinoco ilustrado* suggests that equally important to shaping Gumilla's theory of knowledge were his years of experience with Orinoco region cultures' new and different content and methods. The impact of Gumilla's Orinoco encounters with Amerindian ways of knowing on his own theory of knowledge exemplifies how Jesuit appropriation and adaptation of peripheral knowledge greatly influenced the Hispanic Enlightenment.

The Spanish Jesuit missionary Joseph de Acosta stands out among Gumilla's influences. Gumilla's Jesuit university education included humanistic study, with exposure to Acosta's *Historia natural y moral de las Indias* and works of other great writers and thinkers. This Jesuit training and formation commenced in Spain but was completed in the New Kingdom of Granada. José del Rey has detailed the curriculum at the Javeriana University in Bogotá from its founding in 1623.[70] By the second half of the seventeenth century its Jesuit educational program officially subscribed to eclecticism, especially for natural philosophy. Of course, Acosta had already fostered an eclectic methodology in the sixteenth century when he strove to understand the New World by moving beyond the generalizations of received authorities, and he became the first modern to combine natural and moral history.[71] Yet at the same time Gumilla frames *El Orinoco ilustrado* within the venerated tradition of this renowned Jesuit forefather, his blend of natural and human details from the Orinoco River region advances beyond Acosta's firsthand observation template for natural history writing. While Acosta recognized the limits of scholastic syllogisms, Gumilla's open criticism of this "speculative argumentation" reflects an eighteenth-century Spanish Baconianism. That is, his preference for close observation, experiments, and induction went further than Acosta's methodology. Moving well beyond the classic descriptive natural history and speculation about causes for natural phenomena practiced by those who follow Aristotle, Gumilla's investigation of visible effects and invisible causes incorporates contemporary debates about, for example, particle theory and blood circulation.

Gumilla's natural philosophy was clearly inspired by contemporary Catholic eclectic intellectuals in Spain. Along with Piquer and Mayans y Siscar, the Jesuit Tomás Vicente Tosca (1651–1723) led the eclectic movement in Valencia, where Gumilla was born. Tosca was a mathematician whose corpuscular theory mediated Cartesian mechanism with atomism to create a theory of particles. Gumilla was indebted to this Valencian Jesuit; as he writes at one point: "en sentir del Padre Doctor Tosca" (*El Orinoco ilustrado,* 76) [in the opinion of the Father Doctor Tosca]. According to Mayans y Siscar, Tosca practiced "libertad filosófica" [philosophical freedom] and was a great "amigo de elegir de cada secta filosófica lo que le parecía major"[72] [friend of choosing from each philosophical sect that which seemed best to him]. Like these Valencian eclectics, Gumilla supported the freedom to choose among philosophical systems. He even encouraged his readers to follow the eclectic philosophy, and writes, "Y en fin, cada cual de los instruídos [. . .] eligiendo el sistema que más le cuadrare de los muchos que han propuesto los sabios modernos. [. . .]" (393) [And thus, each one of the well-educated . . . choosing the system that suits him from the many that modern learned people have proposed . . .]. Not surprisingly, this phrasing recalls Piquer's description of "the style of the eclectic philosophers" as one where natural philosophers adopt "whatever seems most suitable to explain the principles of nature."[73] Gumilla's understanding is clearly reminiscent of Piquer's Jesuit-influenced textbook definition of eclecticism:

> [P]or Philosophia Eclectica se entiende un modo de philosophar, en que el entendimiento no se dedica, ni se empeña, a seguir à un solo Philosopho, formando Systema de su Secta, sino que toma de todos aquello, que en cada uno de ellos le parece verdadero.[74]
>
> [What is meant by eclectic philosophy is a way of philosophizing in which the intellect does not dedicate or lock itself into following only one philosopher, thereby forming a system out of the tenets of just one sect, but rather it takes from all philosophies that which in each seems to be true.]

As we saw above, Piquer's writings represent Catholic intellectuals' continued links between knowledge and faith during the Hispanic Enlightenment. In the first half of the eighteenth century, Piquer, Gumilla, and other philosophers give pious justifications for their eclectic approaches. Their reason for exploring advances in modern systems is simultaneously to improve interpretation of the world and of its Creator. For them, the advancement of knowledge remains connected not only to God's glory, but also to his will.

Gumilla's eclectic Catholic formation justified quests for knowledge by maintaining that both knowing and unknowing (or skeptic awareness of what must remain unknown) increases piety. That is, both the pious purpose of knowledge and the pious acceptance of the limits of human understanding found utility in modern investigations into nature. Gumilla reminds his readers that at the same time that it is God's will that humans humbly aspire to know more about his divine works, nobody should seek to "know more than is appropriate": "Es cierto que Dios quiere que investiguemos las obras de su poder; pero quiere que sea con reverencia y humildad: *Non plus sapere, quam oportet sapere,* etc. [. . .]" (312) [It is true that God wants us to investigate the works in his power; but he wants it to be with reverence and humility: *not to know more, than is appropriate,* etc. . . .] In a footnote he underscores this Latin statement on human limits by citing

Romans 12:3.[75] It is God's will that there be inexplicable works that can only be fully understood by him: "*Ut non inveniat homo opus, quod operatus est Deus ab initio usque ad finem*" (311) ["*So that man does not discover/acquire the works of God from the beginning to the end*"]. Enlightened Catholic skeptics throughout Mediterranean Europe echoed this combination of knowledge accumulation and its humble limits. For example, Gumilla's contemporary, Muratori (1672–1750), a great admirer of the Jesuits who compiled one of the best-known contemporary accounts of their enormously successful Paraguay missions, philosophized that whether successful or frustrated, all searches for hidden causes piously advance human knowledge of God.[76] As noted earlier, Muratori supported Bacon's inductive path as the epitome of "good taste" for scientific inquiries. Like Gumilla, though, above all else Muratori praised investigations into the mysteries of nature for their ability to increase human knowledge, wonder, and praise of the all-knowing "Author of nature." Muratori designates these pious increases in "knowing, admiring, and blessing" as the "true fruit" of the limits God puts on human understanding:

> El verdadero fruto que debiera sacarse al ver la cortedad de nuestras fuerzas, quando intentamos descifrar las causas, modos, y fines de tantas maravillosas hechuras, como la naturaleza esconde á nuestro alcance, habia de ser el de conocer, admirar, y bendecir al Autor de la naturaleza, esto es, á aquella Mente, y poder infinito que sabe, y puede hacer tantas cosas superiores á nuestro entendimiento.[77]
>
> [The true fruit that one should get upon understanding the lack of our abilities when we try to decipher the many causes, means, and ends of so many marvelous creations as nature hides from our reach has to be the one of knowing, wondering at, and praising the Author of nature, that is, that Intellect, and infinite power that knows, and can do so many things superior to our intellect.]

Gumilla clearly expresses this understanding of Catholic intellectuals that the exploration of nature and subsequent intellectual discussions either about inexplicable wonders or about new discoveries are all part of God's plan: "Puso Dios el mundo a vista de los hombres, y lo entregó en manos de sus disputas, discursos y averiguaciones" (311) [God put the world in man's sight and handed it over to his disputes, reasoning, and investigations]. The Jesuit's answer to a rhetorical question about why God does this ties his own hunt for knowledge along the Orinoco with the diverse studies of nature that will provide facts and "certain knowledge" about all peoples and lands on earth. He repeats this opening affirmation with a sententious Latin proverb:

> ¿Y para qué? Parece que el fin que tendría Su Majestad sería para que el hombre, con su industria y estudio, consiguiese una noticia de las verdades naturales que resultan de la variedad de los mixtos, de las propiedades de los animales y de las virtudes de las hierbas y adquiriese una cierta ciencia de las provincias y naciones de que se compone el orbe de la tierra: *Mundum tradidit Deus disputationi eorum.* (311)

> [And why? It seems that the purpose His Majesty might have had was that man, with his industry and study, might thereby succeed in getting an idea about the natural truths that result from the variety of the properties of the animals and of the virtues of the herbs and that he might acquire a certain knowledge of the provinces and nations of which the earth is comprised: *God handed the world over to their dispute.*]

This clear expression of Gumilla's theory of knowledge demonstrates how he pairs discovering new "secrets of nature" with recognizing the limits of human investigation into natural things. And he follows this up by articulating how both knowing and the inability to know inevitably serve to praise God:

> [N]o prohibió Dios a los hombres el que trabajen en esta seria y curiosa averiguación de las cosas naturales. Antes bien, liberal y graciosamente, no sólo nos dió la facultad, sí que también nos entregó Su Majestad enteramente, *tradidit Deus,* todo el orbe terráqueo, para que, averiguando en lo factible sus naturales secretos, alabemos al Criador de todo por aquellas noticias que alcanzamos, y veneremos su infinito poder y sabiduría por aquello mismo que no percibimos, y confesando nuestra ignorancia nos humillemos. (312)

> [God did not prohibit men from working on this serious and curious investigation of natural things. On the contrary, liberally and graciously, not only did he give us the ability, but also His Majesty entirely handed the whole world over, *God handed it over,* so that investigating as much as feasible his secrets of nature, we would praise the Creator of everything for that knowledge that we achieve, and we would revere his infinite powers and knowledge for that which we cannot perceive, and admitting our ignorance we would humble ourselves.]

Throughout *El Orinoco ilustrado,* Gumilla punctuates such "curious investigation" of Orinoco things with humble praise of God again and again. The particular chapter that contains this quote concludes by inviting his readers to reaffirm with him skeptical limits of human understanding alongside the utility of all attempts to understand:

> Pero concluyamos este largo capítulo venerando rendidamente la sabia y oculta providencia del Altísimo y humillémonos al considerar que, con tener

> a la vista muchas de sus obras patentes, es tanta la pequeñez de nuestro alcance, que no las entendemos; y así, pasemos a buscar raíz de otras más fáciles de percibir, no menos curiosas, y en gran parte útiles. (323)

> [But let us conclude this long chapter revering with surrender the wise and hidden providence of the Almighty and humble ourselves contemplating that, even having in our sight so many of his obvious works, the smallness of our reach is so great that we cannot understand them; thus, we turn to seeking the root of others easier to perceive, no less curious, and for the most part useful.]

All of these passages illustrate the pious skepticism practiced by Spanish Baconians.

Gumilla is also an excellent example of the impact New World discoveries and Jesuit missionaries' intercultural discourse had on Spanish Baconianism and eclecticism. The rest of this chapter and all the next chapter examine aspects of Gumilla's theory of knowledge that were influenced both by his university studies and by his particular experiences in the Orinoco. On the one hand, Gumilla seems to perpetuate the diffusion model described by George Basalla in that Jesuit intellectual and spiritual enlightenment is dispersed from the European center to an American periphery. On the other hand, his text evidences the reciprocity of knowledge between America and Europe. Gumilla's Jesuit discourse underscores the flow of knowledge from Europe to America as well as the flow of striking new data from the Orinoco River region back to Spain. And Gumilla's incorporation of local paradigms outside traditional Western concepts of science and enlightenment exemplify a key piece in the redefinition of modernity in Spanish America. The Hispanic Enlightenment's bilateral flow of knowledge reveals itself in *El Orinoco ilustrado* through examples that link to the Catholic faith both advances in knowledge and awareness of the limits of human understanding. Throughout his text, Gumilla sets himself up as a mediator between the Orinoco River region and Europe and between God and man: not only between God and Catholic intellectuals, but more importantly between God and his Amerindian charges. His goal for domesticating Orinoco peoples participates in Eurocentric evangelical goals and enlightened reforms. In the process, he expands the natural and human objects of study available to Spain. Gumilla studies the region with the tools born of his Jesuit training, within an eclectic tradition that has mediated not only the European philosophies central to the evolution of Western science, but also more ritualistic, intersubjective ways of knowing witnessed in the peripheries of mission culture. As the examples at the close of this chapter and especially in the next chapter suggest, Gumilla's intercultural conversations not only expanded the data pool about the Orinoco River region, but actually altered his theory of knowledge.

Gumilla's Spiritual Scientific Enlightenment

Gumilla crafts metaphorical ties between nature and man within his enlightened discourse on the advancement of knowledge. The examples of the Orinoco turtles and the mimosa plant fit within the natural history tradition of converting inexplicable wonders into Christian allegories. Yet Gumilla's lengthy plant discourse combines moral exegesis of nature with a speculative scientific explanation that supersedes both Jesuit Neoplatonism and the tradition of natural and moral history codified by Acosta. In the case of the mimosa plant, Gumilla frames his blend of moral and natural philosophy within contemporary debates about particle theory. After examining these two examples of wonderful Orinoco plants and animals, we will turn to two examples of intercultural discourse that correct Amerindian errors about the sun and about the lunar eclipse. Gumilla's narrative of Christian persuasion of the Betoye, Lolaca, and Atabaca tribes using scientific demonstrations with a glass lens and a physical model to explain the lunar eclipse also functions to prescribe Jesuit strategies for leading Amerindians to God by way of European knowledge.

Gumilla's consideration of turtle behavior is a manifestation of how Jesuit natural philosophers cast inexplicable wonders as part of God's plan. José Eusebio Nieremberg's *Historia naturae* (1635) articulated the Jesuit Neoplatonic tradition for reading the book of nature. Two types of moral readings of nature characterize Nieremberg's important source text for Gumilla.[78] First, God intends for humans to decode nature and reveal more about the divine. Because of this, Gumilla seeks to understand the Orinoco turtle's innate sense of direction. Yet secondly, also within Jesuit Neoplatonism, inexplicable wonders, mysteries, and marvels in the book of nature provide useful Christian metaphors to increase praise of God. Allegorical uses for nature to heighten appreciation of God's benevolence are also prescribed in other key source texts for Gumilla's form of Christian persuasion.[79] After reporting on his frustrated efforts to comprehend turtles' innate sense of direction, Gumilla extracts a moral lesson, punctuated by his Christian exclamation. The Jesuit relates various experiments, but says that no matter which way he pointed these Orinoco turtles, their natural inclinations led them directly back to the river:

> [T]omaban el rumbo derechamente al agua, obligándome a ir con ellas, alabando la providencia admirable del Criador, que a cada una de sus criaturas da la innata inclinación a su centro, y modo connatural de llegar a él. ¡Gran reprehensión nuestra, que, aun alentados de los eternos premios y amenazados con imponderables castigos, apenas acertamos a tomar la senda derecha de nuestro último fin y centro de la Bienaventuranza para que Dios nos crió! (232)

[They took the direction right back to the water, forcing me to go with them, praising the wonderful providence of the Creator, who gives to each one of his creatures the innate inclination to its center, and the natural way to reach it. Great warning to us, who, even when encouraged by eternal rewards and threatened with unthinkable punishments, can hardly find the right path to our ultimate end and center of the Bliss for which God created us!]

Gumilla's inability to explain this observable phenomenon clearly stupefies him: "lo que me causó más admiración" (232) [that which caused me much astonishment]. However, he takes advantage of the limits of his human perception to inspire praise of God's wonderful providence. This mystery of nature also allows Gumilla to shift briefly into a moral philosophical application of this Orinoco phenomena. In the passage just quoted, his purposeful shift from declarative to exclamatory discourse rhetorically underscores the moral statement, which could easily be directed at Europeans or Amerindians who ignore or resist their natural inclination for Christianity: God will direct their path.

A very different exemplum that blends Christian moral philosophy with a modern discourse of natural philosophy is Gumilla's allegorical use of the "timid" or "modest" mimosa plant, a mystery of nature called "virgin plant" by missionaries in other Jesuit provinces but still commonly referred to as the *vergonzosa* (bashful) plant along the Orinoco.[80] The moral value of this remarkable plant stems from its leaves' mysterious contraction to avoid being touched. Gumilla first lists the vergonzosa plant as a spiritual remedy, then considers its physical and scientific aspects. Much like the natural example used by Christ in Matthew 6:26 and reproduced by Gumilla in his text—" 'Mirad, atended a los lirios y azucenas del campo y tomad enseñanza de su hermosura y de su candor'—dijo Cristo Nuestro Señor" (Gumilla, *El Orinoco ilustrado,* 444) ["Look, pay attention to the tiger lilies and the white lilies of the field and acquire doctrine from their beauty and innocence," said Christ our Lord]—Gumilla inserts a moral aside of fatherly advice amid a series of practical plant remedies. First he invites readers to pause with him, redirecting their attention away from the obviously spectacular beauty and utility of what he has been discussing to instead look down at the soil to examine the humble plant hiding at their feet: "Apartemos la vista de la hermosura de las plantas y arboledas. Fijémosla un rato en el suelo de estos dilatados campos. [. . .] La primera que ocurre a los pies y a la vista en aquellos terrenos, por vulgar, es la *vergonzosa,* en la cual no se ha conocido virtud alguna" (443) [Let us remove from our sight the beauty of the plants and trees. Let us look awhile on the ground of this extensive countryside. . . . The first thing that happens upon our feet and sight in those lands, by the lay term, is the *bashful one,* in which no virtue whatsoever has been recognized].[81] If the vergonzosa plant provides no obvious botanical virtue like the flora he has just discussed, Gumilla explains, it offers an important moral metaphor, which the Jesuit emphasizes with his rhetorical question:

> ¿[Q]ué más virtud que la lección práctica que da del modo con que se deben portar las mujeres, y especialmente las doncellas, que aún por eso en muchos de aquellos países la llaman la *doncella?* Bien pueden los físicos prevenir sus admiraciones para lo que voy a decir. (443)
>
> [What more virtue than the practical lesson that gives the manners with which women should comport themselves, and especially the virgins? For this reason in many of those countries they still call this plant the *doncella* {maiden}? The physicists can well avoid their wonder about what I am about to say.]

Gumilla applies human traits to this plant (personification, or prosopopoeia) and creates a moral allegory that ties "gentle plants" with "gentle beauties."[82] Gumilla's moral lesson includes the remonstration of how in the blink of an eye—"en un cerrar y abrir de ojos" (444) [in a closing and opening of the eyes]—both the vergonzosa plant and the young maiden can be deflowered and disgraced. Gumilla recommends that mothers and teachers look to the vergonzosa as a natural mirror of modesty when morally instructing youths: "Mírense en el espejo de esta vergonzosa hierba" (444) [Look in the mirror of this bashful plant]. Christ's words about the lilies of the field model how this conversation might proceed, and Gumilla employs similarly simple and sincere language to gently command with exhortation and admonition: "[P]ueden y deben exhortar a sus hijas y discípulas, diciéndolas: 'Venid, observad, atended y aprended de esta hierba vergonzosa; reparad que en cuanto la tocan, se da por muerta, desfallece, se desmaya y se marchita' " (444–45) [They can and should exhort their daughters and disciples, telling them: "Come, observe, heed and learn from this bashful plant; note that as soon as she is touched, she gives herself for dead, she weakens, faints and wilts"]. Gumilla evokes the curious actions of the mimosa plant noted by missionaries and naturalists around the world as a lesson in moral conduct.

Gumilla's lesson, however, goes beyond moral philosophy and into modern physical science. Gumilla declares the vergonzosa a "prodigio de la naturaleza" (444) [miracle of nature]. Instead of merely touting the metaphorical value of this wonder, Gumilla draws upon recent advances in natural philosophy in an attempt at a rational explanation of the vergonzosa plant's physical evasion. At first he describes this bashful plant's previously unexplained movement away from human touch, a strange natural phenomenon that has also earned it a look-but-don't-touch nickname: "*Mírame y no me toques*" (444) [*Look at me and don't touch me*]. He then intensifies its mystery by citing one famous contemporary historian's report on the Philippines' "virgin plant" that seems to flee even the careful gaze of naturalists: "[A]segura que luego que alguno toca aquella mata dobla sus cogollos y los esconde en el agua, como si se corriera y avergonzara, no sólo de sentir el ajeno contacto, sino aun de ser mirada con cuidado" (445) [He assures us that as soon as someone touches that bush it folds its shoots and hides them in the water, as if it runs and is put to shame, not only from feeling the outside contact, but even from just being looked at carefully]. Soon after, however, he posits the natural philosophical reason behind his moral philosophical discourse. The personified vergonzosa's "modesty" can be explained by a corpuscular interaction that spurs a rush of fluids toward its roots and causes a "prudent" retraction of leaves.

Gumilla consciously intends for his example of the vergonzosa plant to provide not only a mirror of moral conduct but also a demonstration of his atomistic understanding of the interaction between matter and energy. As a scholar and literatus, he is well aware of this interaction between natural and moral philosophy. He maintains his personifying language yet supports his physical theory by citing Noël Regnault, the Jesuit author of a recent dialogue explaining a "new system of physics":[83]

> La causa y raíz física de la instantánea mutación, discurro yo consiste en que aquel contacto extrínseco, con efluvios que introduce, inmuta el flujo natural de los sucos que la raíz remite hasta los últimos cogollos, y hace retroceder el curso corriente de los flúidos con que se mantiene la lozanía de la mata; y tomando su retirada hacia las raíces, el desmayo de los cogollos y el encogimiento de las hojas es un efecto que necesariamente se sigue a la sustracción del necesario pábulo: como se ve, el desmayo que la falta de alimentos causa en los vivientes sensitivos. [. . .] entremos en los jardines del Rey Cristianísimo con el Padre Regnault, pongamos los ojos en la mata llamada sensitiva; pero nadie alargue la mano para tocarla. [. . .] (445)

> [The cause and physical root of the instantaneous mutation, I think, consists of that extrinsic contact, with the effluvia that it introduces, which affects the natural flow of the juices that the root remits to the furthest shoots, and makes the running course of the fluids with which the bush maintains its lushness recoil; and taking its retreat toward the roots, the fainting of the shoots and shyness of the leaves is an effect that necessarily follows the subtraction of the necessary fuel: as can be seen with the fainting that the lack of nourishment causes in sentient creatures. . . . [L]et us enter into most Christian King's gardens with Father Regnault, let's set our eyes upon the bush called sensitive; but nobody stretch out his hand to touch it. . . .]

Under the rubric of Regnault's exploration of particle theory, Gumilla reasons that not only touching but mere careful examination of the plant can spur its contraction, because the vergonzosa will flee from the invisible particles emitted from the human hand: "corrida y espantada de solos los efluvios que la mano curiosa despide, antes de tocarla" (445) [embarrassed and scared of just the effluvia that the curious hand shoots out, before touching it].[84] Gumilla's lengthy section about a personified plant running scared from effluvia, placed within a series of brief and practical botanical remedies, exemplifies his innovative blend of traditional Christian morality and symbolic readings of the book of nature with modern natural philosophical disquisitions.

For Jesuits and other Catholic intellectuals the "true fruit" of human knowledge of nature, either natural philosophical or metaphorical, is to stimulate Christian piety. *El Orinoco ilustrado* clearly functions to increase knowledge of God for its Christian reader by increasing natural knowledge of the Orinoco River region. Yet its pages also contain several conversion narratives that function as recommendations to Jesuits on how to employ European science for evangelical purposes. The next two examples reveal intercultural discourse, because Gumilla's scientific demonstrations are directed at Christian persuasion and model Jesuit missionary strategies for leading Amerindians from superstitious beliefs to knowledge of God. In the first example, using a glass lens Gumilla dispels mistaken beliefs about the sun and leads the Betoyes to "rational knowledge." The second example models steps taken to move the Lolaca and Atabaca tribes to scientific understanding about the eclipse and, consequently, to pious knowledge of God's power over nature.

Gumilla is well aware of how conversations about nature transform Orinoco Amerindians' perceptions of the world around them and might be used to spread Christian enlightenment throughout the region. By expanding upon the Neoplatonic tradition of natural history writing as providing symbols to interpret, Gumilla positions himself as a mediator between God and Orinoco Amerindians—"yo de parte de Dios les enseñaba" (282) [I taught them on God's behalf]—as well as an important source for increasing the Jesuits' current success rate for their conversion. Gumilla describes the journey of one Orinoco tribe from false superstitions based on their own "received authorities" to Christian knowledge as a pathway to rationality. To support this he re-creates testimony from grateful Betoye tribe neophytes:

> Pero por el mismo caso que reinan las tinieblas en los entendimientos de aquellas gentes, cuando al abrir los ojos de la razón perciben la luz de las verdades eternas, les da mayor golpe la novedad, y se reconoce por los efectos que entonces derrama Dios a manos llenas su misericordia, según la mayor o menor

disposición de los neófitos, entre los cuales vemos y advertimos la mutación que en ellos hace la diestra del Todopoderoso. Y aun los mismos indios, al cotejar su vida racional y cristiana con su antiguo desconcierto, se regocijan, se admiran y dicen repetidas veces a sus Misioneros: "*Diosó fausucajú, Babicá, ujuma afoca, ubadolandó maydaitú*"; esto es: "Dios te lo pagará, Padre, pues por tu medio vivimos ya racionalmente." (283)

[But in the same way that when darkness reigns in the intellects of those peoples and upon opening their eyes to reason they perceive the light of eternal truths, the novelty gives them a greater blow, and it is clear from the result that God pours out his mercy to full hands, and it is received according to the greater or lesser gift of the neophytes. In them we see and realize the change that the right hand of the Almighty performs. And even these very same Indians, upon comparing their rational and Christian lives with their previous confusion, rejoice, marvel and say repeated times to their missionaries: "*Diosó fausucajú, Babicá, ujuma afoca, ubadolandó maydaitú*"; that is, "God will reward you, Father, because through you we now live rationally."]

Yet even after the Betoye tribe's conversion to Catholicism, Gumilla found it necessary to open their eyes and appeal to their tactile sense to dispel a lingering error about the sun and bring them in line with the Catholic catechism, whose teaching ironically seems to have further confused them. As Gumilla explains,

Los betoyes decían antes de su conversión que el sol era Dios; y en su lengua, al sol y a dios llaman Theos, voz griega [. . .] en la nación betoye hubo que vencer algo, porque pusimos en el catecismo esta pregunta: "*¿Theodá Diosoque?*" ¿El sol es Dios?, y al punto respondían que sí. La respuesta que se les enseña es: "*Ebamucá futuit ajajé Diosó abulú, ebadú, tuluebacanutó.*" "No es, porque es fuego que Dios crió para alumbrarnos." (281)

[The Betoyes used to say before their conversion that the sun was God; and in their language, they call both sun and god "Theos," a Greek word . . . in the Betoye nation it was necessary to overcome something, because we put into the catechism this question: '*¿Theodá Diosoque?*' The sun is God? and they immediately responded yes. The answer that is taught to them is: '*Ebamucá futuit ajajé Diosó abulú, ebadú, tuluebacanutó.*" "No it is not, because it is fire that God created to illuminate us."]

To illustrate this Catholic dogma, Gumilla demonstrates how the sun is fire by burning the hand of an important soldier with a glass lens:

[M]e valí de la mecánica de un lente o cristal de bastantes grados, y junta toda la gente en la plaza, cogí la mano del capitán más capaz, llamado Tunucúa. Preguntéle *si el sol era Dios*, luego respondió que sí; entonces en voz alta, que oyeron todos, dije: "*Day dianu obay refolajuy? Theodá futuit ajaduca, may mafarra.*"

(*¿Cuándo acabaréis de creerme? Ya os tengo dicho que el sol no es sino fuego.*) Y diciendo y haciendo, interpuse la lente entre el sol y el brazo del dicho capitán, y al punto el rayo solar le quemó. Clamó luego él con voz amarga, diciendo: "*Tugaday; tugaday; futuit ajacudacá.*" (*Es verdad, es verdad: fuego es el sol.*) Corrían en tropel los hombres y mujeres a ver el efecto del sol y de la lente; veían la quemadura, y el capitán les explicaba con eficacia la operación, que miraban con espanto correlativo a su nativa ignorancia. (281–82)

[I made use of the mechanics of a lens or glass with enough power, and gathering all the people in the square, I grabbed the hand of the most capable captain, called Tunucúa. I asked hime *if the sun was God,* and he responded yes; then in a loud voice, so that everyone could hear, I said: "*Day dianu obay refolajuy? Theodá futuit ajaduca, may mafarra.*" (*When will you finally believe me? I have already told you that the sun is nothing but fire.*) And saying this, I interposed the lens between the sun and said captain's arm, and immediately the solar ray burned him. He then cried out with a bitter voice, saying: "*Tugaday; tugaday; futuit ajacudacá.*" (*It is true, it is true: the sun is fire.*) The men and women ran in a mad rush to see the effect of the sun and the lens; they saw the burn, and the captain explained to them with effectiveness the operation, which they viewed with a fear correlative to their native ignorance.]

After repeating the lens experiment on a tribal elder, the crowds clamored to experience the burning of the sun for themselves. Left to their own devices, they all concluded that the sun was not God, but instead fire created by God: "Todos querían experimentar (aunque a costa suya) si el sol era fuego o no. [. . .] El efecto de esta maniobra fué cual se deseaba, porque de allí en adelante ningún betoye dijo jamás que el sol era Dios. Luego respondía que el sol era fuego" (282) [They all wanted to experience (even at their own expense) if the sun was fire or not. . . . The effect of this maneuver was all one could desire, because from then on not one Betoye ever said that the sun was God. Afterward they responded that the sun was fire]. Within the Betoye tribe conversion narrative, Gumilla underscores the value of European scientific knowledge not only for their initial conversion, but also for dispelling persistent errors that impede orthodox faith.

One chapter in *El Orinoco ilustrado,* "Turbación, llantos, azotes y otros efectos raros que causa el eclipse de la luna en aquellas gentiles" (457) [Confusion, weeping, scourging and other strange effects that the lunar eclipse causes in those pagans], narrates Jesuit strategies for leading Amerindians from the darkness to the light, both intellectually and spiritually. This chapter compares the mistaken beliefs about the eclipse of several different Orinoco region tribes to other Jesuits' experiences enlightening pagans from the Philippines to China about the true nature of the lunar eclipse. Within Gumilla's narrative about a physical model used to dispel popular errors about the eclipse, the Jesuit employs a simultaneously spiritual and intellectual meaning of enlightenment as "seeing the light" of Christianity by way of experiments and scientific elucidations.

The steps in Gumilla's narrative are directed toward enlightening the Orinoco Amerindians about the earth's occasional positioning exactly between the sun and the moon and then leading them from this darkness of ignorance through the European knowledge of the solar system's workings to knowledge of God. The chapter as a whole exemplifies the Jesuit strategy of using explanations of formerly mysterious wonders for evangelical purposes. Gumilla narrates how he persuaded the Lolaca and Atabaca tribes that an eclipse will not kill the moon. To complete this first step, he fashions a makeshift astronomy lab and demonstrates the true cause of darkness, using a mirror to represent the moon, an orange to serve as the earth, and a flame to signify sun. This model successfully enlightens them:

> [S]aqué un espejo, una vela encendida y una naranja, y llamando a los principales les expliqué con los términos más groseros que pude hallar cómo la privación de luz de la luna no era por enfermedad, porque ella no es una cosa viva, sino porque no tiene otra luz sino la que recibe del sol, poca o mucha, según el aspecto con que el sol la mira; y que llegándose a interponer el orbe terráqueo entre el sol y la luna, durante el tiempo de la interposición no recibía luz si era total, y recibía poca luz, si era interposición parcial. Esto mismo les hice ver con la demostración de la vela y su luz refleja del espejo, interponiendo la naranja entre la luz de la vela y la del espejo. Percibieron algunos de los principales la explicación, y dando grandes palmadas en los muslos, gastaron mucho tiempo en explicar a sus gentes la causa del eclipse, con tan buen éxito, que en adelante no hubo lágrimas, ni gritos, ni ceremonia alguna en los eclipses que se siguieron. (458)

> [I took out a mirror, a lighted candle and an orange, and calling all the principals I explained to them in the most crude terms that I could find how the deprivation of the light of the moon was not because of sickness, because it is not a living thing, but rather because it has no other light except that which it receives from the sun, a little or a lot, according to the angle with which the sun faces it; and that upon the earth interposing itself between the sun and the moon, during the time of its interposition the moon did not receive any light if it was total, and it received little light, if it was a partial interposition. This very same thing I made them see with a demonstration of the candle and the mirror, putting the orange between the light of the candles and that of the mirror. Some of the principals observed the explanation, and giving great slaps on their thighs, they took a good while in explaining to their people the cause of the eclipse, with such success, that from then on there were no tears, nor screams, nor any ceremonies about the eclipses that followed.]

After completing this first step, the fearful lunar eclipse was no longer a mysterious wonder. Through Gumilla's mediation the Amerindians now perceived the reason it happened as it did.

Immediately following this narration of successful abandonment of Amerindian superstition and adoption of European cosmology in the chapter on the lunar eclipse, Gumilla enacts the second step by arousing Amerindian curiosity, expanding their imagination, and leading them from ignorance to knowledge of God:

> [E]stán en una suma ignorancia de todo [. . .] les causa notable gusto saber aquello que jamás habían imaginado; y como de estas conversaciones de las criaturas luego se pasa a tratar del Criador de ellas, se les va embebiendo insensiblemente y con gusto el conocimiento del Criador de todo. (458)
>
> [They are in a great ignorance of everything . . . it causes them noteworthy pleasure to know that which they had never imagined before; and since these conversations about creatures then pass on to treating the Creator of them, without realizing it and with pleasure they go about soaking up knowledge of the Creator of everything.]

The recourse to curiosity in the service of Christian conversion was not a Jesuit invention. The church father Aquinas had revered wonder as "the best way to grab the attention of the soul" and lead it to knowledge of God.[85] St. Ignatius Loyola, the founder of the Society of Jesus, had included this ancient strategy of simultaneous emotional and intellectual appeals to natural curiosity among the "Jesuit ways of proceeding," a method Acosta employed repeatedly in his *Historia natural y moral*.[86]

Gumilla prescribes this two-step technique for Jesuits and, surprisingly, for his European readers. Like Acosta and other Jesuits before him, Gumilla prescribes this to missionaries: they should appeal to natural human curiosity in order to direct man's attention and wonder to God. Earlier in *El Orinoco ilustrado* he had articulated these virtues of curiosity and wonder for inspiring human desire for knowledge that can in turn increase Christian piety. As he put it, "De aquí se excita la curiosidad, o la admiración, y el deseo de saber" (312) [From this curiosity or wonder is aroused and thus the desire for knowledge]. In his chapter on the eclipse, Gumilla explicitly recommends this ancient conversion strategy: "[Y] éste es el medio por donde los misioneros mejor captan la atención de aquellos bárbaros" (458–59) [And this is the route by which missionaries best capture the attention of those barbarians]. Gumilla offers the mysterious eclipse as a cross-culturally relevant example for evangelism. Just as the Chinese had been enlightened by Jesuits, so could Orinoco River region natives be enlightened: "[L]a luz del santo Evangelio [. . .] les ha aclarado también los entendimientos para percibir mejor el curso de los planetas o el movimiento de los astros y la novedad de los fenómenos" (462) [The light of the holy Gospel . . . has also clarified for them their intellects to be able to perceive better the course of the planets or the movements of the stars and the novelty of phenomena"].[87] While the particular experiment Gumilla narrates was for the benefit of "las naciones lolaca y atabaca" (457) [the Lolaca and Atabaca nations], a similar process could be used to enlighten the Salivas (459), the unconverted in the Philippines (460), the Guayana nation on the coast (460), and the Otomac tribe, whose superstitions about the eclipse Gumilla compares to those of the Moors: "[E]stos y otros tales son los partos de aquella nativa ignorancia, bien semejantes a las demostraciones bárbaras que hacen los moros durante el eclipse de la luna [. . .] nacido de la falsa tradición de que la luna está enojada o enferma" (461–62) [These and other similar ones are born of that native ignorance, quite similar to the barbaric displays that the Moors make during the eclipse of the moon . . . born of a false tradition that the moon is mad or sick]. All of these cultures can benefit from Jesuit mediation of the truth about eclipses.

The lesson of the eclipse was not only for Amerindians and Jesuits, but also for Europeans. When Gumilla performs his experiment with the orange, flame, and mirror to dispel erroneous inherited beliefs among the Amerindians, he is also modeling how knowledge gleaned from observation, experience, and induction can supersede received wisdom for Europeans as well. On one level, the misunderstood eclipse serves to underscore how Jesuits should subtly take advantage of the connections between revealing the light of natural knowledge and leading pagans to the light of God. But Gumilla's narration of Amerindians' departure from the darkness of mistaken beliefs passed on from generation to generation also warns European readers against being blinded by the errors of ancient authorities. Thus, the eclipse chapter is also directed at his European readers, who are warned that only "incapable" minds hold fast to mistaken beliefs when experiments challenge previous knowledge. Gumilla plants this particular seed of cross-cultural comparison when he laments at the start of the eclipse narrative, "Así deliran aquellas gentes; no hay asunto tan arduo como querer quitar un error derivado de padres a hijos entre gente incapaz" (458) [In this way those peoples speak nonsense; there is no matter so arduous as wanting to rule out an error passed down from parents to children among incompetent people]. He returns to the comparison after surveying erroneous beliefs and the superstitious practices of several tribes and immediately before praising the Chinese for having abandonded their former errors and superstitions and advanced in their knowledge of astronomy.

In a statement that ties the light of intellectual knowledge with spiritual enlightenment, Gumilla generalizes his comments about human *genios* and *ingenios:* "Tal como éste es el genio humano cuando le falta cultivo, carece de la luz que dan las ciencias y de la sobrenatural con que los alumbra nuestra santa fe" (462) [Such is human intellect when it lacks cultivation; it then lacks the light that the sciences give and the supernatural light with which our holy faith illuminates them]. Europeans, if lacking cultivation, are no better than those along the Orinoco who lack the light of natural knowledge and the light of Catholicism. By asserting the connections between religion and knowledge, this statement warns against European resistance to the cultivation of modern methods for investigating nature. Gumilla's eclectic theory of knowledge does not require acceptance of things not observed firsthand. Like Feijoo in Spain, Gumilla supported modern investigation of nature to challenge popular traditions and superstitions. By voicing the Amerindians' reason for believing that the moon can die during the eclipse, Gumilla gently suggests to his European readers that previous authorities might not always know best: " 'No hemos visto ni uno ni otro—respondieron—pero así nos lo han contado nuestros mayores y ellos muy bien lo sabrían' " (458) ["We have not seen one or the other," they responded, "but in this way our elders have told us and they very well know"]. As he had suggested earlier, true enlightenment depends on curious wits unafraid to see and know for themselves: "Gran rayo de luz es éste si quisieran abrir los ojos para recibirlo aquellos vivos ingenios temeriamente soberbios" (312) [This is a great ray of light if those lively, fearfully proud wits want to open their eyes to receive it].

Gumilla's Baconianism

Thanks to Gumilla's firsthand experience and his mediation of Amerindian sources, *El Orinoco ilustrado* multiplies both the quantity and the kind of data available to Europe. In his expansion of the pool of New World topics, Gumilla demonstrates Spanish Baconianism on several levels. On a general level, he executes a Baconian renovation of natural history studies and cabinets by participating in Bacon's program calling for separate "histories" (here synonymous with "observations") that would contribute to a comprehensive natural history of the world. Like Bacon, Gumilla places a high value not just on the display but also on the scientific study of curiosities, exploring and cataloguing topics according to ordinary and extraordinary states of nature. Gumilla clearly shares Bacon's opinion that, instead of stupefying humans, wonder encourages the empirical investigation of nature. As Bacon put it, "[A] mind awakened by wonders would reject syllogisms based solely on the familiar and commonplace."[88] Like Bacon, Gumilla's textual natural history cabinet converts items that might formerly have been viewed as mere curiosities into objects worthy of inquiry and analysis. The next section will consider Gumilla's report on the strikingly new and different humans he encountered as a missionary. In the previous chapter of this book we saw how Gumilla catalogued some of New Granada's potentially profitable natural remedies and agricultural promises. Here we will see how he catalogues humans. These ethnographic observations model the direct observation of particular cases and evaluation of outside authorities and opinions that typify Gumilla's eclecticism. This inductive path values firsthand experience as the most effective means to mediate existing knowledge and philosophies and arrive at general conclusions.

Another general debt to Bacon could lie in how Gumilla frames his Orinoco data collection as a heroic hunt for "certain knowledge." Gumilla joins other contemporary Spanish Baconians such as Feijoo who echo Bacon's heroic "hunt" metaphor for the investigation of previously unknown data.[89] Although careful observation of nature in the New World did not commence in the seventeenth century with the Baconian reforms, Bacon reinvigorated the heroic template of New World "discoverers" and "conquerors" for more recent scientific discoveries. In separate aphorisms of his *Novum Organum,* Bacon praises Columbus and Acosta's early empirical efforts: Columbus for the "reasons for his confidence that he could find new lands and continents beyond those known already; reasons which, although rejected at first, were later proved by experiment . . . and became the causes and starting points of very great things" and the Jesuit for his "careful research" advancing knowledge of South America (especially Peru).[90]

Gumilla likewise cites Columbus and Acosta as exemplary heroes whose discoveries of new lands and new peoples became the starting point of great intellectual, spiritual, and material riches. *El Orinoco ilustrado* refers to Gumilla's Jesuit forefather when comparing his data with or at times challenging the data collected in Acosta's *Historia natural y moral.* Gumilla also gives Columbus due credit for earlier centuries' paradisiacal perception of the Orinoco River. As William Eamon explains, after Bacon this heroic hunt metaphor "was used increasingly as natural philosophers attempted to elucidate and to vindicate experimentalism. The French savant Pierre Gassendi (1592–1655) described scientific discovery as a way of 'sagaciously' (*sagaciter*) examining nature, looking for clues or signs (*media*) that will lead the searcher to the hidden aspects of nature."[91] Not unlike Gassendi and in keeping with Spanish Baconian skepticism, Gumilla reminds his readers that some aspects of nature are meant to remain hidden; therefore he upholds the pious limits of human understanding within this heroic hunt.

At the same time that God handed over the world as a "wonderful machine" to be analyzed and debated, no hero is meant to understand everything about it. In a lengthy passage reaffirming his theory of knowledge, Gumilla clarifies that by God's will

> el hombre, con su industria y estudio [. . .] adquiriese una cierta ciencia de las provincias y naciones de que se compone el orbe de la tierra: *Mundum tradidit Deus disputationi eorum.* Ocupación muy loable y digna de la atención, aplicación y estudio de los más insignes héroes en los siglos pasados, a que dan realce los del presente. Mas veis aquí que no fué ésta la intención ni el fin total que tuvo la inexcrutable providencia del Criador, sino el que expresa el divino texto: *Ut non inveniat homo opus, quod operatus est Deus ab initio usque ad finem,* para que ninguno de los mortales se alabe de que averiguó, halló y supo los arcanos secretos de la maravillosa máquina de este mundo (*El Orinoco ilustrado,* 311)

> [man, with his industry and study . . . might acquire a certain knowledge of the provinces and nations of which the earth is comprised of: *God handed the world over to their dispute.* It is an occupation very praiseworthy and worthy of the attention, application, and study of the most distinguished heroes of the past centuries, which those of the present enhance. But you see here that this was not the intention or complete purpose that the providence of the Creator had, but rather that which the divine text expresses: *So that man does not discover the works of God from the beginning to the end,* so that none of the mortals can boast that he has investigated, found out, and discovered the mysterious secrets of the marvelous machine of this world.]

Gumilla's consideration of ancient scripture piously validates this heroic metaphor alongside a new scientific hunt paradigm.

Gumilla's scientific analysis of curiosities reveals his general debt to Bacon. In Bacon's renovating vision, natural history objects of study should be displayed and also analyzed with rigorous scientific methodology.[92] Whereas sixteenth-century cabinets had continued the tradition of Plinian natural history and Aristotelian natural philosophy, by the time Gumilla referred to his natural and civil history as a textual cabinet of curiosities many private galleries and cabinets across Europe as well as the Americas had been transformed into public museums symbolizing the "new philosophies."[93] Paula Findlen reminds us that after Bacon's seventeenth-century reforms, royal cabinets and museums had "become a symbol of the 'new' science, incorporated into scientific organizations such as the Royal Society in England, the Paris Academy of Sciences, and later in the Institute for Sciences in Bologna."[94] Baconian reforms encouraged the exploration of three states of nature (namely, the ordinary, perverse, and artistic) such that, as William Eamon puts it, "all the bizarre objects and rarities that had fascinated and delighted visitors to the Renaissance curiosity-cabinets became urgently relevant to the Baconian scientific enterprise . . . an essential part of the Baconian natural histories."[95] Thus, Gumilla wrote *El Orinoco ilustrado* at a time when cabinets of curiosity had grown into museums and these museums and laboratories were founded by learned and scientific societies and served as an aid to both public and private teaching of "modern science."[96]

Gumilla ascribes the same pious purposes to studying human diversity as he does to studying plants and animals: "variedad hermosa, que es reparable espectáculo para los ojos y noble origen de aquellos pensamientos, que de las criaturas deben pasar a quedarse absortos y anegados en el golfo inmenso de la omnipotencia del Criador de todas las cosas" (82) [beautiful variety, that is a restful spectacle for the eyes and a noble origin of those thoughts, that from creatures they should pass to remaining amazed and flooded in the immense gulf of the omnipotence of the Creator of all things]. In various metacommentaries, Gumilla purposely demonstrates the applicability of eclectic scientific methodology to the study of humans. Above all, Gumilla's ethnographic studies demonstrate a paradigmatic Jesuit eclecticism: he bases his conclusions on his firsthand observations while respectfully engaging classical, patristic, and contemporary authorities. For example, in a section subtitled "Contraposición de las opiniones moderna y antigua acerca del origen del color etiópico" (91–102) [Contrast of the modern and ancient opinions with regard to the origin of Ethiopian {i.e., black} color], his philosophizing clearly follows a path similar to that he follows when he discusses plants and animals. Immediately preceding an extended argument reconciling ancient and modern ideas about the power of the female mind over the human fetus, Gumilla clearly articulates the eclectic methodology that has served him well in other parts of *El Orinoco ilustrado:*

> Me veo obligado a dejar esta sentencia moderna, y a seguir la antigua y comunísimo; y por cuanto los argumentos antiguos se dan por ineficaces y de los casos específicos que se alegan se dice que no tienen la certidumbre necesaria, procuraré dar fuerza y eficacia a los argumentos y alegar casos de hechos innegables y específicos corroborados con muchos testigos, y abonados [. . .] (91)
>
> [I find myself obligated to leave behind this modern maxim, and follow the ancient and most common one; and since the ancient arguments are taken as ineffective and with specific cases that are put forth it is said that they do not have the necessary certainty, I will try to give force and efficacy to the arguments and put forth cases of undeniable facts and specifics corroborated by many witnesses, and guaranteed. . . .]

By self-consciously taking into account previous authorities, current opinions, and his own inductions based on particular cases, as well as inviting philosophically inclined readers to choose whichever approach has the most appeal—"escoja el erudito la que más le gustare" (91) [may the learned choose that which most pleases him]—Gumilla models his eclectic theory of knowledge.

Gumilla's references to his eclectic philosophy also announce the path along which he will lead readers to their agreement with his conclusions, as the following statement of methodology clarifies:

> La sentencia moderna duda, y no decide; pero tampoco asiente a la sentencia antigua [. . .] y en atención a que la experiencia es madre de la mejor y más cierta filosofía, de un solo caso de hecho, cierto y notorio, deduciré la razón de dudar; daré mi parecer, lo corroboraré con razones filosóficas (desatando de paso los argumentos contrarios) y concluiré confirmando la opinión, con otro caso de hecho cierto y notorio. (91–92)
>
> [The modern maxim has doubts, and does not convince; but neither does the ancient maxim agree . . . and given that experience is the mother of the best and most certain philosophy, from only one case in point, certain and well known, I will infer the reason for doubting; I will give my opinion, I will corroborate it with philosophical reasoning (refuting along the way the contrary arguments) and I will conclude by confirming my opinion with another case in point certain and well known.]

Within a discourse on skin pigmentation among Orinoco Region Amerindians and blacks, Gumilla applies to humans the same categories of nature he has applied to plants; he includes rarities and exceptions among his more ordinary physical descriptions. This kind of cultural data enters the realm of the latest philosophical inquiries in the emerging disciplines of anthropology and ethnology, combining the "new science" of Bacon's three states of nature with Giambattista Vico's three stages for human development. When hypothesizing reasons for some Orinoco Amerindians' continued resistance to European civilizing efforts, he evokes the *Sciencia nuova* (1725–44) of Vico, who philosophizes that the human mind develops from primitive mentation to rational, abstract thinking over time in three stages. Gumilla's discourse shares much with Vico's "new science" for studying humanity in that in their efforts to modernize the "science" of history both took inspiration from recent developments in natural philosophy while maintaining biblical chronologies. Gumilla suggests that the persistent problem of Amerindian resistance along the Orinoco is rooted in the region's lack of Inca or Aztec dominions, which would have unified and domesticated diverse tribes before Europeans arrived. With this, Gumilla eliminates the second of a familiar three-stage paradigm for human development that had been articulated by earlier philosophers such as El Inca Garcilaso de la Vega for South America had been but more recently generalized by Vico.[97]

Gumilla's eclectic exploration of Orinoco region peoples obeys the imperative to engage in a dialogue with other philosophers even as it appeals to human curiosity. For example, he appeals to curiosity by describing the Amerindian features common to the Orinoco in the same order one would describe a Renaissance portrait: starting with their thick, black hair, he passes to their beardless faces, "bellísimos" [most beautiful] eyes, flat noses, well-proportioned lips, and "enviable" teeth, which he compares to the purest ivory. Within this ecphrasis, however, he appeals to the received authority of two French humanists with regard to the cause of white hairs: "El cabello en todos, sin excepción alguna, es negro, grueso, laso y largo, con el apreciable privilegio de que necesita de largo peso de años para ponerse cano, argumento nuevo que robora la opinión antigua [. . .] (82) [The hair of all, without any exception, is black, thick, languid and long, with the noticeable privilege that it needs the long burden of years to turn white, a new argument that corroborates the ancient opinion . . .]. At this point, a footnote references specific pages in works by Julius Caesar Scaliger (1484–1558) and Bernard de la Monnoye (1641–1728), two renowned sources. As we will see, by applying Baconian categories of nature and inductive reasoning to his study of Orinoco region Amerindians and blacks, Gumilla participates in an eighteenth-century application of methodologies originally intended for the natural sciences to ethnographic topics (today included in the social sciences).

Gumilla's discussion of the humans inhabiting the Orinoco River region simultaneously appeals to his readers' curiosity and demonstrates his Jesuit eclecticism. In a chapter entitled "De los indios en general y de los que habitan en los terrenos del Orinoco y de sus vertientes en particular" [On the Indians in general and those that live in the lands along the Orinoco and its banks in particular], alongside the imperative to engage in dialogue with other authorities, Gumilla argues for the authority granted by close examination of particulars. This evokes Baconian methodology, which was eclectic in its own right. On the issue of miscegenation, Gumilla asserts his firsthand observations and mediates standard, accepted terminology with popular denominations. In the two lengthiest subdivisions of this chapter, Gumilla mediates his careful observation of specific cases with both ancient and modern authorities, comparing the authorities' opinions with his own experience. Instead of deducing particulars about the Orinoco region tribes from previous authorities' generalizations about New World Amerindians, Gumilla's Baconian methodology casts his particular study of Orinoco peoples as a stepping-stone toward study of South American Amerindians in general. He proposes as a tool his survey of the Orinoco nations' particular body types, facial structure, skin color, and customs. This tool can serve as a structuring gateway for examining all Amerindians:

> No es razón entrar en una noble y curiosa fábrica, sin fijar algo la vista en su frontispicio y fachada, que es de ordinario índice de la interior arquitectura; y así, antes de poner a la vista la capacidad, propiedades e inclinaciones, usos y costumbres de los indios americanos, daremos un bosquejo del talle, aire, aspecto y color de aquellas gentes de Orinoco y sus vertientes. (81–82)
>
> [It makes no sense to enter into a noble and curious structure, without fixing somewhat the eyes on its frontispiece and facade, which is ordinarily an index of the interior architecture; and so, before setting the eyes on the capabilities, properties and inclinations, ways and customs of the American Indians, we will give an outline of the physique, appearance, aspect, and color of those peoples of the Orinoco and its branches.]

This chapter justifies Gumilla's Baconian approach from the particular to the general and elevates his contributions to the studies of humans among modern "historiadores geográficos" (90) [geographical historians].

Gumilla supersedes the tradition of curious cultural displays and applies his eclectic scientific methodology to the emergent discipline of ethnography. He reconciles ancient and modern theories of human pigment variations and maternal impression while providing taxonomies of skin color. Just as with his study of plants and animals, Gumilla suggests that his time in New Granada revealed many particular cases that allow him to make a general hypothesis about humans: by introducing European blood both Amerindians and blacks are at most four steps away from being white. However, Gumilla warns, unions with non-Europeans delay or even regress within the four steps. This, explains Gumilla, is why so few blacks actually reach white status: "Ya se ve que si esta pochuela [*sic*] se casa con mulato propio, la prole vuelve a retroceder; y si se casa con un negro, se atrasará mucho más: y de estos atrasos depende el que pocos de ellos lleguen a puros blancos; pero algunos realmente llegan" (86) [It is already seen that if this entirely white woman marries a typical mulatto, the offspring falls back again; and if she marries a black man, she will fall much further behind: and on these delays depends the fact that few of them become purely white; but some of them actually do arrive].

The Jesuit's human taxonomy makes use of accepted racial classifications such as "quadroon," but also takes into account "vulgar" denominations: a mixture of formal and informal terminology. In both cases the whitening of Amerindians and blacks depends on continued unions with Europeans that progress mathematically from wholly black to half to fourth to eighth to "entirely white." Amerindian women whiten in the following way: *india, mestiza, cuarterona, ochavona,* and *puchuela,* or "entirely white" (84–85). African women follow an equivalent path: *negra, mulata, cuarterona, ochavona,* and *puchuela,* or "entirely white" (86).[98] Gumilla describes Amerindians who depart from the four-step path with popular classifications like *tente en el aire* (hold steady in the air), as in neither whitening or regressing, or "*salta atrás;* porque en lugar de adelantar algo, se atrasa o vuelve atrás, de grado superior a inferior" (86) [*jump backward;* because instead of advancing something, it delays or turns back, from superior to inferior degree].[99] These descriptions are echoed by Spanish and Spanish American contemporaries as well as Charles Marie de La Condamine during this eighteenth-century birth of ethnographic vocabulary.

Gumilla suggests that his missionary experience allowed him to observe years of "experiments" in miscegenation, and he bolsters his authority by evoking the Royal Society of London. He compares his personal observations about Amerindian skin color to a recent study by the Royal Society that followed Baconian methodology to test a technique for predicting the eventual color of newborn babies:

> Al nacer aquellos niños, son blancos por algunos días, lo que sucede también a los negrillos; y es digno de saberse que así como los hijos de los negros nacen con su pinta negra en las extremidades de las uñas (Academia Real de las Ciencias, año 1702, pág. 32), como muestra de lo que luego serán, así también nacen los indiecillos con una mancha hacia la parte posterior de la cintura de color oscuro. [. . .] Esta seña o mancha es cierta, y cosa que tengo vista y examinada repetidas veces. (84)

> [Upon the birth of those [Indian] children, they are white for some days, which also happens to the little black babies; and it is worthy of knowing that just as the children of the blacks are born with their black dot on the extremity of the fingernails (Royal Academy of Sciences, year 1702, page 32), as a sign of what they will later be, the little baby Indians are born with a dark stain toward the back of the waist. . . . This sign or stain is certain, and something I have seen and examined repeated times.]

This quote appears in a section on interracial marriages, where Gumilla also underscores the power of missionaries to forge Christian unions. Throughout this chapter, Gumilla cites his own eyewitness authority, and either gently corrects the writings of historians and philosophers or frames his references to previous authorities as mere confirmations of what he has seen.

Within his discussion of causes for different skin colors, Gumilla depends on induction from particular cases. Two of the most fascinating cases upon which Gumilla based his inferences include his generalizations about the force of the imagination to shape a developing fetus. These curious human phenomena occurred in Cartagena de Indias, where Gumilla served as rector of the local Jesuit college. Gumilla frames the first anecdote about a married slave couple's albino black babies as proof of the biblical origins of all the world's nations and skin colors from Noah's three sons, namely, Shem, Ham, and Japheth. The second anecdote vividly describes a spotted baby (neither fully black nor fully white) born of another married slave. In both anecdotes the Jesuit engages Bacon's classifications of the three states of nature: the ordinary, the perverse, or the artistic.[100] These striking cases evoke Bacon by discussing "irregular births" that demonstrate God's power to create playful irregularities in nature.

In both cases, Gumilla's paradigmatic eclecticism reveals itself in the Jesuit's combination of outside authorities with firsthand observation. In the first anecdote, Gumilla cites the case of eight births from the marriage of two black slaves, in which successive offspring alternated between black and white, and he describes this as "un posible de muy extraordinaria contingencia, tal, que en rarísimo caso se reduce a acto" (93) [a possibility of very extraordinary circumstance, such that in extremely rare cases it becomes reality]. However, before making his inference he investigates other cases of "negros blancos" [white blacks] or "negros albinos" [albino blacks]:

> Fuera de esto, negros de Angola, que yo examiné sobre ello en Cartagena, me aseguraron que allá en su patria también nacen algunos de dichos albinos. [. . .] De este hecho y hechos infiero esta consecuencia: luego después de la dispersión de las gentes pudieron nacer de padres blancos hijos negros, y casados entre sí, ir poblando países que hasta hoy poseen y llenarlos de negros a fuerza de tiempo. [. . .] ¿Es posible que de Sem, Cham y Jafhet se han originado todas las naciones que hoy pueblan la faz de la tierra? Sí, porque éste no es negocio de tres, ni de cuatro siglos, sino de muchos millares de años y generaciones. (93–94)
>
> [Aside from this, blacks from Angola that I examined about it in Cartagena assured me that also there in their homeland are born some of these said albinos. . . . From this fact and facts I infer this consequence: after the dispersion of peoples, black children could be born from white parents; and married among themselves, they go about populating countries that up to today possess and are filled with blacks over time. . . . Is it possible that from Shem, Ham, and Japheth have originated all the nations that today populate the face of the earth? Yes, because this is not a transaction of three, nor of four centuries, but rather of many thousands of years and generations.]

In his second anecdote, Gumilla couples a description of the spotted baby's skin, which combined white and black pigmentation, with a narration of the wonder-filled reactions that both the human child and her portrait (*copia*) inspired throughout New Granada as well as across the ocean in London.[101] To emphasize the pious worth of extraordinary natural or human phenomena, Gumilla hints at the appropriate conclusion to such a spectacle. He says of the noblemen of Cartagena who gathered to view the spotted baby: "Todos se volvían atónitos y alabando al Criador, que, siendo siempre admirable en sus obras, suele también jugar en la tierra con las hechuras de sus poderosas manos" (101) [Everyone became amazed and praised the Creator, who, being always wonderful in his works, also is in the habit of playing in the earth with the creations of his powerful hands].

However, before expanding on this second case, Gumilla speaks of the evidence that led him to form a general hypothesis about the force of human "fantasy," or imagination. He applies various ancient and modern sources to the topic of maternal impression, then adds what he "saw, observed, and reflected upon" before arriving at his conclusion: "[Y] entonces fué cuando yo tuve, finalmente, para mí por indubitable la conclusión que aquí he propuesto de la eficacia natural de la imaginativa" (101) [And then it was that I took as undeniable the conclusion that here I have proposed on the natural effectiveness of the imaginative]. First, his pathway to knowledge provides a laundry list of church fathers and ancient authorities who support the concept of imprinting a developing fetus. Next, Gumilla offers a series of replies to possible contrary arguments that demonstrate his awareness of current opinions.[102] He opens these *ante occupatio* with phrases like "Bien sé que a esta réplica responden. [. . .] Tal vez dirán (y no falta quien lo afirma)" (97–98). [I well know that to this reply they respond. . . . Perhaps they will say (and there will not be missing someone to affirm it)]. Finally, he wraps up this lengthy subsection (what he actually refers to as a "short dissertation") of his chapter on Orinoco Amerindians as follows:

> Como es evidencia en el caso de hecho con que me ofrecí concluir esta corta disertación, y es como se sigue, sin quitar ni añadir un ápice de lo que vi, observé y reflexioné. El año de 1738, estando a mi cargo el Colegio de la Compañía de Jesús [. . .] en Cartagena de Indias, salí a una enfermería. [. . .] Hallé, entre otros, a una negra casada. [. . .] Levantó la negra la mantilla, y vi (mas no sé si vi, hasta que salí de la suspensión con que me embargó la novedad), vi, en fin, una criatura cual creo que jamás han visto los siglos. [. . .] Volví repetidas veces con otros Padres de aquel Colegio a contemplar y admirar esta maravilla [. . .] tomé a la criatura en mis brazos para observar más y más la variedad dicha de sus colores. . . ." (99–102)
>
> [As is demonstrated by the case in point with which I will to conclude this short dissertation, and it is as follows, without removing nor adding an iota to what I saw, observed, and reflected on. In the year 1738, being under my charge the College of the Company of Jesus . . . in Cartagena de Indies, I

went out to an infirmary. . . . I found, among other things, a married black woman. . . . The black woman lifted up her shawl, and I saw (but I do not know if I saw, until I recovered from the amazement with which the novelty overcame me), anyway, I saw a baby I think the centuries have never seen. . . . I returned time after time with other fathers of that college to contemplate and wonder at this marvel . . . I took the baby in my arms to observe more and more the said variety of its colors. . . .]

This dissertation ends with a reaffirmation of careful observation in tandem with outside authorities, as Gumilla self-consciously applies his eclectic methodology to the study of humans.

Gumilla's maternal impression dissertation concludes with a definitive hypothesis. However, before generalizing about this particular case of the spotted baby, he corroborates his own authority with that of additional eyewitnesses. The anecdote that confirms his beliefs includes a surprising revelation, which the Jesuit re-creates with flair. While examining the spotted baby, narrates Gumilla, a small lapdog jumps up on the skirts of the black mother. After comparing the spotted pattern of the dog with the baby, Gumilla questions the slave and finds out that well before the birth of her spotted child this young mother had raised her spotted canine companion from infancy, and that her beloved lapdog was never far from sight. Before definitively forming his hypothesis about this particular case of maternal impression, however, Gumilla confirms his thoughts on the matter with other eyewitnesses. The Jesuit reconstructs their investigation and testimony in a vivid narrative:

[S]altó a las faldas de la negra una perrilla de color blanco y negro; empecé a cotejar en general aquellas pintas con las de la criatura y, hallando notable correspondencia de unas con otras, las fuí cotejando parte por parte, unas con otras; y, en fin, hallé una total uniformidad entre unas y otras, no sólo en la forma, figura y color, sino en lo respectivo al lugar en que estaban colocados los colores. [. . .] Y así creí, y creo, que la continua vista, el afecto con que la miraba y los muchos ratos que jugaba con ella, fueron causa suficiente para dibujar toda aquella variedad de colores de la perilla en su fantasía e imprimirlos después en la configuración natural de su hija en la matriz. Este pensamiento comuniqué a sólo dos sujetos del dicho Colegio de Cartagena, y ambos hicieron el mismo cálculo y cotejo de colores y manchas de la perra y de la niña, y la total correspondencia y uniformidad. Los convenció totalmente y obligó a creer ejecutada allí la fuerza de la imaginación de la madre. (102)

[A little dog of black and white color jumped into the lap of the black woman; I began to compare in general those dots with those of the baby and, finding a noteworthy correspondence of some with others, I went about comparing part by part, some with others; and, anyway, I found a complete uniformity between some and the others, not only in the form, figure, and color, but also with regard to the place in which the colors were placed. [. . .] And so I believed, and I believe, that the constant view of the dog, the affection with which she gazed at it, and the many moments that she played with it were sufficient cause to sketch all that variety of colors of the lapdog in her fantasy and fix them afterward in the natural configuration of the daughter in the womb. I conveyed this thought to only two individuals from said College of Cartagena, and they both did the same calculation and comparison of colors and stains of the dog and the girl, and they also found a complete correspondence and uniformity. This convinced them totally and obliged them to believe executed therein the force of the mother's imagination.]

The total persuasion of these eyewitnesses based on their own careful examination of both spotted creatures rhetorically obliges readers of Gumilla's story to believe this case of "tan peregrino y singularísimo juguete de la Naturaleza" (100) [such a peculiar and extraordinary plaything of nature].

With both this singular case of the spotted baby and the previous case of a mother's alternating series of black and white (albino black) children, Gumilla explains how he formed his opinion on the cause of these playful irregularities of nature by reconciling ancient and modern testimonies regarding the force of a mother's imagination to imprint an unborn fetus with his first-hand observations in Cartagena. As a concluding reference, Gumilla cites Father Feijoo as another modern authority about the ongoing debates regarding maternal impression: "Y porque éste basta, no añado otro parto de nuestro tiempo, digno de saberse, no por ser raro ni inaudito, sino por ser moderno. Hallarálo el curioso en el tomo último de las *Obras* del Rvdo. P. M. Feijoo, que es el primero de sus *Cartas eruditas, Carta* II, pág. 73" (102). [And since this is enough, I am not adding another birth from our times, worthy of knowing about, not because it is strange or unheard of, but rather because it is modern. The curious can find it in the last tome of the *Works* of the most reverend Father M. Feijoo, which is the first of his *Cartas eruditas,* Letter II, page 73).] Following an eclectic Baconian pathway to knowledge, Gumilla's dissertation also reaffirms the weight of his specific observed case by evoking and affirming this preeminent Spanish Baconian's authority.

In summary, Gumilla's ethnographic discussions in *El Orinoco ilustrado* serve as excellent examples of his Baconian methodology and eclecticism, just as do his studies of plants and animals. With regard to the examples detailing New Granada "playthings of nature," purely scholastic philosophers would have simply accepted the ancient theory of maternal impression. Gumilla instead validates the theory for modern times by mediating previous and contemporary authorities with his own authority evoked through both his being an eyewitness to this strange human phenomena as well as being rector of Cartagena's Jesuit college. With the publication of *El Orinoco ilustrado* came immediate recognition of Gumilla's worth as a source of authoritative, firsthand knowledge about the region.[103] Gumilla references Feijoo as a valuable modern source. Yet with the appearance of *El Orinoco ilustrado,* this transfer of knowledge reverses and Gumilla becomes Feijoo's primary source for knowledge about Orinoco Amerindian ethnography. Feijoo was hailed as the "new Spanish Bacon" and is still today credited for spreading knowledge of modern European scientific ideas throughout Spain and the New World. Feijoo paid tribute to his Jesuit sources, however, and in his *Cartas eruditas y curiosas,* he recognizes Gumilla's successful mediation of Amerindian knowledge.[104] For example, in one article discussing the medical value of French snuffs, Feijoo compares them with Orinoco purgative powders and refers his readers to Gumilla's documentation of Amerindian practices.[105] In another article discussing shamanistic culture, Feijoo praises Gumilla's "truthful" reporting and speaks of his authority as an eyewitness. He also underscores the Jesuit's solid character and methodology before quoting extensively from *El Orinoco ilustrado.*

Gumilla's taxonomies of skin color and disquisitions on maternal impression reveal how he applied the same eclectic approach to his study of humans as to his examination of plants and animals. The Jesuit blended his own observations and induction with deductions based on previous authorities, following an experimental approach very much influenced by Spanish Baconianism, yet open to other methodologies. Born of his Jesuit training and evolved by its application within the Orinoco River region, Gumilla's eclectic theory of knowledge provides an important example of Jesuit mediation with Catholicism of both natural philosophical developments and Amerindian knowledge. Increased attention to the Society of Jesus as a multinational, multidisciplinary scientific institution founded by Spaniards, headquartered in Rome, and with access to local knowledge across the globe can expand definitions of an Hispanic Enlightenment that did not divorce science from religion and did not always deny knowledge produced outside of European paradigms. Jesuit epistemology depended on a unique methodological combination of both missionary naturalists and university professors. The former made observations and formed hypotheses on-site (in the peripheries), mediating human and environmental sources for new knowledge with their academic training in Catholic theology, moral, and natural philosophy. The latter were scholars with formal libraries and laboratories who taught and published in Jesuit centers of learning.

Scholars are beginning to explore how much the Baconian hunt for particular, empirical truths might be indebted to Acosta.[106] Future research may conclude that eighteenth-century Spanish Baconianism and the increased emphasis on experimentalism owed more to the Jesuits than to Bacon himself. Certainly the Jesuit tradition of missionary data collection and mediation with Amerindian sources expanded European methodologies and contributed to Spain's shift toward direct observation and inductive reasoning. Far from being "dry and uncreative opponents to progress," eighteenth-century Jesuit natural philosophers incorporated new methods for the advancement of natural knowledge. Thanks to the Society of Jesus's international, evangelical mission to spread Christianity, their eclecticism was influenced by the mediation of local knowledge paradigms. Jesuit eclectics like Gumilla justified scientific advances as contributing to both European and Amerindian knowledge of God. Gumilla's simultaneously scientific and religious discourse exemplifies how missionary firsthand experience and experiments in the field superseded the scholastic heuristic and effected mutual exchanges of knowledge to enlighten Amerindian and European souls and intellects.

4

¡Oh Monstruo, Oh Bestia!: Pathways to Knowledge in *El Orinoco ilustrado*

AS WE SAW IN CHAPTER 3, GUMILLA'S MEDIATION OF CATHOLIC dogma, European scientific discourse, and new information about the Orinoco River region's flora, fauna, and people clearly contributed to the eclectic approach that allowed him to expand Europe's knowledge base. Chapter 4 considers nine examples from *El Orinoco ilustrado* that demonstrate Gumilla's Jesuit eclecticism, Baconian empirical methods, and reconciliation of his own firsthand experience with information from various authorities. It also suggests ways Gumilla's intercultural discourse with Amerindians affected his simultaneously religious and scientific theory of knowledge. This book's understanding of intercultural discourse underscores Jesuit appropriation of indigenous knowledge. However, textual evidence of Gumilla's incorporation and adaptation of Amerindian epistemologies has implications for understanding the evolution of enlightened thought in Spain and South America. On one level, Gumilla's text celebrates the benefits that dispelling Amerindian superstitions or correcting their "cruel customs" had for his evangelical agenda. Yet, on another, Gumilla's rhetoric of wonder reveals the subtle ways that aspects of Amerindian popular wisdom and folklore altered his theory of knowledge.[1] *El Orinoco ilustrado* reveals pathways to knowledge that challenge the Eurocentric discourse of the Enlightenment.

Both Gumilla's Jesuit education and his exposure to Orinoco Amerindian folk knowledge influenced the rhetorical strategies he employs to present scientific data in *El Orinoco ilustrado.* As noted previously, Gumilla builds upon his Jesuit authority (ethos) and persuades his readers by blending reason (logos) and emotion (pathos). This persuasion depends on a rhetoric of wonder that combines several literary strategies for amplification with internal and external proofs. Gumilla privileges emotion-evoking rhetorical mechanisms that are outlined in the handbooks of sacred oratory standard to Jesuit training. And the strange and new Orinoco plants, animals, and Amerindian nations complement this stirring rhetoric, as Gumilla reproduces local folklore to amplify his text as well as the wonder it induces. The pathways from wonder to knowledge in *El Orinoco ilustrado* are rhetorical.

This chapter will focus on Gumilla's recourse to a vivid presentation of anecdotes and analogies, which are two simultaneously evocative and persuasive techniques for rhetorical amplification, in order to explore how he captures his readers' imaginations and leads them on an inductive path to explanations of several mysteries of Orinoco River region nature and culture. The different pathways to knowledge revealed in the Jesuit's Christianization, demonization, correction, and validation of Amerindian folklore and popular wisdom all take Gumilla's readers on a rhetorical journey from wonder to knowledge that depends on their persuasion more than on readers' perception or certainty. After a brief consideration of some classical and Jesuit roots for Gumilla's rhetorical pathway from wonder to knowledge, we will look at nine examples that illustrate the Jesuit's main approaches to Amerindian knowledge. In the first example, Gumilla appropriates and Christianizes the myth of El Dorado. With the next anecdote, he demonizes the poison curare. The third through fifth examples demonstrate how Gumilla simultaneously learns from and corrects Amerindian medical practices. In the sixth, he corrects European misunderstandings of the Orinoco region war drums and applies an eclectic approach to acoustics. The last three examples take up the second half of this chapter. With the seventh through ninth examples, Gumilla's legitimization of mysteries surrounding anacondas and crocodiles, two iconographic Orinoco reptiles, offers models for how, instead of rejecting strange Amerindian knowledge as irrational, Gumilla validates fantastic natural phenomena and popular beliefs through pathways to knowledge that mix religious and scientific discourses in an attempt to make sense of autochthonous Orinoco region opinions and cultural practices for his eighteenth-century Christian readers. Taken together, the nine examples illustrate four seemingly contradictory types of rhetorical paths (Christianization, demonization, correction, and validation) that provide alternatives to notions of knowledge in Enlightenment studies.

Like the Jesuit Joseph de Acosta, and like Catholic intellectuals in Spain during the first half of the eighteenth century, Gumilla drew on classical roots to frame his scientific pursuits as a vehicle for praising God and better understanding Catholic truths. In Gumilla's theory of knowledge, human understanding does not remove wonder, as wonder should both inspire and result from the production of knowledge. For Catholic intellectuals, wonder (*admiratio*) at nature serves as impetus for knowledge (*scientia*), which encourages knowledge of and sacred awe toward God (*Admiratio*).[2] Therefore, by providing rational explanations of mysterious natural phenomena, knowledge heightens both the passion of wonder and human Christian devotion. Jesuits combined reason and passion in pursuits of science that remained tied to the Catholic faith and evangelization. In this wonder-to-knowledge paradigm the passion of wonder awakens man's natural thirst for knowledge, or as Gumilla puts it, "De aquí se excita la curiosidad, o la admiración, y el deseo de saber"[3] ["From here is stimulated curiosity, or wonder, and thus the desire for knowledge]. By the time Gumilla evoked this paradigm, the role of wonder as the starting point for philosophical pursuits of knowledge had been recognized in Western philosophy for over two thousand years. Contemporary articulations in Bacon, who called wonder the "seed of knowledge," as well as in Hobbes, Descartes, Malebranche, and Vico were based on Plato and Aristotle and church fathers like Aquinas, who cemented the tradition of referencing the opening words of Aristotle's *Metaphysics* ("All men by nature desire to know") as well as the oft-repeated *admiratio-scientia* paradigm: "For it is owing to [the earliest philosophers'] wonder that men both now begin and at first began to philosophize."[4]

El Orinoco ilustrado incorporates Gumilla's Orinoco River region experience into pathways to knowledge that appropriated and assimilated instead of rejected Amerindian folklore. Rather than denying seemingly fantastic anecdotes or local beliefs, Gumilla integrates them into a wonder-provoking production of knowledge about his Jesuit province directed at European readers. A key source for Gumilla's valorization of wonder as the starting point of pathways to knowledge even about seemingly fantastic phenomena was the Jesuit Imperial College professor and royal apothecary Juan Eusebio Nieremberg, whose 1635 textbook, *Historia naturae* Gumilla cites.[5] Most of Gumilla's contemporary missionary brothers around the globe had also read Nieremberg's textbooks and absorbed his philosophy on the value of wonder sparked by extraordinary natural history topics for inspiring the production of credible knowledge about them.[6] Gumilla includes Amerindian knowledge with his topics in a manner indicative of his Jesuit eclecticism, and *El Orinoco ilustrado* employs a scientific discourse that combines reason and emotions and provides alternatives to the hegemonic narrative of knowledge accumulation during the Enlightenment.

In the following examples, Gumilla deploys four types of rhetorical pathways to knowledge that process and legitimize it through anecdotes and analogies. Both of these strategies support inductive reasoning and depart from Aristotelian causes and effects and deductive syllogisms. Particular analogies, which use figurative language such as similes and metaphors, are employed to create the chain of reasoning required by induction. Gumilla turns to these types of rhetorical amplification in order to persuade his readers by appealing to their senses and reason. The anecdotes and allegories bridge Amerindian folklore and European science.

They also capture and transport his readers' imagination with a rhetoric of wonder that validates Orinoco region legends with scientific demonstrations. As we will see, Gumilla creates pathways from wonder to knowledge that blend metaphorical and literal language in a simultaneously religious and scientific discourse. *El Orinoco ilustrado* reconciles scholastic authorities, Amerindian popular wisdom, and seventeenth- and eighteenth-century advances in natural philosophy.

Christianizing El Dorado

Since *El Orinoco ilustrado* first appeared in print, Gumilla's stated belief in El Dorado has been pointed to as evidence of the influence of Amerindian lore on the imagination of an overly credulous Jesuit. Misunderstandings about why Gumilla professes faith in the existence of El Dorado have prevented readers from appreciating the rhetorical function of this New World trope in his text. Like others before him, Gumilla engages European wonder at the mythical El Dorado. However, the Jesuit applies an evangelical purpose to the quest for El Dorado, maintaining that European explorers' constant search for El Dorado's mineral riches was divinely inspired. As Gumilla explains, God helped Amerindian souls bait Christians by painting images of gold, and this helped to advance evangelization.

Amerindians provided diverse anecdotes about El Dorado. In contrast to the current reality he affirms, Gumilla designates these early reports as mere fables spun by tribes. "Las primeras [. . .] eran a fin de apartar de sí a los españoles" (*El Orinoco ilustrado,* 270) ["The first ones . . . had the purpose of moving the Spaniards away from themselves]. Anxious to be rid of their European "guests," who seemed inclined to overstay their welcome, Amerindians tricked them by capturing their imaginations with wonderful stories of gold and then sending them away on fruitless searches:

> [E]ste Dorado el que estaba ideado en la mente [. . .] y nada menos ofrecían los indios que iban conquistando; porque éstos, viendo que lo que más apreciaban aquellos forasteros era el oro, pintaban con muy vivos colores la copia de oro del país que les parecía más a propósito para estar más libres de sus huéspedes. (264)

> [This Dorado was the one that was conceived in the mind . . . and the Indians that they were conquering offered nothing less; because these, seeing that what the foreigners most appreciated was gold, painted with very vivid colors the portrait of gold in the country that seemed to them the most convenient for being more free of their guests.]

Thus commenced a long-standing tradition of treasure hunters inspired by El Dorado, a city that very early existed in the European imaginary as a place and civilization, a fabled mountain of gold, a lake, or a cacique covered in gold. Gumilla's list of those who searched in vain from the Amazon to the Orinoco River region includes Pizarro, Quesada, Orellana, Berrío, and Ralegh.[7] Yet, as Gumilla insists, even though they were not missionaries trying to convert Amerindians, these European treasure-hunters were also fulfilling God's will, because God made Spaniards believe fantastic Amerindian anecdotes about El Dorado to inspire their discovery of—and, more importantly, spiritual enlightenment of—unknown regions:

> Permitía Dios que los españoles creyesen tan seriamente dichas noticias, para que se descubriesen más y más provincias, donde rayase la luz del Santo Evangelio, como por su bondad rayó, creció y llegó a claro y perfecto día, mediante la predicación de muchos varones apostólicos que reputaron el oro por lodo a vista de la preciosidad de tan innumerables almas. (264)
>
> [God allowed the Spaniards to believe so seriously said notices, so that they would discover more and more provinces, where the light of the Holy Gospel would break through, just as by his kindness it broke through, grew, and became a clear and perfect day by way of the preaching of many apostolic men who esteemed the gold as sludge in light of the loveliness of such innumerable souls.]

Gumilla believes that not only do El Dorado and its mountains of gold exist, but they have a providential purpose. This Christianization of the Amerindian myth of *El Dorado* effectively converts it into a reality. He writes that the time has come to cease quests for so many mythical El Dorados: "Pero recojamos ya las noticias del célebre Dorado o cuidad de Manoa, separando al mismo tiempo las cosas fabulosas de las probables, reteniendo éstas y despreciando aquéllas" (263) [But let us collect already the notices of the famous Dorado or city of Manoa, separating at the same time the mythical things from the provable, keeping these {the likely} and rejecting those {the fantastic}]. As the Jesuit will conclude, one real El Dorado does exist in New Granada, but it still patiently awaits discovery and evangelization.

Thus, Gumilla commences several pages proving the existence of El Dorado within his Jesuit province. In the process, he both supports and challenges earlier European reports as well as legitimizes Orinoco Amerindian lore, including testimony from one Christianized Amerindian named Agustín that Gumilla spoke with at a Jesuit mission—"Misión nuestra de Guanapalo, en el río Meta" (267) [Our mission of Guanapalo, in the Meta River]—who spent fifteen years in this real El Dorado as a captive. According to Gumilla, when compared with reports from a sixteenth-century Spanish expedition, Agustín's story has merit by "juntando la declaración del indio Agustín, que fue tantos años esclavo en la ciudad capital de *El Dorado* [. . .] con lo que vieron, padecieron y declaran [Felipe de] Utre y sus treinta y nueve soldados" (269) [joining together the statement of evidence by the Indian Agustin, who was so many years a slave in the capital city of El Dorado . . . with what Felipe de Utre and his thirty-nine soldiers saw, suffered, and testified]. Before concluding that El Dorado definitively exists within the territory of New Granada, Gumilla evokes the authority of a contemporary Jesuit missionary:

> [D]igo que estos testigos y circunstancias, juntas con el dictamen constante del Padre José Cabarte, fundado en su larga experiencia de misionero, en casi cuarenta años de tratar y trabajar entre aquellas naciones, por donde fué el derrotero de Utre, este agregado de cosas constituye un fundamento grave a favor de la existencia de *El Dorado,* y una probabilidad no despreciable [. . .] a vista de estos sólidos fundamentos. (270)

> [I declare that these witnesses and circumstances, together with the steadfast opinion of Father José Cabarte, based on his long experience as missionary, in almost forty years of being acquainted with and working among those nations, around where Utre's course was, together constitute serious evidence in favor of the existence of El Dorado, and make it existence even a probability . . . in light of these solid bases.]

Gumilla consciously Christianizes the trope of El Dorado and clearly states his reasons for narrating Amerindian lore. While the real El Dorado awaits discovery and promises great mineral wealth, more importantly it offers a spiritual gold mine with Amerindian souls as the real treasure. He underscores his conversion of a mythical El Dorado into a Christian reality by predicting future Jesuit success in this city, analogous to current glories in New Mexico and California. His urgent call to evangelize El Dorado likens its potential to the latest successful conversion projects:

> Todo lo cual he querido apuntar, porque tal vez con el tiempo moverá Dios nuestro Señor algún corazón magnánimo a descubrir aquellas provincias y se abrirá puerta para que entre en ellas la luz del Evangelio, con la felicidad con que nuevamente ha entrado cerca del Nuevo Méjico [. . .] con las Californias [. . .] ignorados hasta el año de 1739 de este siglo. No repugna que algún día conste lo mismo y se publiquen las mismas o semejantes noticias, ya verificadas, del famoso Dorado y de sus gentes. ¡Ojalá sea cuanto antes, para bien y salud eterna de aquellas almas! (271)

[I have wanted to note everything, because perhaps with time God our Lord will move some magnanimous heart to discover those provinces and a door will be opened so that the light of the Gospel enters into them, with the happiness with which {the light} has entered near New Mexico again . . . with the Californias . . . ignored until the year of 1739 of this century. No doubt someday the same or similar news will be published, already verified, about the famous Dorado and its peoples. May God grant that it happens soon, for the well-being and eternal health of those souls!]

Far from falling victim to an active imagination, Gumilla's pathway to knowledge about El Dorado rhetorically confirms both abundant gold and spiritual treasures awaiting European discovery and converts the wonder of preexisting Amerindian anecdotes into a Christian reality.

Curare: "The Devil Made Them Do It"

In contrast to the providential aspect of his Christianization of El Dorado, Gumilla demonizes curare. Instead of presenting this non-Western knowledge as valuable for medical procedures, Gumilla frames its discovery as dictated by the devil.[8] And while the Jesuit continually underscores his eyewitness status as a key component of his authority about Orinoco River region data, his chapter on curare takes a very different approach. In "Del mortal veneno llamado curare: raro modo de fabricarlo, y de su instantánea actividad" [On the deadly poison called curare: strange means of making it, and on its instaneous activity] Gumilla asserts his second-hand mediation of Amerindian knowledge about this poison and its "strange fabrication" in an effort to distance himself from its production: "[A]unque he tenido muchas veces el curare en mis manos, no soy testigo ocular de la referida maniobra" (366) [Although I have had curare many times in my hands, I am not an eyewitness to the referred operation]. From the curare root's discovery through the production of its poison, to its application as a weapon against the tribe's Amerindian enemies, Jesuit missionaries, Spanish soldiers, and settlers, "¿quién lo creerá, sino confesando que todo ello, desde el hallazgo de la raíz hasta el fin, fué dictado por el demonio? Yo así me lo persuado" (365–66) [who will believe it, unless confessing that all that, from the discovery of the root until the end, was dictated by the devil?][9] Far from mitigating European wonder at the mysteries of curare, Gumilla designates it as diabolical knowledge, yet at the same time heightens his readers' desire for knowledge about this lethal Orinoco botanical product: "¡Oh prodigio grande de las causas ocultas que ignoramos!" (361) [Oh great wonder of the hidden causes we are unaware of!] As opposed to the single paragraph he allots to the safe, civilized healing balm *aceite de María* (Virgin's oil), Gumilla dedicates an entire chapter to the horrific curare. Here his blend of anecdotes that demonize Amerindian cultural practices alongside analogies that can be drawn from human and animal experiments epitomizes this alternative pathway to knowledge.

Although Gumilla includes several mysterious processes of nature in *El Orinoco ilustrado,* he normally frames them as reasons to praise God and acknowledge the limits placed on human understanding. In the curare chapter, however, Gumilla demonizes both the poison and the guardians of its secret recipe. This is a secret about which Europeans should be warned more than informed. As Gumilla tells it, God hid the curare root in festering swamps,[10] but the devil revealed it to the wicked Caverre tribe: "la más inhumana, bruta y carnicera de cuantas mantiene el Orinoco es la maestra. . . . Sola esta nación se tiene el secreto y lo fabrica y logra la renta pingüe del resto de todas aquellas naciones que por sí o por terceras personas concurren a la compra del curare" (360) [the most inhumane, brutal and bloody of all the Orinoco supports is the craftsman. . . . Only this nation has the secret and makes it and earns substantial income from the rest of all those nations that either by themselves or through third parties come together for the buying of curare]. The Jesuit steeps his anecdotes about curare production in pagan rites and social hierarchy and plays up the horrors of "la malignidad del curare" (367) [the evil nature of the curare]. By framing the entire chapter as a warning about how God never meant for humans to know about this root while he himself purposely presents facts based on the Caverre tribe's previously unknown, dangerous knowledge about how to fabricate—"este inaudito y fatal veneno" (363) [this unheard-of and fatal poison]—Gumilla leads his readers on a pathway from wonder to knowledge that demonizes Amerindian cultural practices.

Curare has unfortunately been revealed and therefore should be studied. In an effort to comprehend the horrifyingly instantaneous effect of the poison, Gumilla experiments with animals, applying his theories about blood circulation. If men are inflicted even with minor wounds, the arrow tips poisoned by curare cause an instantaneous, bloodcurdling death. As Gumilla exclaims: "[H]erido el hombre levemente con una punta de flecha de curare, aunque no haga más rasguño que el que hiciera un alfiler, se le cuaja toda la sangre y muere tan instantáneamente que apenas puede decir tres veces Jesús" (361) [When a man is slightly injured with the tip of a curare arrow, even if it does not make more of a scrape than a pin would make, all his blood coagulates in him and he dies so instantaneously that he can only just say Jesus three times]. To try to understand this diabolical effect of the curare on the bloodstream, Gumilla experiments first on monkeys, then on other animals: "El mismo instantáneo efecto reconocí después en los tigres, antes, leones y otras muchas fieras y aves" (363) [Afterward I examined the same instantaneous effect on tigers, elk, lions, and many other wild beasts and birds]. But its mysteries cannot be explained except by analogy, so Gumilla compares the coagulation of blood with the freezing of liquids. This scientific method allows for Baconian experimentalism and induction.

Yet at the same time that Gumilla applies his individual, empirical process to advance understanding of the poison's efficacy, he maintains his negative judgment of the Amerindian cultural practices known about curare. The Jesuit finds obvious reasons to condemn curare, since it is a weapon that could halt the civilization of the Orinoco River region. Still, Gumilla's condemnation ultimately seems more influenced by the ruthless community ritual of its fabrication. In an anecdote that re-creates the vivid, performative aspect of the production of curare, *El Orinoco ilustrado* reveals what Walter Mignolo and Anibal Quijano have described as an "intersubjective dimension in the production of knowledge,"[11] here manifest in the indigenous knowledge produced within the Caverre tribe's tradition, which includes "affective actions as a different kind of rationality":[12]

> [B]uscan para esta faena la vieja más inútil de la población, y cuando ésta cae muerta a violencias del vaho de las ollas, como de ordinario acontece, luego sustituyen otra vieja del mismo calibre en su lugar. [. . .] [L]a pobre anciana amasando su muerte [. . .] exprime con todas aquellas pocas fuerzas que su edad le permite [. . .] y a poco rato de hervir las ollas, ya atosigada, cae muerta; y entra la segunda, que a veces escapa y a veces no [. . .] y así le mandan a la triste anciana que prosiga su peligro próximo de muerte. (364–65)

> [They go and get for this job the most useless old woman of the population, and when this one violently falls dead from the vapors from the cauldrons, as usually happens, then they substitute another old woman of the same caliber in her place. . . . The first poor old hag cooking up her death . . . squeezes [out the essence of curare] with what little strength her age permits her . . . and soon after the pots boil, already overwhelmed, she falls dead; and the second one enters, who sometimes escapes and sometimes does not . . . and in this way they order about the poor old hag who continues in her dangerous proximity to death.]

Gumilla condemns this callous cultural practice of forcing expendable old women to distill the poison at the same time he marvels that "todo esto sea invención de la nación más tosca y bárbara del río Orinoco" (365) [all this is an invention of the most rustic and barbarous nation of the Orinoco River]. If only the production of curare and knowledge about the poison followed a European scientific social hierarchy instead, the Jesuit seems to suggest, then negative opinions might be altered: "maniobra se ejecutara por uno de nuestros científicos, con las vasijas competentes y con las reglas de la facultad" (360) [if this maneuver were carried out by one of our scientists, with adequate jars and university norms]. In contrast to his negative judgment of the mysterious group, performative way of knowing enacted on the frontiers of civilization, Gumilla praises the laboratories found in European and South American cities. The placement of the Amerindian "laboratory" site deep in forest swamps and the emphasis on curare's fabrication by the lowest members in the Caverre tribe's social hierarchy (ancient "hags") influence the Jesuit's determination that this knowledge is demonic.

This demonization of curare contrasts greatly with Gumilla's praise of Jesuit pharmacists who converted Amerindian folk remedies into medically and economically useful data for Europeans. When Gumilla refers to a Jesuit chemistry laboratory in Bogotá that offers up results from experiments and distillation, this site of knowledge production illustrates the other extreme. He openly praises the skills of his contemporary, "hermano Juan de Agullón, boticario, médico y excelente químico del colegio máximo de mi provincial de Santa Fe" (399) [Brother Juan de Agullón, druggist, doctor, and excellent chemist of the Jesuit college of my province of Santa Fe]. When, for example, Gumilla sent Agullón some healing powders concocted from dried snakes to test in his chemical lab, he was satisfied by the Jesuit's authority and confirmation of Amerindian knowledge through European scientific methodology, a process worthy of emulation: "[M]e aseguró que tenía de ello repetidas experiencias. A un hombre que era buen religioso y por otra parte científico no es razón negarle se autoridad" (399) [He assured me that he had made repeated experiments of it. There is no reason to negate the authority of a man who was a good member of the order and a good scientist]. This last sentence rhetorically legitimizes both Agullón's and Gumilla's ethos as well as appropriated Amerindian knowledge that has been mediated or distilled by Jesuits.

Gumilla's demonization of the Caverre tribe's group knowledge also contrasts greatly with his praise of another Amerindian individual's revelation of a different poison in a chapter entitled "De otros venenos fatales: su actividad, la cautela con que los dan, y cómo los descubrí" [On other fatal poisons: Their activity, the cunning with which they are given, and how I discovered them]. Thanks to the revelation of an individual's nondemonic empirical knowledge, in these pages that immediately follow the curare anecdote Gumilla offers a more acceptable pathway that validates instead of demonizes Amerindian knowledge. His anecdote conveying how he gained knowledge about another secret production of poison, this time made from ants, contrasts greatly with the mysterious, secondhand knowledge about curare. Instead of framing knowledge as diabolical, Gumilla frames this entire chapter as a service to evangelical progress. Here cultural knowledge is passed on by one new Christian Amerindian source who first saves Gumilla from being stung by poison ants and then shares the process of harvesting ant poison:

> Yo, enamorado de sus bellos colores y de su nunca visto modo de caminar en su especie, estaba divertido [. . .] cuando llegó un indio de buena ley, que no lo son todos, y dando un grito formidable, me dijo en tono asustado: "*¡Day Jebacá, Babí, alabuquí, ajaducá!*" "¿Qué haces, Padre, que ésas están llenas de veneno?" Apartéme luego y me puse a examinar al indio, el cual, no reservando el secreto, como acostumbran casi todos, dijo: "Estas hormigas son muy bravas y muy ponzoñosas [. . .] con ellas basta y sobra para sacar cantidad de veneno con que matar mucha gente." "¿Cómo las cogen y cómo sacan su veneno?", repliqué yo. [. . .] Hasta aquí la declaración del indio, para mi cierta e indubitable [. . .] por lo cual creí y creo que aquel indio me dijo cándida y sinceramente la verdad en la declaración que llevo referida. La cual, sin faltarle al secreto, sirvió y sirve grandemente a todos los misioneros, y se pone aquí para que sirva a los venideros. (370–71)

> [Enamored by their beautiful colors and their species' means of walking, never seen before, I was entertained . . . when an Indian of good faith, which they are not all, arrived and, giving a formidable shout, said in a frightened tone: "*¡Day Jebacá, Babí, alabuquí, ajaducá!*" "What are you doing, Father, for these {ants} are full of poison?" Then I withdrew and began to question the Indian, who, without holding back the secret, as almost all of them usually do, said: "These ants are very fierce and very poisonous . . . in them there is enough poison to kill many people." "How do you catch them and how is their poison removed?" I replied. . . . Up to here the evidence of the Indian, in my opinion, was true and indubitable . . . for this reason I believed and still believe that that Indian told me sincerely and candidly the truth in the statement that I have referred to. Which, without lacking the secret, served and still greatly serves all the missionaries, and it is put here so that it will serve all those to come.]

This anecdote reproduces Gumilla's own pathway from wonder to knowledge about poisonous ants. The fact that he attributes all barbarous, bloodthirsty Caverres knowledge of the curare root to the devil and warns that it might defeat the progress of evangelization and colonization, yet praises the individual knowledge of a civilized Amerindian for his "sincere and candid declaration" about the ant poison and even suggests that this witness's revelation of a secret will aid the advancement of Christianity, can be tied to the group/performative versus individual/empirical production of knowledge and Gumilla's assessment of Amerindian social hierarchy.

Modernizing Amerindian Medicine

Gumilla's third pathway to knowledge deliberately corrects and modernizes Amerindian cures and treatments, as well as reveals their effect on the Jesuit's medical procedures. As we saw in chapter 2, Gumilla clearly maps several valuable pharmaceutical products in *El Orinoco ilustrado*. In those examples, instead of rejecting "particular, local, regional knowledge" for falling outside of European paradigms, Gumilla presented autochthonous products as valuable resources.[13] These reports on Amerindian medicines and practices in the Orinoco River region were included throughout his text as well as in an alphabetical index of exotic specimens investigated by Jesuits such as the "Caña agria, útil para muchos remedios" [Sour cane, useful for many remedies], "Piñones americanos, son purga eficaz" [American pine seeds, they are an effective purgative], "Hueso último de la cola de armadillo, es contra el dolor de oídos" [Last bone of the armadillo's tail, it works against earaches], and "Piedra de la iguana, es contra mal de orina" [Iguana stone, it works against urinary illness].[14] Many valuable remedies were simply appropriated without change for European consumers.

In other cases, Gumilla presents local knowledge within anecdotes about his efforts to correct and improve upon local cures and medical practices. Despite this corrective discourse, Amerindian knowledge obviously influences his own. For example, Gumilla's anecdote about his experience with the ringworm rash intends to challenge popular wisdom handed down from generation to generation in the Orinoco River region. But instead of simply presenting his improved remedy for this "culebrilla," (410) [little snake] he narrates the different cures employed by an "indio silvestre" (410) [wild Indian] and then by a "vieja mestiza" (410) [old mixed-blood woman]. The pathway to knowledge consists of a rhetorical series of corrections and improvements, and the improvements include his own remedy. In this anecdote, first the mixed-blood healer corrects the "wild" Amerindian's skin burning strategy. Next Gumilla experiments with her complicated mixture of fruit and powders before presenting his improved remedy:

> [El indio silvestre] me consoló diciendo: "*Rabicá, fajijú, futuit fu, rufay fafolejú,*" que a la letra fue decirme: "*Padre mío, tú mueres sin falta; no hay más remedio que dejarte quemar.*" [. . .] [D]urante la cura vino a visitarme una vieja mestiza, quiero decir que era medio india y medio mulata; ella se preciaba de médica y se lastimó mucho del rústico remedio que me aplicó el indio y añadió que ella de sus mayores había aprendido que para matar la tal culebrilla [. . .] añadió la vieja que tenía por experiencia que [etc.]. . . . Después experimenté que con sola la untura del limón tibio repetida basta para atajar esta rara enfermedad. (410–11)

> [The wild Indian consoled me saying: *"Rabicá, fajijú, futuit fu, rufay fafolejú,"* which to the letter was telling me: *"My Father, you die without fault; there is no other remedy than to let yourself get burned."* . . . During the cure an old mestiza came to visit me, I mean to say that she was half Indian and half mulatto; she fancied herself a doctor, and the rustic remedy that the Indian applied to me pained her a lot. She added that she had learned from her elders how to kill such a ringworm . . . the old woman added that she had it from experience that (etc.). Afterward I experimented and found that just the repeated application of the ointment from the tepid lemon was enough to halt this strange sickness.]

Finally, Gumilla corrects the common error of Orinoco peoples who believe that the ringworm is a live animal: "[L]o que yo no acabo de creer es que sea animal vivo, como lo afirman aqullas gentes" (411) [What I cannot get myself to believe is that this is a live animal, as those people affirm"].[15] He has corrected the local cure but also changed his own European cure.

While improving upon existing Amerindian (and European) treatments, Gumilla introduces entirely new techniques to the Orinoco region that improve medical procedures. As he explains, during his years as a missionary many have benefited from his medical knowledge. In the example of how to remove impacted chiggers (*niguas*), Gumilla uses his healing techniques on Amerindians, Spaniards, and blacks:

> De este modo me he visto siempre libre de niguas desde que supe el secreto, y por mi aviso se han librado cuantos lo han sabido y se librarán cuantos usaren lo que aquí referido. [. . .] Esto es tan cierto que con mis manos he curado a muchos indios, negros y blancos con sola la referida diligencia. (408–9)

> [In this way I have seen myself free of chiggers ever since I found out the secret, and because of my notice as many as have found out have been freed and as many as use what I have here related will be freed. . . . It is very certain that with my own hands I have cured many Indians, blacks, and whites with only this referred care.]

As he explains, one of his book's purposes is to teach civilized medical practices to missionaries and "aquellos pobres ignorantes indios" (400) [those poor ignorant Indians].[16] To this end he will correct the "bárbaro, cruel y necio" (400) [barbarous, cruel, and foolish] erroneous cures employed by Amerindians by replacing them with proper European cures.

One dramatic anecdote that demonstrates Gumilla's correction of erroneous cures includes his healing of a poor boy who faced certain death at the hands of his people. It also demonstrates ways that Gumilla moves beyond deductive reasoning about causes to instead hypothesize medical cures based on experiments and induction. Before the Jesuit taught them a particular remedy, Amerindians struck by stingrays often died from an infected wound. One of Gumilla's anecdotes narrates how, when a wounded boy was brought before him, he observed no blood flowing from where the ray had stung him. In an analogy that recalls his application of knowledge about blood circulation to the effects of the curare poison, Gumilla at first suggests that the blood either coagulates or recedes to escape its natural contrary (the stingray's poison). Then he makes careful observations of the wound and carries out two experiments to thin the blood. Gumilla completes an inductive process before declaring how every victim can be saved from infection:

> [M]e excitó a hacer dos experimentos, que son los que hoy se practican ya en todas aquellas Misiones contra las cotidianas heridas de rayas, contra las cuales los indios no habían hallado otro remedio que morir después de encancerada la herida. (414)
>
> [I was stimulated to carry out two experiments, which are those that today are still practiced in all those missions for the everyday stingray wound, against which the Indians had not found any other remedy except dying after the wound became ulcerated].[17]

Gumilla centers this anecdote on an innocent child who has yet to learn adult strategies for avoiding the stingray's tail and the bony spine that wounds human flesh. He brings the boy into his own home and heals him within three days. Gumilla parlays the success of his first experiment into a second one and, based on its equal outcome, he makes a general hypothesis based on firsthand observations:

> [D]e modo que se infiere que lo cálido del ajo pone flúida la sangre coagulada con el frío del veneno. [. . .] Este experimento me dió motivo para el segundo, y fué llenar la herida hecha por la dicha púa de raya con raspadura de nuez moscada, y surtió el mismo efecto y con las mismas circunstancias dichas ya en el experimento primero. (414)
>
> [So it can be inferred that the warmth of the garlic liquefies the blood coagulated with the cold of the poison. . . . This experiment gave me reason for the second one, and that was to fill the wound made by said stingray barb with scrapings of nutmeg. It gave rise to the same effect and under with the same conditions as the first experiment.]

These generalizations drawn from particulars exemplify the pathway to knowledge taken by Spanish Baconistas. The Jesuit models an inductive process for his Amerindian charges and for his European readers whereby maxims are concluded from experience, experiments, and the study of particular, empirical truths. First with the ringworm rash and chiggers, and then with the wounds from the stingray, Gumilla improves upon Amerindian remedies by employing modern scientific methods. However, both appropriation and correction of Amerindian knowledge influences his own.

Orinoco War Drums

The sixth example of Orinoco region war drums offers a perfect demonstration of how Jesuit eclecticism mediated European physics and American cultural knowledge. Gumilla leads his readers on a pathway from initial horror to rational knowledge about drums that includes understanding of Amerindian cultural practices, Orinoco region geography, and a disquistion on physics. He divides this atomistic discourse into four areas: production, dissemination, reflection, and amplification.[18] And again employing anecdotes to serve his rhetoric of wonder, Gumilla invites readers to imagine the first time that Europeans heard the terrible noise that Orinoco drums can make. The anecdotes transport these new arrivals—and, by extension, his readers—from initial horror, through knowledge of local culture as well as European physical theories of acoustics, to a scientific explanation of sound production and propagation. Gumilla adds analogies to his anecdotes and legitimizes Amerindian cultural practices while demonstrating his scientific eclecticism.

Gumilla opens the anecdote about his own and others' initial wonder at the horrifying sounds of the Orinoco war drum with a few other strategies for amplification that grab the attention of European readers and contribute to the pathos of the anecdote: "quién los podrá ponderar? Y ya ponderados, ¿quién en Europa lo querrá creer?" (*El Orinoco ilustrado,* 344). [Who will ponder the noise and echo? And once pondered, who in Europe will want to believe?][19] The uses of the rhetorical question (*interrogatio*), various forms of *ponderar,* as well as the purposeful manipulation of the consonants *p, q,* and *r* (another rhetorical figure often called homoioptoton) demonstrate techniques recommended for provoking emotions in Jesuit handbooks. At first the drums' thunderous noises and terrible echoes fill new arrivals with a dread analogous to that caused by a great storm: "de modo que lo que percibe el oído es un continuado trueno con sus altibajos ya más, ya menos intensos, que es cosa muy notable y que causa mucho pavor y asombro a los forasteros" (352) [in such a way that what the ear perceives is a continuous thunder with its ups and downs—now more, now less, intense—which is a very noteworthy thing and it causes much fear and wonder in foreigners]. The war drums' sound is amplified by a confusion of echoes bouncing off, in, and around hills, boulders, mountain valleys, and trees, which increases their fear

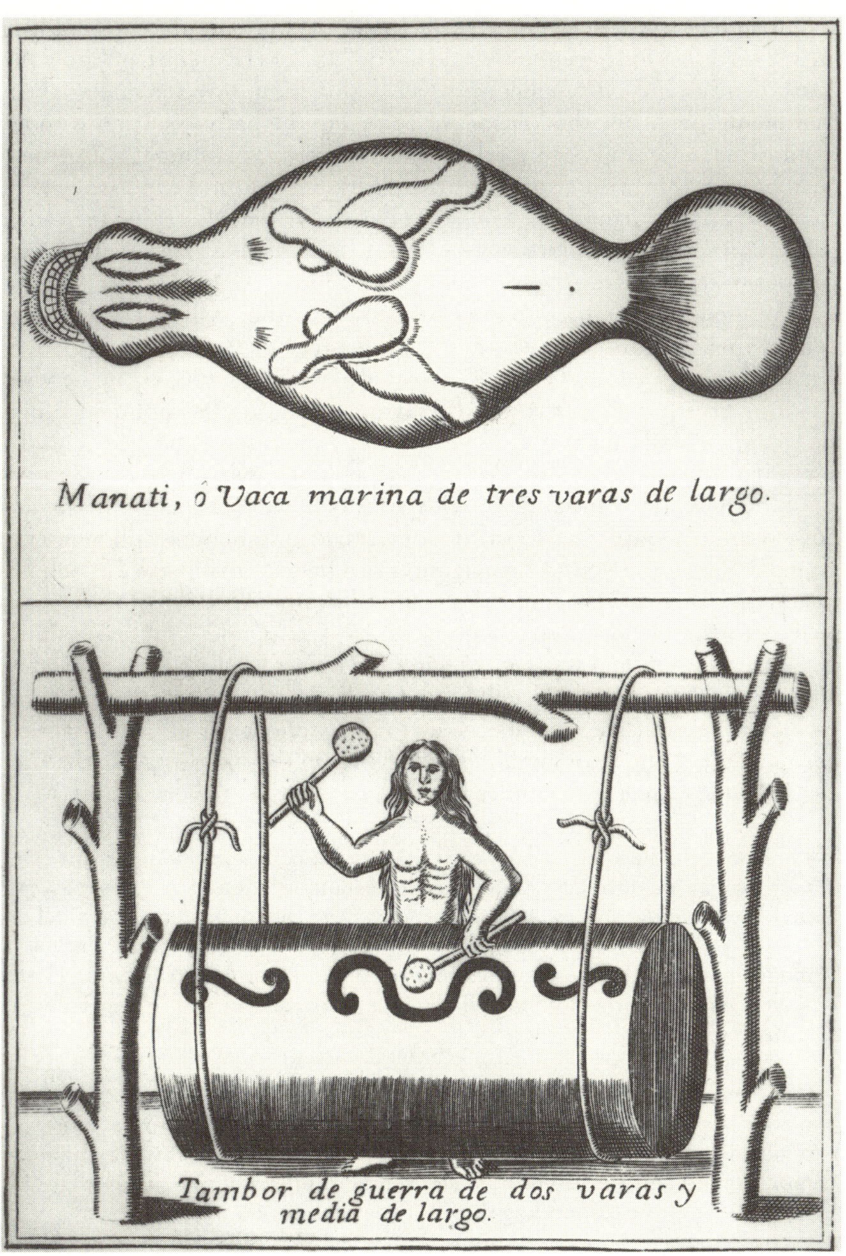

Manati, o Vaca marina de tres varas de largo [Manatee, or marine cow three yards long]. Tambor de Guerra de dos varas y media de largo [War Drum two and one-half yards long]. The John Carter Brown Library at Brown University.

and wonder: "[S]i bien conocen la causa de tan singular novedad, la misma novedad los hace temblar de miedo" (352) [Even if they know the cause of such an extraordinary novelty, this same novelty makes them tremble in fear]. For this reason, to complete the pathway from wonder to knowledge Gumilla cannot depend on merely explaining it with European physics. He must also re-create his Orinoco cultural experience with Amerindian war drums. He would not have to mediate knowledge of such cultural practices if readers came to the region: "[S]i le pica la curiosidad, con pasar al río Orinoco podrá salir de sus dudas" (344) [If their curiosity is piqued, by coming into the Orinoco River they will be able to leave behind their doubts]. However, since they are not there, Gumilla offers himself as their eyes and ears: "[Y]o refiero ingenuamente lo que he visto y oído, y presto que es fiero y extravagante el ruido y estrépito de aquellas cajas" (344) [I ingenuously refer what I have seen and heard, and I agree that the sound and racket of those drums is wild and extravagant].

To help readers unfamiliar with Orinoco war drums' extravagant noise, Gumilla employs several analogies to known acoustic phenomena. Gumilla likens the sound propagation of Orinoco drums to French artillery fire, their amplifying echoes to the doubling effect of mirrors as well as to Port Charenton's famous tripling echo, and Orinoco geography's frightening acoustics to the architectural effect of some churches where preachers' booming, echoing voices terrify congregations.[20] Next, he moves beyond such comparisons to instructions for carrying out experiments that allow for analogies that foster understanding, like the similes of sound waves and ripples in the water or rays from the sun:

> Este modo de filosofar consta por el siguiente experimento: tóquese una campana o una caja de guerra junto al mismo estanque, o junto a una ventana por donde el rayo del sol descubra los átomos, y se verá así cómo el agua del estanque y los átomos que se descubren al rayo del sol se conmueven y a su modo corresponden a los golpes sonoros de uno y otro instrumento, en que se ven los efectos de la vibración con que las partes del aire se impelen unas a otras. (348–49)
>
> [This means of philosophizing consists of the following experiment: ring a bell or beat a war drum next to the same tank, or next to a window through which a sunbeam reveals the atoms, and it will be seen in this way how the tank's water and the atoms revealed by the sunbeam move and in that way correspond to the resounding blows of one or the other instrument. That way are seen the effects of the vibration with which the air particles propel themselves against one another.]

These explanatory analogies are based on European science, but are presented with evocative rhetorical strategies that combine emotion and reason to persuade readers.

When discussing sound production and propagation through the air by expanding circular waves, Gumilla refers directly to several modern physicists and their own experiments in Europe. Gumilla's atomistic understanding of hearing and sound is similar to that of his fellow Valencian Jesuit Tosca, who shares understanding about particles with Gassendi and Newton. Sound vibration in air is considered analogous to the movement of atoms. Gumilla combines his personal experience with external proofs: references include the Jesuit Regnault's "repetidos experimentos" (349) [repeated experiments], "el experimento del Padre Grimaldi" (349) [Father Grimaldi's experiment], and Father Mersenne's *Harmonie Universelle,* as well as a report on echo experiments for the French Academy of Sciences.[21] Gumilla asserts that he has avoided speculative ideas and scholastic generalizations about sound. Instead of following a systematic pathway to knowledge about acoustics, his disquisition on sound waves promotes induction from several experiments: "[N]o es idea especulativa, ni argumento fundado en formalidades metafísicas, sino una serie de experiencias que concurren a evidenciar la certidumbre de mi experimento" (351) [It is not a speculative notion nor an argument founded in metaphysical formalities, but rather a series of experiments that combine to prove the certainty of my experiment]. His study addresses the interaction of matter and energy, of vibrations, and their movement in waves through the air to create sound. Gumilla announces that these findings are supported by "good philosophy":

> [V]oy a evidenciar la certidumbre del sonido del tambor caverre de Orinoco por buena filosofía, deducida de experimentos físicos, cuya solidez conocerá el que tuviere alguna tintura de filósofo, y el que no la tuviere no se disgustará de ver los fundamentos y los experimentos con que pruebo y confirmo mi proposición. (348)

> [I am going to prove the certainty of the sound of the Caverre drum of the Orinoco by good philosophy, inferred from physical experiments, whose soundness anyone who has even a tincture of philosopher will recognize, and whoever does not have it will not be displeased to see the basis and the experiments with which I prove and confirm my proposition.]

By good philosophy he means an understanding of sound that privileges individual experiments over Aristotelian common experience. Such statements are concordant with early eighteenth-century natural philosophy and what was considered "modern physics" in his day.[22] Not only does Gumilla's treatment of Orinoco war drums demonstrate his eclecticism, but it explains Amerindian cultural practices. While leading European readers on a pathway to knowledge that departs from negative wonder at a terrifying acoustic phenomena and arrives at a modern disquisition on sound waves, Gumilla shares his geographical and cultural knowledge of the Orinoco River region with readers.

¡Oh Monstruo, Oh Bestia! Facing the Orinoco Anaconda

The seventh example follows the fourth type of pathway to knowledge by demonstrating how Gumilla validates the seemingly fantastic phenomenon of the fatal attraction caused by an invisible chain emitted in the anaconda's pestilent breath. Those who have seen this amazing attraction have no need for proof: "[E]n orden a los americanos, digo que la experiencia que tienen de la atracción del buío les da suficiente luz y fundamento para confirmarse" (389) [With regards to the Americans, I say that the experience they have with the attraction of the *buío* gives them sufficient light and grounds to be confirmed].[23] Like other Europeans and Americans who live in the region, Gumilla accepts popular opinion: "Confieso ingenuamente que he tirado a deslindar y averiguar de raíz esta noticia, por el mismo caso que se repute por vulgar y común, y apurada bien la material y atestiguada con sujetos fidedignos que por su ocupación pasan su vida en los campos, es para mí indubitable la verdad del hecho" (387) [I ingenuously confess that I have tried to clarify and investigate the root of this idea, taking into account that it is deemed vulgar and common; since the material is well investigated and borne witness to by reliable witnesses who because of their occupations spend their lives in the country, the truth of the matter is, in my opinion, indubitable]. In a chapter entitled "De las culebras venenosas de aquellos países" [On the poisonous snakes of those lands] Gumilla legitimizes these beliefs about the anaconda by placing them within a discourse of European scientific methodology, reconciling Spanish Catholic orthodoxy with Amerindian popular wisdom and the evolved empiricism of seventeenth- and eighteenth-century natural philosophers. He employs an eclectic, pious inquiry into the inner workings of nature to explain the diabolical powers of the anaconda.

Here again, Gumilla's anecdotes and analogies serve first to induce and then mitigate negative wonder. The central location of his snake chapter (number 14 of twenty-seven chapters in the second part of *El Orinoco ilustrado*) and its length (tripled for the definitive 1745 edition) underscore the centrality of Gumilla's moral and scientific use of the giant boa in his epistemology. For this pathway to knowledge, the Jesuit legitimizes Amerindian folklore by incorporating it into a Christian, eclectic paradigm of natural philosophy. That is, the Jesuit describes and analyzes this snake within the context of both Christian and Amerindian experience. As we will see, instead of dismissing as fantastic the natural phenomenon of the anaconda's ability to pull in its prey, Gumilla seeks to legitimize South American folklore by comparing it with North American and European reports and philosophizing, and he does so with tools from his eclectic Jesuit education. Gumilla purposely preys upon his readers' natural curiosity. His pathway from initial horror toward reasoned explanations for the snake's fascinating faculties acknowledges the allegorized descriptions of monstrous snakes found in ancient treatises, early modern treatises, and previous Jesuit natural histories. Underscoring the continued value of wonder to the advancement of knowledge during the Hispanic Enlightenment, this pathway engages ideas about magnetism and atomism in natural philosophy while also incorporating Amerindian popular knowledge.

Gumilla's choice to portray the giant anaconda as a monster engages over two thousand years of horror at man-eating "monstrous serpents."[24] From the earliest years of European travel to South America, these "monstrous boas" inhabited their writings as mythical creatures.[25] At the same time Gumilla mentions "reliable subjects" who have seen and therefore believe in this monster, he also directs his rhetoric of wonder at exploiting the anaconda's continued mysterious status. To this end, he opens his discussion by playing up its mystique both within and outside of the Orinoco River region. As he claims, despite a clear abundance of anacondas, many still doubt their existence: "allá mismo, donde se crían y abundan los buíos, hay personas de toda forma que niegan su existencia, hasta que entrándolas al espanto por los ojos hasta el corazón se desengañan" (382) [Right there, where the buíos are raised and abound, there are all sorts of people who deny its existence, until the astonishment enters them through their eyes and reaches their hearts and they face the facts]. Gumilla's anaconda anecdotes invoke this mystique when re-creating his own encounters, as well as those of other Jesuits and unfortunate natives, to vicariously bring his readers face-to-face with anacondas.[26] The pathway to knowledge about these Orinoco monsters is rhetorical: the journey from wonder to knowledge of the facts Gumilla has gathered depends on visualization through vivid literary presentation.

Gumilla well understood that very few of his readers would ever experience the kind of face-to-face encounter he described, where initial shock inevitably leads to the affirmation of such a monstrous snake's existence. However, since humans tend to believe what they can see for themselves, Gumilla strives to create virtual encounters, to heighten readers' emotional reaction and capture them in a magnetic pull that moves them from curiosity to knowledge. Since readers must experience the anaconda vicariously, Gumilla emphasizes the authority of his eyewitness experiences and uses his rhetorical skills to bring them before everyone's eyes. He turns yet again to Jesuit strategies for sacred rhetoric that invite readers to experience dramatic visualization and travel along a pathway from wonder to knowledge. Gumilla was fully aware of Latin ties between vision and wonder, the "mir" in "admiratio" being based on several Latin words for miracles and wonders and on the verb for "looking at with wonder." From the opening lines of his treatment of the anaconda, the Jesuit emphasizes the visual nature of this path: "El primer horrible serpentón que *se nos pone a la vista,* por hallarse con gran frecuencia [. . .] es el buío [. . .] sólo el verle da notable espanto" (376; emphasis mine) [The first horrible snake that is *put before our eyes,* because it is found with great frequency . . . is the buío . . . just seeing it gives a noteworthy fright]. Here Gumilla rhetorically constructs an increase of wonder at seeing such a horror.

Handbooks for sacred rhetoric prescribe wonder for Christian persuasion. Just as Jesuit missionaries exploit wonder to evangelize Amerindians, Gumilla employs his rhetoric of wonder to move and ultimately persuade European readers. In *El Orinoco ilustrado,* he directs the wonder typically evoked when converting Amerindians to Catholicism at converting readers' doubts about the region into belief. For example, Gumilla's invitation to contemplate the anaconda makes use of the same rhetorical strategies of sermons that induce heated contemplation of hell. Through one of the most popular figures for amplification (*exclamatio*), readers are incited to imagine the cold sweat of any victim facing such an "insatiable monster": "¡[Q]ué congoja, qué sudores fríos, qué angustias fatales no sofocarán el ánimo del pobre que contra su voluntad se ve llevar a la tremenda boca de aquella bestia carnicera e insaciable monstruo!" (377) [What distress, what cold sweats, what horrible anguish will smother the spirit of the poor person who against his will finds himself led into the tremendous mouth of that bloody beast and insatiable monster!] The next phrases craft an analogy between the anaconda and the devil. Both are "infernal serpents" that draw unfortunates in against their will. Both require people to wake up and "open their eyes" before it is too late:

> Gran similitud es la de este apretado lance, para que abran los ojos, suden y se acongojen los que, halagados de la serpiente infernal, se dejan llevar de su vaho y atractivo, sin reparar que el paradero es la boca de un infierno inacabable, que ya tiene abierta su garganta para tragarlos sin remedio. . . . (377–78)

> [This is a great resemblance to this dangerous situation, so that they might open their eyes, sweat, and grieve over those who, flattered by the infernal serpent, allow themselves to be led by its vapor and attraction, without notic-

ing that the end point is the mouth of an unending hell, that already has its throat open to swallow them without fail. . . .]

Here Gumilla employs tried-and-true rhetorical strategies for evangelization that make moral use of similarities between the anaconda's fatal attraction and the devil's temptations. Gumilla moves and excites his readers, channeling emotion at this moral lesson to Christian knowledge. Yet if, on one hand, the Jesuit priest's application of sermonic strategies engages the Christian symbolic weight of snakes and fulfills readers' expectations for allegorical asides regarding snake/devil temptation, on the other hand his vivid presentation moves well beyond traditional biblical resonances and into recent advances in natural philosophy. Gumilla's chapter on snakes models his methodology of reports based on internal proofs (his observations) combined with argumentation from external proofs (witnesses, comparisons, and contrasts). Gumilla's external authorities include the sixteenth-century missionary Joseph de Acosta's *Natural and Moral History* as well as more immediate Orinoco region writings by Bishop Lucas Fernández de Piedrahita and Procurator Matías de Tapia. Still, his anaconda discussion supersedes these predecessors.

For example, at the same time that Gumilla humbly defers to the authority of the seventeenth-century Amazon River trailblazers Christoval de Acuña and Manuel Rodríguez, he points out that they neglect to mention any boas, take little advantage of their firsthand Amazon observations, and instead merely echo Pliny's curious description of two-headed snakes. He establishes their Jesuit ethos, however, to heighten his own: "El Padre Manuel Rodríguez [. . .] sin duda no tuvo de ellas las demás noticias que yo averigüé despacio y a todo seguro y pondré aquí, no sólo para la curiosidad, sino también para la utilidad del bien común" (398) [Father Manuel Rodríguez . . . without a doubt did not have the rest of the notices that I investigated slowly, found all reliable, and put here, not only for the curiosity, but also for the usefulness of the public good]. Gumilla's anaconda anecdotes spark readers' curiosity, then quickly pass to useful knowledge directed at the public good, and these contrastive comparisons support his argument. He also makes comparisons of similarity to the findings of ancient and modern authorities. After providing analogies to Augustine's passages on the lodestone[27] as well as Lucretius's *De rerum natura,* Gumilla turns to Sanctorius's *De Estática Medicina* to discuss respiration and insensible perspiration. These highly esteemed natural philosophers detailed oft-repeated "experiments." By accepting received knowledge that his experience has not contradicted while also rejecting ancient ideas such as Aristotle's denial of the existence of the vacuum, Gumilla demonstrates his eclecticism.[28] Gumilla's acceptance of Old World scholastic and modern authorities alongside New World realities currently investigated by Jesuits is in line with contemporary philosophers' belief about animals emanating a mysterious fluid that had magnetic properties.[29] Gumilla's philosophizing about the vacuum, magnetism, and the anaconda effluvia, are, as he notes, "buena filosofía" for his time.[30]

In addition to the above renowned authorities and Catholics, Gumilla cites a key non-Catholic intellectual for his discussion on snakes: Sir Hans Sloan, then the president of one of the oldest scientific societies in Europe. In 1727 Sloan had succeeded Sir Isaac Newton as president of the Royal Society of London for the Promotion of Natural Knowledge. Gumilla cites Sloan's "Conjectures on the Charming or Fascinating Power Attributed to the Rattle-Snake: Grounded on Credible Accounts, Experiments and Observations."[31] These diverse scientific sources are in line with Gumilla's eclectic philosophy. Thus, in addition to mining the visualization strategies of sacred rhetoric, Gumilla appeals to his readers with an eclectic explanation of the anaconda's "cierto y notorio" (377) [certain and notorious] breath. Gumilla mixes church fathers, scholastic authorities, and modern physicists in a discourse that moves beyond an Augustinian miracle of nature (magnetic lodestone) to various theories from "los físicos modernos" (384) [modern physicists] that evoke the vacuums created not only by natural phenomena (tornadoes and whirlpools) but also by the recent invention of the air pump. To Gumilla, the reality of the attractive force of the anaconda is far from an "out of this world" idea: "Fuera de que no hay para qué extrañar tanto esta operación del buío, ni es razón mirarla como entusiasmo ideado en el otro mundo. [. . .] [E]s para mí indubitable la verdad del hecho" (387) [Beyond that there is no reason to be so surprised by this operation of the buío; there is no reason to view it as an enthusiasm conceived in the other world . . . The truth of the matter is in my opinion indubitable]. Moving beyond his original statement that "el atraer con el vaho es cierto y notorio" (377) [the attracting with the vapor is certain and notorious] Gumilla dedicates several pages to explaining possible hidden causes behind the anaconda's fatal attraction. His explanation depends on orthodox Catholic traditions while simultaneously applying empirical methods to a seemingly fabulous natural phenomenon.

Even more interesting than Gumilla's recourse to well-established scientific authorities, however, are the analogies, anecdotes, and other rhetorical mechanisms employed to make readers vicariously experience the wonder of face-to-face anaconda encounters. Gumilla's reliance on particular observations before making generalizations is repeatedly shown in the several subsections of his snake chapter.[32] Early on, he invites readers to "get their hands dirty" by examining each operation alongside him in the field, and also to face the snake's wide-open mouth and "noxious jaws": "[M]anos a la obra, haciéndonos presentes al culebrón buío, que abierta la boca y dilatadas sus pestíferas fauces, tiene la puntería puesta [. . .]" (384) [Let's get down to work, making ourselves present to the buío snake that, with his open mouth and his noxious jaws extended, has his aim set . . .]. Gumilla describes the inductive pathway to knowledge through analogies in another metacommentary on his eclectic methodology. When investigating various possible causes for the anaconda's attractive power, Gumilla refers to the path (*vereda*) that through analogies brings face-to-face (*careando*) both similar and opposite causes and effects. That is, both experiments of similitude and arguments *a contrariis* will allow readers to make inferences from visible and invisible things: "*careando* entre sí causas y efectos contrarios [. . .] ésta es una de las *veredas* que se pueden tomar, para buscar la raíz incógnita de un efecto tal cual es la atracción actual del buío, dónde reside y en qué consiste esta virtud atrayente [. . .] fijemos la vista" (390; emphasis mine) [*bringing face-to-face* contrary causes and effects . . . this is one of the *paths* that can be taken, to seek out the unknown root of an effect such as is the present attraction of the buío, wherein it lies and of what this attractive virtue is composed . . . let us set our gaze]. This attractive force can be compared to the invisible effects of magnetism, but vibration is the most plausible cause for Gumilla:

> [A]unque imploremos el auxilio de los físicos modernos y de sus mejores microscopios, no hallaremos en este monstruo más armas ofensivas que la vibración y la atracción del ambiente inficionado con la ponzoña que exhala su hálito. Esta vibración de efluvios malignos, y la atracción de que de ellos resulta, comprende todo el nervio de la dificultad para cuya cabal solución debemos examinar de una raíz una y otra operación, cada una de por sí. (384)

> [Even if we entreated the aid of the modern physicists and their best microscopes, we will not find in this monster more offensive weapons than the vibration and attraction of the air infected with the venom that its vapor gives off. This vibration of malicious effluvia, and the attraction that results from them, comprises all the force of the difficulty for whose exact solution we should examine entirely one operation after another, each one by itself.]

In addition to analogies, the seeing-is-believing validation for Gumilla's snake chapter depends on vivid anecdotes "seen" by the readers. A pivotal example of this pathway to knowledge occurs with his reenactment of the walk during which he and another missionary entertained a Jesuit secretary with stories of the anaconda. Gumilla's strategy to persuade readers through a descriptive episode is echoed in the strategy both he and his Jesuit superior employed when trying to convince Father Anisón of the reality of this monster. However, this Orinoco visitor needed to see to believe. As this anecdote unfolds, Gumilla leads his readers from anticipation of the anaconda's inevitable appearance to the moment of Anisón's horrified realization:

> Digo que caminando [. . .] a fin de aliviar el fastidio del camino, iba yo refiriendo al Padre Secretario Carlos de Anisón la figura, vaho venenoso y daños de los buíos; no daba asenso y apelaba al Padre Provincial que también había sido misionero, y práctico del terreno y como tal tiraba a convencer al Padre Secretario; pero éste se mostró incrédulo hasta que poco después vio por sus ojos en una laguna un buío feroz, que acababa de atraer para sí una garza y se la comenzaba a engullir, teniendo ella las alas abiertas al uno y otro lado de la boca del culebrón; de que se infería, al pasar volando, la atrajo, siendo los pies los que primero llegaron a la fatal boca. Aquí fue donde aturdido exclamó el Padre Anisón, diciendo: "¡Oh monstruo! ¡Oh bestia! ¡Oh, y qué horror!" (381–82)
>
> [I say that walking . . . in order to alleviate the boredom of the trail, I was relating to Father Secretary Carlos de Anisón the shape, poisonous vapor, and dangerousness of the buíos; he would not give assent, and so I appealed to the provincial father, who also had been a missionary and pilot of the lands, and thereby tried to pull the father secretary in and convince him; but the latter appeared incredulous until a little later he saw with his own eyes in a lagoon a ferocious buío. It had just finished attracting to itself a heron and was starting to swallow it; the bird's wings were open on both sides of the mouth of the snake. From this it was inferred that when the bird was flying by, the snake attracted it, and the bird's feet were the first things to arrive in the deadly mouth. At this the stunned Father Anisón cried out, saying: "Oh monster! Oh amazing beast! Oh, and what horror!"]

Gumilla punctuates this anecdote with screams of horror. He also employs a common rhetorical strategy for amplification (*epiphonema*) by citing a Latin quote from Horace's *Ars poetica* that underscores both vision's impact for wonder and the value of eyewitness experiences. Gumilla reduces Horace's verse to: "Segnius irritant animos demisa per aurem. Quam quae sunt oculis subiecta fidelibus" (382) [The things that are sent down by the ear excite the souls more slowly than those which have been captured by the trustworthy eyes.][33] Gumilla's amplification makes explicit his visual rhetorical strategies for leading his readers from wonder to knowledge.

The rhetorical function of the anecdote detailing Father Anisón's embodiment of Horace's verse is to persuade readers by making them vicarious spectators at the scene of Gumilla's and Anisón's observations. Additionally, this rhetorical path of narrating initial disbelief, a wonder-provoking encounter, and finally a startled exclamation of full knowledge embodies the wonder-to-knowledge strategies employed throughout *El*

Orinoco ilustrado. Gumilla's method for "pulling in" or "attracting" Father Anisón and then convincing him included description and testimony as well as a seeing-and-believing experience that rhetorically establishes yet another eyewitness to the mysterious properties of the anaconda's effluvia. Certainly the image re-created with Father Anisón's encounter—that of a heron pulled feet-first, wings flapping, into the anaconda's jaws—recalls this classic line from Pliny the Elder's *Natural History:* "[T]here be serpents that catch and deuoure the fowles of the aire, be they neuer so swift winged, and soare they neuer so high."[34] However, Gumilla's curious blend of striking monster anecdotes and reasoned, scientific discourse builds upon Pliny's one-sentence affirmation of snakes' power to pull in birds. This is because Gumilla's scientific explanations do not participate in a process that must invalidate "old" discoveries with a chronological series of "new" ones. Yet if Father Anisón comes face-to-face with the anaconda swallowing a freshly captured heron, seeing and believing cannot completely finish the pathway to knowledge about the anaconda. Gumilla also wants his readers to understand this attractive power. To this end he devotes several pages to helping his readers face the facts and get to the root of it through scientific inquiry.[35] Finally, by way of conclusion to his explanation of the fascinating faculties of the anaconda, Gumilla invites his readers to "draw back the curtain to reveal" the next Orinoco spectacle, which he will again explain by leading his readers along the path from wonder to knowledge.

> Y, pues queda largamente establecida la existencia del buío, la acción y vibración de sus nocivos efluvios, la fuerza atractiva de ellos y apuntadas varias sendas para la inteligencia de su virtud atrayente, ya es hora de correr otra cortina y poner a la vista otros espectáculos que llaman con la curiosidad la atención de unos y la admiración de otros. (393)
>
> [And, since the existence of the buío, the action and vibration, and the attractive force of its harmful effluvia are well established, and various pathways for understanding their attractive virtue indicated, it is about time to draw back another curtain and bring before the eyes other spectacles that call with curiosity the attention of some and the wonder of others.]

Thus, Gumilla's scientific persuasion depends on wonder-provoking vivid descriptions, the virtual eyewitnessing of his readers, and his firsthand observation and external authorities' testimony.

Gumilla's natural-history treatment of the monstrous South American boa caused a stir.[36] His unique combination of firsthand knowledge of the Orinoco, Amerindian folklore, and a Jesuit education allowed him to augment conventional allegorical discourses on snakes with reasoned scientific argumentation. Gumilla's pathway proved so convincing that Feijoo, whose writings famously strive to dispel all hints of superstition and myth, would affirm the validity of Gumilla's claims. In one of the letters in his *Cartas eruditas y curiosas* (1750), Feijoo underlines the terror inspired by Gumilla's explanation of the anaconda's fatal attraction:

> El P. Gumilla dice, que el Buío le atrae. . . . [E]ste horrible Serpentón, que verosímilmente es el más formidable que hay en toda la naturaleza [. . .] [que] no pudiendo, por su lentísimo movimiento, alcanzar al hombre [. . .] tiene otro modo muy singular de apresarle, que es disparar hacia él un vaho de tal actividad que no sólo le impide la fuga, mas le precisa al movimiento opuesto, con que aunque reluctante, y congojado, se va a meter en las fauces del monstruo.[37]

> [Father Gumilla says, that the buío attracts man. . . . This horrible serpent, which truly is the most wonderful that there is in all of nature . . . not being able to, because of its very slow movement, catch up with man . . . has another very extraordinary way to capture him, and that is by hurling toward him a vapor of such activity that not only does it impede man's flight, but it forces the opposite movement, so that even reluctantly, and with anguish, he finds himself introduced into the jaws of the monster.]

Although twentieth-century herpetologists no longer accept Gumilla's reasoning about the magnetic properties of the anaconda's breath, the fact that Feijoo and others did demonstrates the validity his pathway to knowledge in his time.[38] Belief in the anaconda's capacity for fatal attraction provides an excellent example of the mediation of Amerindian popular wisdom, European science, and Catholicism employed throughout *El Orinoco ilustrado* for journeys to natural and cultural knowledge.

"On caimans, crocodiles, and on the newly discovered virtue of their fangs"

Another monstrous creature featured in *El Orinoco ilustrado* is the crocodile, "monstruo infernal" (419), whose portrait we saw in chapter 1 of this book. Gumilla dedicates an entire chapter, "De los caimanes y cocodrilos, y de la virtud nuevamente descubierta de sus colmillos" [On the caimans and crocodiles, and on the newly discovered virtue of their fangs], to both paint a vivid description of this mythical "dragón de cuatro pies horribles, espantoso en tierra y formidable en el agua" (419) [dragon with four horrible feet, frightening on land and dreadful in water] and also to legitimize Amerindian folklore.[39] Two final examples of Gumilla's validation of Orinoco-region beliefs through scientific mediation include his discussion of crocodile stones and fangs. After relaying the Otomac nation's explanation of the large quantity of rocks found in the caiman's stomach, he legitimizes their folklore in three paragraphs that mock the deductive path of scholastic syllogisms and validate his own and Amerindians' observations of particular cases. Specifically, Gumilla refutes the deductive logic of a supposedly indisputable syllogism that has been put forth by those who doubt that caimans and crocodiles swallow ballast.[40] First he negates the major premise of the syllogism—namely, that caimans and crocodiles are fish. Next he evokes his Orinoco experience to negate the minor premise—namely, that God has given fish all the nimbleness needed to swim, sink, and rise in water: "la menor, que hallo falsificada en la América" (420) [the minor, which I find falsified in America]. Rejecting this deductive logic, Gumilla refutes the syllogism's conclusion that the caiman need not swallow stones to sink into the river.

Instead of persuading through syllogisms, Gumilla provides a series of analogies to other animal instincts. For example, he proves that God gave caimans the instinct to swallow ballast by comparing it to the crane's God-given instinct to rest the weight of their heads on the backs of other birds to stay in the air during flight: "[E]n este mismo sentido le dió Dios al caimán lo que ha menester para hundirse en el río, dándole instinto para tragar las piedras que necesita para ello" (421) [In this same sense God gave to the caiman all that was necessary to submerge itself in the river, giving to it the instinct to swallow the stones that it needs for it]. The pathway to knowledge about the caiman's behavior blends folkloric tales with scientific explanation. The analogies used to confirm that crocodiles swallow rocks also include Gumilla's survey of several Amerindian nations' beliefs. They all have their theories about the stones, but induction based on the Jesuit's observations of crocodiles leads Gumilla to deem the Otomacs' explanation the most reasonable:

> Procuré averiguar este secreto y las causas de este lastre, y hallé que cada nación de indios tiene su opinión en la materia. [. . .] El parecer que más me cuadró es el de los indios otomacos. [. . .] Dicen aquellos indios que, cuando va creciendo el caimán, va reconociendo dificultad en dejarse aplomar al fondo del río. [. . .] de lo que se infiere que cuanto más crece, de más piedras necesita para su lastre y contrapeso. (420)

> [I tried to find out this secret and the causes of this ballast, and I found that each nation of Indians has its own opinion on the matter. . . . The opinion that most squared with my own was that of the Otomac Indians. . . . Those Indians say that, as the caiman is growing, it realizes the difficulty of keeping itself at the bottom of the river. . . . from which it is inferred that the more it grows, the more stones it needs for its ballast and counterweight.]

To confirm the Otomacs' assertion as "not far off course," Gumilla offers an analogy to ships and then concludes by asserting its validity for both of the Indies:

> [I]nclino a los indios otomacos: no van muy fuera del camino diciendo que el caimán engulle piedras para lastre, arbitrio de que usan los marineros para que, hundido con proporción el navío, navegue con la seguridad que no tuviera sin lastre: de modo que así como cuanto mayor es la embarcación requiere más lastre, así cuanto más crece el caimán más piedras tiene en el buche, y así es materia de hecho indubitable, no sólo por haberlo visto yo, como ya dije, sino porque es notorio en donde quiera que hay caimanes y cocodrilos, así en las Indias Occidentales como en las Orientales (422).

> [I opt for the idea of the Otomac Indians: they do not go far off course by saying that the caiman swallows stones for ballast. Sailors use ballest so that, the boat submerging with proportion, it sails with a security it would not have without ballast. The bigger the boat, the more ballast it requires, and the bigger the caiman, the more stones it has in its belly, and this is an indubitable matter of fact, not only because I have seen it, as I already said, but also because it is well known wherever there are caimans and crocodiles, both in the West and East Indies.]

Finally, Gumilla refers to external sources, including a Spanish soldier and one of his most valued resources, monsieur Salmón.[41] Through this rhetorical path Gumilla moves from exciting Amerindian anecdotes to persuasive scientific analogies, using the latter to mediate the former and ultimately establishing his scientific opinion.

Another more striking legitimization of local folklore in Gumilla's chapter on the caimans and crocodiles appears in its mediation of anecdotes with experimentation about a newly discovered virtue of crocodile fangs as protection against poisons. An anecdote about one slave bent on poisoning another on a plantation near Caracas relays this exciting discovery, which has recently converted crocodile fangs into valuable commodities:

> [V]enden los colmillos a muy buen precio y se buscan con ansia para enviarlo a personas de estimación, que los reciben y agradecen como un apreciable y rico regalo, a causa de haberse descubierto en la Provincia de Caracas ser dichos colmillo un gran contraveneno. [. . .] El descubrimiento de la virtud del dicho colmillo es moderno, y fué así [. . .] corrió la voz, y con la experiencia el aprecio. (426–27)

> [They sell the fangs at a very good price and they are searched for with eagerness to send to respectable people, who receive and appreciate them as a precious and rich gift, because in the province of Caracas said fangs have been discovered to be a great antivenom. . . . The discovery of the virtue of said fang is modern, and it was in this way . . . word spread, and with the experiment regard circulated.]

The story of how each one of the vindictive slave's attempts was foiled by the other slave's crocodile fangs charm concludes with a notorious case from Panama. The protective capacity of the fangs become famous "con ocasión de no poder matar una enojada y cruel mujer a su marido, para lo cual le había dado varios venenos, se averiguó que no habían tenido fuerza por que . . . él traía siempre consigo un colmillo de caimán. [. . .] El caso fue notorio en la ciudad de Panamá, pasó la noticia a las de Guayaquil y Quito" (477) [When an angry and cruel wife was not able to kill her husband to whom she had given various poisons, it was discovered that the poisons had not had any effect because . . . he always carried with him a crocodile fang. . . . The case was notorious in the city of Panama, and the news reached the cities of Guayaquil and Quito]. For this anecdote, Gumilla evokes the Baconian language of experimentalism and validates the far-fetched belief about the fangs' virtue. Many in Guayaquil and Quito heard this news of the fangs' virtues, so to confirm them various experiments were performed on animals: "[S]e hicieron varios experimentos, dando tósigos a varios animales [. . .]" (427) [Various experiments were performed, giving poisons to various animals . . .]. With this, crocodile fangs became more valuable than the mythical unicorn horn: "el tal colmillo . . . se ha visto ser el antídoto más activo y más universal, como es ya notorio en las tres citadas provincias. [. . .] Sólo lo ya experimentado equivale a más de lo que se afirma del unicornio" (427) [This fang has been seen to be the most active and universal antidote, as is already well known in the three aforementioned provinces. . . . Just with the already experimented it is tantamount to what is affirmed about the unicorn]. And with time, Gumilla promises, skilled druggists will certainly discover even more virtues of crocodile fangs: "y la pericia de los botánicos descubrirá con el tiempo mucho más" (427) [and with time the skills of the druggists will reveal many more]. The crocodile fangs anecdotes evoke wonder and then validate their protective property against poisons with the language of modern empiricism. Particular anecdotes and experiments provide evidence about the fangs, while an inductive chain of analogies confirms the instinct of caimans and crocodiles to swallow stones and sink deep in the river.

To conclude, these nine examples from *El Orinoco ilustrado* reveal rhetorical pathways to knowledge mediating Amerindian knowledge with European paradigms by Christianization, demonization, correction, and validation. Gumilla adheres to the Jesuit tradition of combining rhetorical strategies with scientific methodology. In this, he combines his authority (ethos) as an eclectic, Catholic intellectual with emotion (pathos) and reason (logos) to construct pathways to modern and religious knowledge. We have seen how Gumilla maximizes the emotional appeal of his material by relating anecdotes based on his personal experiences or reports from reliable witnesses, often slaking his readers' thirst for curiosities and seemingly fabulous phenomena by incorporating

Amerindian folklore. He relays experiments and scientific explanations with a rhetoric that persuades Europeans through alternative pathways to knowledge that lead them from wonder at several mysterious secrets of the Orinoco to explanations for them.

El Orinoco ilustrado illustrates Gumilla's construction of scientific authority through firsthand experience and physical experiments, references to ancient and modern authorities, and appropriation and mediation rather than denial of Amerindian knowledge. For this Jesuit, wonder constitutes the key emotion inspired by and inspiring investigation of nature and culture. All pathways to knowledge are also meant to evoke pious wonder at God's magnificence. *El Orinoco ilustrado* offers a representative glimpse into the mind of an early eighteenth-century Jesuit missionary with a simultaneously religious and scientific agenda. Gumilla combined his Bogotá Jesuit education with firsthand knowledge gathered along the Orinoco and successfully traveled beyond an emblematic discourse on nature toward enlightened scientific discoveries. In his natural and civil history, Gumilla presents the Orinoco River region as both wondrous and real. His rhetoric and epistemology depend on passion and reason to pull readers into pathways that lead from initial wonder at nature's marvelous effects to knowledge of hidden causes for them. Further studies that examine Jesuit interactions with non-European peoples can help us understand a simultaneously spiritual and intellectual Hispanic Enlightenment that reconciles European Enlightenment science, traditionally seen as secular, with devout Catholicism and takes into account other kinds of knowledge that missionaries encountered in the peripheries.

Conclusion: The Legacy of Joseph Gumilla's *El Orinoco ilustrado*

WESTERN EUROPE HAS TRADITIONALLY DEFINED THE ENLIGHTENment as a period of critical reform brought about by the secularization and rationalization of culture, and by the application of mathematics and experimentalism in the "new" sciences. However, this model leaves behind Catholic Spain and her colonies, which were sites of mediation between European paradigms and different ways of knowing.[1] Instead of perpetuating time-honored generalizations, the current book has argued for elucidating individual cultural constructions in order to reject universalizing characterizations of the scientific revolution and to expand Eurocentric notions of the Enlightenment. Jesuit interactions with New World knowledge facilitated a blend of modern empiricism and new scientific theories from Europe with local cultures, anecdotal evidence, and Amerindian knowledge gathered in missions. Scholars examining the role of missionary naturalists and Jesuit science are reconsidering the development of a modern scientific identity in Spain and South America.[2] My analysis of *El Orinoco ilustrado* recuperates a text previously studied separately from Enlightenment discourse and joins recent studies of Hispanic modernity that point the way for challenging restrictive conceptions of what enlightenment entails.

Throughout the eighteenth century, *El Orinoco ilustrado* was widely read and highly regarded as the key source of information about the Orinoco River region. Yet Gumilla's status as Catholic missionary, combined with his use of wonder-provoking strategies, which successfully attracted readers from diverse fields of knowledge to his lively prose, did cause some to question his credibility. Even today, the Jesuit's ornate style, replete with astonishing anecdotes about the marvels of the Orinoco region, limits critics' abilities to appreciate the legacy of Gumilla's religious and scientific contributions to the Enlightenment. Gumilla was influenced not only by ancient and Jesuit rhetorical traditions but also by Amerindian folklore, knowledge, and firsthand encounters with astonishing plants, animals, and humans. His use of wonder and blend of pathos with logos was at odds with developments in philosophical history writing and secular scientific narratives, a paradigm shift that contributed to the commonplace of "unenlightened" Spain and Spanish America and their supposed resistance to modernity.[3] Ralph Bauer and Jorge Cañizares-Esguerra have discussed the Protestant philosophy of history that wrote off Catholic cultures. Their work exposes the "Eurocentric bias in contemporary historiography that still treats Spain and Spanish America as outcasts from the history of modernity."[4]

Instead of classifying Gumilla's natural history as "pre-" or "unmodern," we have seen how its combination of sentiment and reason for the accumulation, enumeration, and dissemination of knowledge is the hallmark of alternative pathways to modernity and Enlightenment. Wonder continued to serve as a key strategy for Jesuits during the Hispanic Enlightenment, benefiting epistemology, commerce, and evangelism. Gumilla's use of wonder exemplifies the simultaneity of religious and scientific motivations for enlightenment during the Age of Reason. My book provides a model for vindicating peripheral areas traditionally left behind in historical narratives of progress and a march toward modernity limited by geographical, chronological, or religious boundaries. By articulating the lasting legacy of Gumilla's rhetoric of wonder for diverse readers, this conclusion continues to refute stereotypes of a delayed entrance into modernity for Spain and Spanish America and to broaden perceptions about the Enlightenment in general.

El Orinoco ilustrado has enjoyed a remarkable print run. Its five European editions in the eighteenth and nineteenth centuries and nine twentieth-century editions, including two published in Spain and the rest in Venezuela and Colombia, are evidence of its wide readership from the eighteenth century through the present. The 1741 and expanded 1745 editions of *El Orinoco ilustrado* enjoyed immediate popularity inside and outside the Jesuit community, from Bogotá, Quito, and Santo Domingo to Barcelona, Madrid, and Rome. Before it was translated into French, several periodicals such as the *Mémoires de Trévoux* (1748) and the *Journal Étranger* (1756) favorably reviewed *El Orinoco ilustrado*. Later, *L'Année Littéraire* (1758) praised the translation's importance for "philosophers and non-philosophers alike."[5] These first editions reached the libraries of cosmopolitan figures of the Enlightenment such as Thomas Jefferson, and entries on the Orinoco in representative late eighteenth-century encyclopedias invariably cited Gumilla.[6]

This conclusion will consider various readings of *El Orinoco ilustrado* that underscore possibilities for wider definitions of the Enlightenment. It focuses on two distinct classes of the readers that Gumilla's text has been enlightening ever since the eighteenth century. An examination first of how *El Orinoco ilustrado* was read by nonscientific American readers and second by scientific European readers suggests alternative routes to the Enlightenment. The first series of examples demonstrates ways that Alejo Carpentier, José Eustasio Rivera, and Jorge Isaacs informed their novels with Gumilla's scientific observations. The second series examines how Charles Marie de La Condamine, Jorge Juan, Antonio de Ulloa, and Alexander von Humboldt borrowed literary strategies for their scientific interpretations of the flora, fauna, and peoples in underexplored areas. The legacy of Joseph Gumilla's *El Orinoco ilustrado,* as revealed in the appropriation not only of specific facts about Orinoco nature and peoples, but also of the wonder-provoking rhetoric with which the Jesuit presented them, indicates a few ways that "peripheral" eighteenth-century Catholic cultures were in fact central to the making of modernity.[7] The novelists' use of scientific data and the earlier scientific travelers' appropriation of Gumilla's literary discourse suggest differing routes to the Enlightenment that have implications for how we understand the production and dissemination of knowledge during the first half of the eighteenth century and beyond.

Gumilla's Legacy for Novelists

Well after the peak of Jesuit mission culture in the eighteenth century, *El Orinoco ilustrado* found mainstream readers during and beyond the nineteenth-century and early twentieth-century vogue of natural history writing.[8] Novelists would take from Gumilla rhetorical strategies that employed vivid descriptions of the region and included tropical tropes such as the abundance of multicolored birds and verdant vegetation. Renowned authors from various literary movements have borrowed literary techniques and at times even lifted entire descriptions from early modern chronicles to add authority to their jungle representations.[9] Literature written about and within Latin America continues this tradition today. Even more interesting, however, than Gumilla's influence on the literary descriptions of Carpentier, Rivera, and Isaacs is how these novelists were enlightened by his facts. The following examples reveal the scientific legacy of *El Orinoco ilustrado* first in the details about Orinoco peoples in *Los pasos perdidos* [*The Lost Steps*], next through facts about Orinoco flora and fauna in *La vorágine* [*The Vortex*], and finally in the echoes of both ethnographic and natural historical details in *María.*

Alejo Carpentier's *Los pasos perdidos* (1953) takes its readers on a myth-like journey into an Orinoco River region seemingly unaffected by time. Here Carpentier has undoubtedly incorporated Gumilla's vivid river and jungle descriptions and wonder-provoking literary strategies.[10] The Cuban-born novelist acknowledges his own debt to colonial sources through his narrator, also a writer. The latter self-consciously evokes the early chronicles that serve as pre-texts for his jungle descriptions: "Donde el cronista se asombraba ante la presencia de árboles gigantescos, he visto árboles gigantes, hijos de aquéllos, nacidos en el mismo lugar, habitados por los mismos pájaros, fulminados por los mismos rayos"[11] [Where the chronicler was amazed before the presence of gigantic trees, I have seen gigantic trees, the sons of those trees, born in the same place, inhabited by the same birds, struck by the same lightning bolts]. Even though Carpentier's narrator gains his own firsthand knowledge of the region, by the end of *Los pasos perdidos* the rise of the river has restored his adoptive Orinoco village to "lost world" status, forcing him to leave the jungle for civilization and thus consult source texts to garner tribal information. Thanks to Gumilla's work with various tribes of the Arawak indigenous language group, the legacy of *El Orinoco ilustrado* reveals itself not only as a pioneering chronicle of the Orinoco River and its flora and fauna, but also as a primary source for ethnographic details. In a passage at the end of chapter 2 of *Los pasos perdidos,* the narrator details the Saliva tribe's funeral ceremonies and mentions the musical instrument first described and illustrated in Gumilla's text: "la famosa jarra con dos embocaduras"[12] [the famous jar with two mouthpieces]. A year before publishing his novel, Carpentier had mentioned Gumilla as a source in newspaper articles describing the Orinoco as a "world of stopped time."[13] In short, far beyond his jungle descriptions, Carpentier borrows from Gumilla's field experience with Orinoco region tribes. In this way, the Jesuit missionary enlightened the novelist with data gleaned from his privileged access to cultures, practices, and human artifacts such as the above-mentioned musical instrument described in part 1, chapter 13 of *El Orinoco ilustrado.*

In contrast to Carpentier's borrowing of ethnographic facts, José Eustasio Rivera borrowed jungle horrors from *El Orinoco ilustrado* for the man-eating jungle he portrayed in *La vorágine* (1924).[14] So, while Gumilla does echo Columbus's theory that the Orinoco River region contains God's earthly paradise and offers several views of amenable places waiting to be mastered by civilization from the safety of watchtowers, he also provides fodder for Rivera's horrific sublime, built around an anthropophagic *locus terribilis* that intensifies as his landmark novel progresses and its characters travel deeper into Amazonia.[15] As legend has it, the same rainforest/vortex that swallowed the protagonist of *La vorágine* first devoured the poet Rivera's earlier literary efforts and later caused the untimely death of this author, who had temporarily escaped the jungle.[16] Rivera certainly focused on the hazards of nature. One jungle horror that he takes from Gumilla are the nearly invisible parasites lurking in swamp water, though he does also echo the Jesuit's remedy by

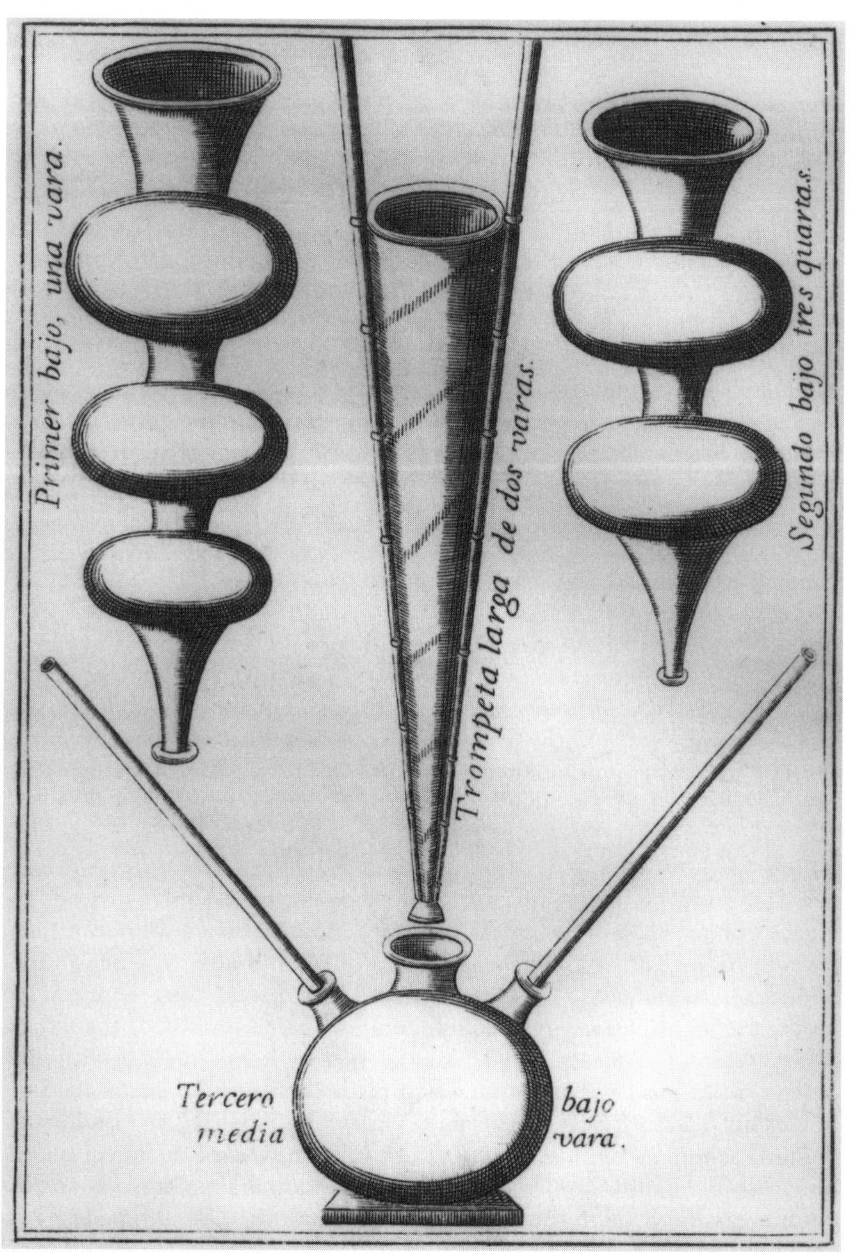

Trompeta larga de dos varas [Trumpet two yards long]. The John Carter Brown Library at Brown University.

providing details on how to strain swamp water two or three times with a handkerchief so it will act as a sieve to remove insects.[17] However, not even Gumilla could find a way to cleanse the waters of dangerous reptiles such as crocodiles and anacondas: the iconic South American boa figures prominently in Latin American literatures from the era of the *Florentine Codex* all the way to Horacio Quiroga and beyond. Part 2, chapter 14 of Gumilla's account provides a great deal of scientific information about giant serpents of the Orinoco River region. Here in "De las culebras venenosas de aquellos países" [On the poisonous snakes of those lands] the Jesuit details the fatally attractive vapors that spring forth when a giant buío opens its jaws wide. Not only does Rivera refer to the anaconda's magnetic breath, but he also appropriates Gumilla's rarely used name for the giant boa: "[P]artiendo una rama, me incliné para barrer con ella las vegetaciones acuátiles, pero don Rafo me detuvo, rápido como el grito de Alicia. Había emergido bostezando para atraparme una serpiente guío, corpulenta como una viga. [. . .]"[18] [Breaking off a limb, I bent down to sweep away the aquatic vegetation with it, but don Rafo stopped me, quickly like Alicia's scream. A *guío* serpent, as thick as a trunk, had emerged yawning in order to trap me. . . .] The novelist explains what the *guío* (buío) is in his regional glossary found at the end of the novel, and Rivera's most recent editor provides an explanatory footnote within the text. *La vorágine* was clearly indebted to particular details from *El Orinoco ilustrado,* especially for its natural history of jungle terrors.

This giant snake with attractive properties also appears in Jorge Isaacs's Colombian national romance, *María* (1867). This novel has long been written off as a servile imitation of French romanticism. Roberto González Echevarría summarizes this categorization of Isaacs's novel as falling "within the sphere of influence of European literature."[19] However, far from limiting himself to echoes of European works of fiction, Isaacs borrows heavily from the autochthonous discourse of natural history in vogue during the years surrounding the South American independence movements. In Isaacs's novel, the protagonist expresses a strong regional patriotism through romantic descriptions of the Cauca Valley region of Antioquia that are also tied to descriptions of the character María. His love of country and of the cousin who has shared his home(land) are equally passionate, and when "exiled" to London for medical school he misses them both. Toward the end of the novel (chapters 57–60), he re-creates his thrilling journey along the Dagua River, including a ride through the jungle, where a majestic and at times threatening nature leads him away from the coast toward his homeland (and fiancée) while at the same time delaying his return to Colombia's interior. It is here that Isaacs echoes scientific details about the ethnography, geography, flora, and fauna from Gumilla that are particular to Colombia. *María* echoes *El Orinoco ilustrado* when it includes details about an underground cooking practice that caused some to believe that regional tribes were earth-eaters, about musical instruments, about virgin forests of palm trees, about unforgettable swarms of pesky mosquitoes, about vampire bats, about folkloric beliefs about crocodile fangs protecting against snakebites, and about the attractive properties of the anaconda.[20] Both Gumilla and Isaacs include some finer points on how to break the anaconda's invisible magnetic force. Isaacs writes: "[A]quella víbora hacía daño de esta manera: mientras la presa que acecha no le pasa a distancia tal que solamente extendida en toda su longitud la culebra, puede alcanzarla, permanece inmóvil, y conseguida esa condición [. . .] la atrae a sí con una fuerza invencible. [. . .] Casos han ocurrido en que cazadores y bogas se salven de ese género de muerte [. . .] arrojándole una ruana sobre la cabeza."[21] [That viper harmed in this way: when the prey that it lies in wait for does not pass it close enough that the snake, even extended, can reach it, it stays still, and achieving this condition . . . the snake attracts the prey toward itself with an invincible force. . . . There have been cases in which hunters and oarsmen have saved themselves from this type of death . . . by hurling a poncho onto its head]. Here and in other sections of his novel Isaacs clearly evokes *El Orinoco ilustrado.*

Gumilla's natural history has had a lasting legacy in part because of the Jesuit's combination of missionary authority (having "been there" to accumulate data) and the rhetoric of wonder to disseminate it. This combination provides a pathway to knowledge by blending empirical fact with wonder-provoking details that appeal to European expectations of New World curiosities. *El Orinoco ilustrado* maintains its authority as an eighteenth-century biological-anthropological field report not only because of its wealth of information but also because of Gumilla's rhetorical skill in creating in readers the sense that they see, hear, and even feel these details almost as if they had "been there" too.[22] Gumilla's text continues to enlighten twentieth- and twenty-first century anthropologists and ethnologists, who often borrow the same facts and details as novelists, further continuing a long-standing tradition of citing Jesuit natural histories published since the early modern period that convey information about Amerindian customs, racial typologies, regional animals, and plant products ranging from chinchona bark quinine to niopo powder snuff.[23] Because Gumilla presents these details within startling anecdotes that are meant to intensify his readers' wonder at the Orinoco River region, the scientific weight of his firsthand knowledge has at times been undervalued. Nonetheless, novelists such as Carpentier, Rivera, and Isaacs availed themselves of Gumilla's scientific data to craft realistic descriptions in their fictional prose. In fact, numerous readers absorbed these scientific facts from the poetic prose in which they were relayed, which resulted in their being enlightened about the region's flora, fauna, and Amerindian tribes.

Gumilla's Legacy for Scientists

Perhaps more surprising than Gumilla's legacy among novelists, anthropologists, and ethnologists is his influence on some of the most famous scientific expeditions to South America.[24] To write their groundbreaking scientific travel accounts, Charles Marie de La Condamine and Alexander von Humboldt gathered data both in the field and from various sources such as *El Orinoco ilustrado*. However, just as important as the details they borrowed from Gumilla's Orinoco region experience is how they presented these and other facts with rhetorical strategies that provoked a sense of wonder in their readers. We clearly see the Jesuit's legacy in these travelers' choices about which sights to bring before their readers' eyes. Even more fascinating is how they chose to enlighten Europe about the still barely known (and therefore exotic) lands of South America. The French explorer La Condamine and his Spanish companions Jorge Juan and Antonio de Ulloa chose some of Gumilla's most unusual descriptions of Orinoco nature, and placed their narrative within a traditionally marvelous framework while also bolstering their scientific apparatus. Humboldt's famed synthesis of art and science—which blends imagination, reason, and emotion—also shares several literary strategies employed by Gumilla's travel narrative while adding the precision of his own form of fieldwork.

La Condamine's quest to resolve one of the greatest scientific questions of the first half of the eighteenth century by measuring the circumference of the earth near the equator also exposed valuable South American secrets and resources to Europe.[25] Two reports—La Condamine's *Relation abrégée d'un voyage* (1745)[26] and Juan and Ulloa's *Relacion historica del viage* (1748)[27]—echo Gumilla's stories that personify Orinoco plants and animals. Juan and Ulloa may question the reliability of local eyewitnesses as to the attractive properties of the anaconda's breath, but they still include four pages of details straight from Gumilla's narrative.[28] For example:

> El aliento, que despide de sí es tan ponzoñoso, que embriagando con él a la Persona, o Animal, que está en el camino por donde lo dirige, lo hace moverse acia ella involuntariamente hasta que teniéndolo cerca se lo traga. Esto dicen, y adelantan, que el modo de librarse en semejante trance es cortando el tal aliento, quando se empieza a sentir, con un otro Cuerpo, que passando violentamente por medio, lo divida, y rompa: lo que executado puede, el que em-

pezaba a padecer, tomar otra senda, y salir del peligro. Todo esto bien considerado tiene mas viso de fabula, que apariencias de realidad, como el mismo ya citado *Mr. de la Condamine* da a entender en su Relacion.[29]

[The breath it discharges from itself is so harmful that, enrapturing with it the person, or animal, that is in the path on which it directs the breath, it forces it to move toward it involuntarily until, having it close, it swallows it. They claim that the way to free oneself from a similar trance is by cutting that breath, when it is first felt, with another body, which by violently passing in between divides and breaks the breath: thus executed, one who had started to suffer can take another route and get away from the danger. All of this when considered well has more the appearance of a fable than of reality, like the already cited Mr. de la Condamine implies in his *Relation.*]

Later in this section the Spaniards again clearly paraphrase Gumilla's narrative. While partially supporting and partially refuting the anaconda's ability to stun its victims, they take full advantage of the "strange but true" presentation of this same phenomenon that Gumilla described:

No debemos oponernos a que pueda ser de tal calidad el efecto de su aliento, que embriague al que lo perciba [. . .] a correspondencia pues de esto no encuentro yo dificultad, en que el aliento de esta Culebra tenga la propiedad, que se le atribuye [. . .] pues perdiendo los sentidos el Animal [. . .] y no quedándole arbitrio para huir, ni libertad para continuar su rumbo; antes bien dexandolo inmóvil, es regular vaya la Culebra con su tardo movimiento acercándose a él, hasta que lo tenga a tiro, para cogerlo, y engullirlo.[30]

[We should not oppose the possibility that the quality of its breath could have such an effect that it enraptures all that perceive it . . . I find no difficulty believing that the breath of this snake would have this property that is attributed to it . . . since upon losing its senses the animal . . . would have no means left to flee, nor freedom to continue on its path; but having left it immobile, it is normal that the snake would slowly approach until it had the animal in range for grabbing and devouring it.]

Even if the Jesuit's tales of cows, herons, and human victims being pulled into the mouth of the buío strike La Condamine, Juan, and Ulloa as "repugnante a la credulidad"[31] [unworthy of belief], they still privilege Gumilla's wonder-provoking anecdotes and rhetorical strategies for this and other New World marvels.

La Condamine, Juan, and Ulloa borrowed a number of memorable stories from *El Orinoco ilustrado;* and while they contested Gumilla, they also imitated him. To assume, as some do, that La Condamine dismissed Gumilla's "views as those of a traveler who was both too skeptical and too credulous"[32] effectively limits any consideration of Gumilla's legacy to the geographical error that the Jesuit had corrected[33] and some seemingly doubtful facts, such as the misunderstood insistence on the existence of El Dorado and the aforementioned fantastic treatment of the anaconda.[34] Not only does this downplay the value of Gumilla's firsthand knowledge of the region, from which La Condamine and others benefited, but it misunderstands Gumilla's nonfactual legacy. The travel writers both confirmed and rebutted Gumilla's scientific knowledge; but more striking is how Gumilla's literary legacy manifests itself, as when the travelers make use of unforgettable "characters" such as Gumilla's personified virgin plant, which flees from human touch or gaze. The travelers' delight at the modesty of the vergonzosa or doncella plant (*mimosa pudica*), whose leaves are sensitive enough to evade men, supersedes other natural historical references to this plant:

> Es tan visible la de esta, que luego que se toca alguna de sus Hojitas, se cierran todas las de aquella Rama, y aprietan unas contra otras con tanta prontitud, que no parece sino que los resortes de todas ellas estuvieron esperando aquel instante con prevención, para jugar todos a un mismo tiempo . . . y así admiraban que en una Yerva huviesse Sentido, y Instinto para manifestar la obediencia a lo que se le mandaba.[35]

> [This one's modesty is so clear that as soon as one of its little leaves is touched, it closes up all the leaves on that branch, and they clench against the others with such speed that it seems as if the springiness of all of them were waiting for that moment so that they all might make their move at the same time . . . and in this way people wonder at the fact that a plant would have the sense and instinct to manifest obedience at that which was commanded of it.]

Here the reference to the virgin plant's instinct for obedience alludes to its rational faculties, to which Gumilla added a moral aside, first commanding his readers to look into this mirror of nature: "[M]írense en el espejo de esta vergonzosa hierba" (*El Orinoco ilustrado,* 444) [Look at yourselves in the mirror of this bashful herb]. He then placed these commandments in the mouth of hypothetical mothers and teachers: " 'Venid, observad, atended y aprended de esta hierba vergonzosa; reparad que en cuanto la tocan, se da por muerta, desfallece, se desmaya y se marchita' " (444–45) ["Come, observe, attend to, and learn from this bashful plant. Note that when she is touched, she gives herself for dead, fainting and wilting"]. Here there is certainly an overlap as to which Orinoco region topics Gumilla, La Condamine, Juan, and Ulloa privilege. However, to underscore the wonder of individual facts, the scientific travelers also borrowed Gumilla's literary strategies for personification and anecdote.[36]

Alexander von Humboldt's appropriation of Gumilla's rhetoric of wonder particularly demonstrates the Jesuit's legacy. Humboldt famously explored the waterways from the Orinoco to the Amazon. Yet over fifty years before the Prussian's more famous journey, Gumilla culminated his own twenty-three-year journey to learn about and catalogue the Orinoco by publishing his natural history in Madrid. Humboldt's narrative of this journey is replete with accounts of the same wonder-provoking tropical biodiversity; he provides vivid descriptions of virulent insects, electric eels, monstrous crocodiles, and both the Christian and cannibal natives who shoot poison arrows from blowguns. Humboldt is said to have invented a "new literary genre": a modern travelogue that effects a happy union between literature and science.[37] However, not only is Humboldt indebted to Gumilla for what and how he presents the Orinoco River region, but his purposeful use of imagination and wonder to lead his readers on pathways to knowledge echoes the Jesuit's epistemology as well.[38] *El Orinoco ilustrado*'s synthesis of literary strategies and scientific elucidations prefigures that in Humboldt's writings, even though Humboldt used scientific measuring instruments for precise calculations and illustrations.[39]

Humboldt gained immediate fame for his extensive opus. He traveled to America, gathered ethnographic and botanical information, measured geographies, then retired to "civilized" Paris and published copious volumes that enlightened both Europeans and Americans of both the Northern Hemisphere and the Southern Hemisphere about the emerging new republics' peoples, politics, nature, and geographies. Humboldt's writings have been posited "not only as the product of a genius working in isolation but also as a summary of the Spanish American Enlightenment."[40] Although regarded as a "rediscoverer" of South America, Humboldt in fact compiled knowledge from Amerindians, criollos, and clerics who continue to be insufficiently cited, as their contributions to modernity have been eclipsed by Humboldt's international fame.[41]

Recently, historians and literary critics have been exploring Humboldt's diverse textual sources. An examination of his scientific writings reveals many individual facts from *El Orinoco ilustrado*.[42] Humboldt even cites specific pages from Gumilla's 1791 edition. But, even more important than the scientific data about the Orinoco, which constitutes a small portion of Humboldt's extensive opus, is the rhetoric he borrows from Gumilla. Both traveling naturalists' vivid descriptions of this "land worthy of a European's curiosity"[43] entertained, educated, and persuaded their readers with wonder-provoking rhetorical constructs. As a Catholic missionary on a spiritual journey, Gumilla constituted a different brand of natural traveler than the Protestant scientist-observer.[44] His agenda served political and religious goals, convincing government representatives to support the Jesuits' Catholic civilizing mission while exciting his fellow Jesuits' imaginations about travel to this distant land. In addition

HISTORIA NATURAL,

CIVIL Y GEOGRÁFICA

DE LAS NACIONES

DEL ORINOCO.

INTRODUCCION

Á LA PRIMERA PARTE.

La historia que voy á emprehender, natural, civil y geográfica del rio Orinoco, comprehenderá Paises, Naciones, Animales y Plantas incógnitas, casi enteramente hasta nuestros dias: para cuya cabal inteligencia se requieren especial claridad y método. Lo uno y lo otro procuraré en quanto pueda: para lo qual no saldré un paso fuera de los límites, que me he propuesto, sino fuere para comprobar la materia que lo requiere, ó para refutar lo que no dice con la verdad de lo que tratare. Y para que con mas

First page of part 1 in the 1791 edition. The John Carter Brown Library at Brown University.

Pag. 1

HISTORIA NATURAL,

CIVIL Y GEOGRÁFICA

DE LAS NACIONES

DEL ORINOCO.

INTRODUCCION

Á LA SEGUNDA PARTE.

Aunque esté bien tendida y fabricada á toda costa y gusto la escalera de un Palacio; con todo, el arte, la conveniencia ó la costumbre han introducido el descanso y plan en su mediania, para tomar resuello, y subir con mas brio ó ménos fatiga lo restante de ella. Es así; pero si no me engaño, creo que los pasos y capítulos con que hemos venido hasta aquí subiendo contra las corrientes del Orinoco, no han sido tan árduos ni fastidiosos, que requieran este descanso ó division de segunda Parte. Fuera de que, de las novecientas leguas que ya por via recta, ya en repetidos semicírculos creemos que corre el Orinoco, tenemos vistas y navegadas quatrocientas y cinqüenta, des-

Tom. II. A de

First page of part 2 in the 1791 edition. The John Carter Brown Library at Brown University.

to affecting their political consciousness, Humboldt's travelogues incited nineteenth-century gentlemen naturalists and ladies of means to repeat his journeys and write their own accounts.[45]

To clarify Gumilla's literary legacy and the implications his influence on Humboldt has for our understanding of the Enlightenment, this last section will examine a few of their shared rhetorical strategies. First, both authors kindled their readers' emotions by recording their own passionate responses to nature's sublimities.[46] Second, they depended on stirring rhetorical strategies, especially vivid description, or ecphrasis, which fixed images in the mind's eye of their readers. In order to do this, Humboldt borrowed from Gumilla the rhetorical technique of displaying individual specimens in a textual natural history cabinet and echoed Gumilla's selection of Orinoco objects (*inventio*), such as palm trees and curare poison, their arrangement (*dispositio*), and a hyperbolic style for their description (*elocutio*).

The first notable rhetorical strategy in Humboldt's text is his re-creation of emotional responses to nature through poetic prose. While Humboldt does not tie the purpose of his natural history descriptions to worshipping God, he does invite his readers to join him in experiencing nature's sublimities; he says, "[F]ollow me in spirit with willing steps to the recesses of the primeval forests."[47] The Jesuit, of course, seeks emotional reactions primarily to invite his readers to direct their wonder to the Creator, and the harmonious nature on display in Gumilla's "Orinoco cosmos" illuminates the mysteries of the divine order. Instead of echoing Gumilla's goal of expanding man's knowledge in order to better know God, Humboldt addresses late eighteenth-century readers' assumptions that an effusive style and an abundance of natural wonders precludes serious scientific intent. Humboldt's second preface (1849) to *Aspects of Nature* summarizes the difficulties involved with "the combination of a literary and of a purely scientific object—the endeavor at once to interest and occupy the imagination, and to enrich the mind with new ideas by the augmentation of knowledge."[48] As his original preface had apologized:

> [S]uch an artistic and literary treatment of subjects of natural history is liable to the difficulties of composition. . . . The unbounded riches of Nature occasion an accumulation of separate images; and accumulation disturbs the repose and the unity of impression which should belong to the picture. Moreover, when addressing the feelings and imagination, a firm hand is needed to guard the style from degenerating into an undesirable species of poetic prose. But I need not here describe more fully dangers which I fear the following pages will show I have not always succeeded in avoiding.[49]

Humboldt appealed to human pleasure, and awe of nature, with vivid descriptions meant to inspire new awareness of little-known regions. This technique was an integral part of Jesuit natural philosophy inquiries as well. Gumilla had already relied on poetic prose to stimulate man's imagination, awaken vivid enjoyment of nature, and increase natural knowledge. Even though Gumilla tied human wonder for nature to Christianity, both men saw scientific value in artistic strategies, and both men invited readers to view representations of Orinoco grandeur and support investigation of its natural phenomena.

The key literary influence of Gumilla for Humboldt is his aforementioned use of ecphrasis as the most important strategy for displaying objects in a rhetorically constructed natural history cabinet. As we saw in chapter 1 of this book, Gumilla preached through rhetorical strategies that use colorful mental images. Gumilla also constructed a mental and textual cabinet of curiosities to showcase diverse Orinoco region flora, fauna, and ethnography. In the introduction to *El Orinoco ilustrado* Gumilla even likened his natural history to a "most curious gallery."[50] If we view this within the context not only of Jesuit natural history cabinets but also the cabinets of amateur naturalists in Europe and Spanish America, we understand that Gumilla meant to evoke the gentleman's cabinet, which boasted spectacular wonders such as stuffed crocodiles or caimans, the enormous shed skin of an anaconda, and other rare fauna, flora, and curious human artifacts from diverse cultures and Amerindian nations. Humboldt also evoked actual cabinets, once even comparing a display of Carib cannibal skulls to their living Orinoco counterparts and marveling: "Having seen only skulls of these Indians in European collections, we were surprised to see that their foreheads were more rounded than we had imagined."[51] At another point, Humboldt rejoiced that "we saw for the first time live animals that we had only previously seen stuffed in European cabinets."[52] In his *Personal Narrative* and *Aspects of Nature,* Humboldt borrows the literary strategy of a rhetorical cabinet display of Orinoco wonders. Naturally, both men choose the most vivid and memorable plants, insects, river creatures, and native men and women, from the bloody Carib cannibals to the peaceful Christianized Saliva tribe.

Humboldt's treatment of the Orinoco palm provides a perfect example of how Gumilla's literary legacy goes beyond simply the selection of which objects to showcase, a process that corresponds rhetorically to the inventio stage. In addition to the subject matter of palms (inventio), Humboldt also borrows Gumilla's arrangement (dispositio) and style (elocutio). The order of things throughout Gumilla's textual natural history cabinet confirms the palm tree's importance. It is both the first and nearly the last plant treated in *El Orinoco ilustrado.*[53] Humboldt follows an identical arrangement in his own textual cabinet: "We will begin with palms, the loftiest and noblest of all vegetable forms, that to which the prize of beauty has been assigned."[54] Another aspect of the palm discourse that Humboldt borrows is the authority granted by eyewitness experience. Gumilla underscores the extensive detail he can offer from personal observation: "[A]lgo de esto se lee en algunos autores que han escrito acerca de los indios, pero no tanto como lo que he visto en los guaraúnos" (*El Orinoco ilustrado,* 131) [Something of this can be read in some authors who have written about the Indians, but not as much as I have seen myself with the Guaraúnos]. Humboldt purposely builds upon Gumilla's authority when adding his own view of the Guarani tribe's tree houses: "The Padre José Gumilla, who twice visited the Guaranis as a missionary, says, indeed, that this people had their habitation in the palmares (palm groves) of the morasses; but he only mentions dwellings raised upon high pillars, and not scaffoldings attached to trees still in a growing state (Gumilla, Historia natural, civil, y geografica de las Naciones situadas en las riveras del Rio Orinoco, nueva imp. 1791, pp. 143, 145, and 163)."[55] Finally, to increase readers' wonder both Gumilla and Humboldt compare the palm tree with the banana tree, a well-established "tree of life" that was praised in Pliny's *Natural History* as well as the Bible.[56] Humboldt explains that "in all parts of the globe the palm form is accompanied by that of Plantains or Bananas,"[57] while (as we saw in chapter 1 of this book) Gumilla's admiration for the banana knows no bounds: "[N]o hay fruta más sana en las Américas ni tan sustancial ni tan sabrosa. [. . .] En fin, los plátanos son el socorro de todo pobre; en la América sirven de pan, de vianda, de bebida, de conserva y de todo, porque quitan a todos el hambre" (436) [There is no healthier nor substantial and delicious fruit in the Americas. . . . In short, bananas are the relief of all the poor; in America bananas serve as bread, viands, drink, preserves, and everything, because they take away hunger from all]. Both authors activate panegyrical strategies that favorably compare the Orinoco palm to this providential gift to the poor: the banana tree. Even the briefest comparison of Humboldt's treatment of the "lofty and noble palm tree" and Gumilla's lengthy palm panegyric reveals striking similarities in how these authors describe and privilege this wonder of nature.

To conclude our discussion of shared literary strategies we turn to perhaps the most famous example of Gumilla enlightening Humboldt. A quick comparison of their narrations about the most famous South American poison, curare, illustrates Gumilla's legacy in Humboldt's more famous synthesis of literary strategies with science despite fundamental differences in their scientific practices. If, on the one hand, Humboldt insists on efforts in the field to re-create the precision of European laboratories, on the other he borrows the anecdotal aside Gumilla first used to spark his readers' curiosity and then interest them in appropriating Amerindian knowledge. Gumilla's version of the Caverre tribe's distillation of the liana root (*bejuco maracure*) made its way from *El Orinoco ilustrado* into several eighteenth-century encyclopedic and scientific journals published in French[58] and then into Humboldt's *Aspects of Nature* and, more extensively, his *Personal Narrative.* Humboldt directly acknowledges his source ("Gumilla asserts that 'this preparation was enveloped in great mystery'") before quoting the Jesuit's story featuring ancient "women who (being otherwise useless) are chosen to watch over this operation."[59] As with the treatment of the palm, Humboldt clearly borrows Gumilla's literary techniques to enlighten his reader. And in this case Humboldt amplifies the Jesuit's story with his own firsthand experience in the mission town of Esmeralda in order to reveal his more meticulous scientific practices. At the same time Humboldt plays up Gumilla's discourse of mysterious Amerindian secrets, he also emphasizes that he saw a male who was not using a cauldron but instead a "chemical laboratory" to fabricate curare. Humboldt's narrative about this "chemist of the place . . . [who] had that tone of pedantry of . . . the pharmacopolists of Europe" and performed "this chemical operation . . . [that] appeared to us extremely simple" contrasts greatly with Gumilla's evocation of witchlike old hags.[60] Humboldt underlines his view that "the greatest order and neatness prevailed in this hut, which was transformed into a chemical laboratory"[61] and speaks of the simple chemist's desire for European knowledge:

> [His] funnel was of all the instruments of the Indian laboratory that of which the poison-master seemed to be most proud. He asked us repeatedly if, *por alla* (*out yonder*, meaning in Europe) we had ever seen anything to be compared to this funnel. . . . [He] seemed flattered by the interest we took in his chemical processes. He found us sufficiently intelligent to lead him to the belief that we knew how to make soap, an art which, next to the preparation of curare, appeared to him one of the finest of human inventions. . . .[62]

Finally, since the Esmeralda chemist's laboratory is grossly inferior, Humboldt proposes to relocate research: "[A]n interesting chemical and physiological investigation remains to be accomplished in Europe on the poisons of the New World, when, by more frequent communications, the *curare de bejuco,* the *curare de raiz,* and the various poisons of the Amazon, Guallaga, and Brazil, can be procured. . . ."[63] Gumilla also praises European laboratories and blends the scientific with the artistic—wonderful stories alongside narrations of experiments—when favorably comparing the results of his Jesuit "in-the-field" science, tested in the Bogotá laboratory, to the knowledge gained in "those European labs" with "the latest microscopes."[64] Gumilla's artistic dissemination of knowledge about the mysterious curare production and a host of other Orinoco region topics clearly influenced Humboldt's more celebrated synthesis of literature and science.

By examining how Gumilla enlightened the scientific travelers La Condamine, Juan, Ulloa, and Humboldt with much more than unfamiliar Orinoco region facts, we can appreciate the legacy of this Jesuit's rhetoric of wonder. They read Gumilla's lively prose, which led them on pathways to scientific knowledge and also influenced the wonder-provoking literary strategies they employed to capture their own readers' imaginations and transport them to far-off lands. From actively involving readers in stirring representations of sublime nature to privileging ecphrasis to create mental images of objects that are displayed in a rhetorically constructed cabinet of natural history, Humboldt's writings in particular reveal the literary legacy of *El Orinoco ilustrado.* Novelists writing about the Orinoco River region also availed themselves of Gumilla's rhetoric of wonder. By exploring how Carpentier, Rivera, and Isaacs enhanced the scientific weight of their fiction with information from *El Orinoco ilustrado,* we can start to construct a wider perspective on the nature of the Enlightenment.

El Orinoco ilustrado continues to enlighten twentieth- and twenty-first century biologists, herpetologists, and pharmacologists, who evoke Gumilla's mediation of Amerindian knowledge and folklore about, for example, the "mythical anaconda" and curare alongside current discussion about such topics.[65] The legacy of Gumilla's text suggests an expanded definition of the Enlightenment. *El Orinoco ilustrado* exemplifies alternative pathways to the Enlightenment that do not conform to traditional, secular definitions and limiting conceptions of scientific authority. Further research into how cultural documents such as Gumilla's that have not usually been examined as Enlightenment texts in fact enlightened many can broaden our understanding of the Enlightenment in the peripheries and help us move beyond sweeping generalizations about Spain and Spanish America's tardy entrance into modernity.

Notes

Introduction

1. Key scholars who contributed to the cultural construct of France, England, Germany, and Holland as hegemonic secular Enlightenment models include Ernst Cassirer, *The Philosophy of the Enlightenment,* trans. Fritz C. A. Koelln and James P. Pettegrove (Princeton, NJ: Princeton University Press, 1951); Frank Manuel, *The Age of Reason* (Ithaca, NY: Cornell University Press, 1951); and Peter Gay, *The Enlightenment: An Interpretation,* 2 vols. (New York: Knopf, 1966–69). Gay's Enlightenment is defined by devotion to modern science alongside "hostility to Christianity," and his two volumes omit Spain and Spanish America (Gay, *Enlightenment,* 1:18).

2. Menéndez y Pelayo summarized the general view shared by the great Spanish philologists of the late nineteenth century and early twentieth century (such as Américo Castro and others). For more on the negative transatlantic legacy of Menéndez y Pelayo's *Historia de los heterodoxos españoles,* see Eduardo Subirats, *Modernidad truncada de América Latina* (Caracas: CIPOST [Centro de Investigaciones Postdoctorales], 2001), 18–19.

3. Recent studies have challenged the image of the Enlightenment as "the Goddess of Reason riding on the shoulders of French revolutionaries." Francisco Sánchez-Blanco, *La Ilustración en España* (Madrid: Akal, 1997), 6. While Wolfgang Röd's revalorization of sensibilities underlining the value of both reason and emotions in Enlightenment philosophy in *Die Philosophie der Neuzeit 2: Von Newton bis Rousseau* (Munich: Beck, 1978) has opened the door to studies like Jessica Riskin's *Science in the Age of Sensibility: The Sentimental Empiricists of the French Enlightenment* (Chicago: University of Chicago Press, 2002), Catholic Spain and Spanish America still have received scant attention in Enlightenment studies.

4. Marcus Hellyer's chapter, "The Jesuits and Their Contemporaries in the *Aufklärung,*" in *Catholic Physics: Jesuit Natural Philosophy in Early Modern Germany* (Notre Dame, IN: University of Notre Dame Press, 2005) provides a model for moving beyond the prejudices in eighteenth-century studies that maintain that the Jesuits "modernized their philosophical curriculum only when pressured by Enlightened reformers" (203). As Hellyer reminds us, "[T]his reduced the complexity of the situation to caricature. . . . [S]imple dichotomies of Enlightened versus obscurantist do not adequately describe the Jesuits' interaction with their contemporaries in the Age of *Aufklärung*" (203).

5. In the second half of the eighteenth century some reports based on missionaries' mediation of Amerindian philosophies and firsthand knowledge of exotic locales were written off by enlightened thinkers in Europe as credulous *cuentos de frailes* (travelers' tales). On this, see Jorge Cañizares-Esguerra, "Nation and Nature: Natural History and the Fashioning of Creole National Identity in Late Colonial Spanish America," in *Cultural Encounters in Atlantic Societies, 1500–1800, International Seminar on the History of the Atlantic World,* Working Paper Series (Cambridge, MA: The Charles Warren Center for

Studies in American History, 1998); and idem, *How to Write the History of the New World* (Stanford, CA: Stanford University Press, 2001). The current book joins recent criticism that instead underscores the value of Jesuit negotiation of indigenous secrets and emerging new philosophies between Europe and the Americas. See, for example, Steven Harris, "Jesuit Scientific Activity in the Overseas Missions, 1570–1773," *ISIS* 96 (2005): 71–79; and Luis Millones and Domingo Ledezma, *El saber de los jesuitas, historias naturales y el Nuevo Mundo* (Madrid: Iberoamericana, 2005).

6. Literary histories have not yet challenged the *eclecticismo, iluminismo,* and *protonacionalismo* thesis of such representative studies as Manfred Kossok, "La Ilustración en América Latina: ¿Mito o Realidad?" *Ibero-Americana Pragensia* 12 (1973): 89–100.

7. Jonathan Carlyon analyzes this contribution to the republic of letters in *Andrés González de Barcia and the Creation of the Colonial Spanish American Library* (Toronto: University of Toronto Press, 2005). See also Ramon Ceñal, "Fuentes jesuíticas francesas de la erudición filosófica de Feijoo," in *El padre Feijoo y su siglo,* ed. Jean-Louis Flecniakoska (Oviedo: University of Oviedo, 1966), 285–314 for Benito Jerónimo Feijoo's absorption of French, English, and German ideas in the *Mémoires de Trévoux.*

8. Miguel Batllori documents this scholarly activity in *La cultura hispano-italiana de los jesuitas expulsos: Españoles, hispano-americanos, filipinos, 1767–1814* (Madrid: Gredos, 1966). See also Manfred Tietz, *Los jesuitas españoles expulsos: Su imagen y su contribución al saber sobre el mundo hispánico en la Europa del siglo XVIII* (Madrid: Iberoamericana, 2001).

9. The full title of Gumilla's natural history is *El Orinoco ilustrado, Historia natural, civil, y geographica, de este gran río, y de sus caudalosas vertientes: govierno, usos, y costumbres de los indios sus habitadores, con nuevas, y utiles noticias de Animales, Arboles, Frutos, Aceytes, Resinas, Yervas, y Raìces medicinales: Y sobre todo, se hallaràn conversiones muy singulares a nuestra Santa Fè, y casos de mucha edificacion* (Madrid: Manuel Fernandez, 1741) [*The Orinoco Enlightened, natural, civil and geographical History of this great river and its abundant branches: government and customs of the indians who inhabit it, with new and useful notices about animals, trees, fruits, oils, resins, herbs and medicinal roots. And, above all can be found singular conversions to our Holy Faith and very edifying cases*]. The second, expanded edition's title is identical except for the addition of "defended": *El Orinoco ilustrado, y defendido, historia natural, civil y geographica de este gran río, y de sus caudalosas vertientes; govierno, usos, y costumbres de los Indios sus habitadores, con nuevas, y utiles noticias de animales, arboles frutos, aceytes, resinas, yervas, y raíces medicinales* (Madrid: Manuel Fernandez, 1745) [*The Orinoco Enlightened and defended, natural, civil and geographical history of this great river and its abundant branches; government and customs of the indians who inhabit it, with new and useful notices about animals, trees, fruits, oils, resins, herbs, and medicinal roots*].

10. What the Jesuit *ratio studiorum* called philosophy or natural sciences included topics such as geometry, logic, physics (in the early modern sense of the word), metaphysics, and natural history. Theology included study of the scriptures, Hebrew, dogmatic theology, ecclesiastical history, canon law, and moral theology. The *studia inferiora* covered Latin and Greek grammar and several courses in rhetoric. For Gumilla's biography and a bibliography of his essential writings, see José del Rey Fajardo, *Misiones Jesuíticas en la Orinoquia (1625–1767),* vol. 1 (San Cristóbal: Universidad Católica del Tachira, 1992), 325–53, 526–32. For general information on Jesuit missions and education, see John W. O'Malley, *The First Jesuits* (Cambridge, MA: Harvard University Press, 1993); idem, *The Jesuits: Cultures, Sciences, and the Arts, 1540–1773* (Toronto: University of Toronto Press, 1999 and Ángel Santos Hernández, *Los Jesuitas en América* (Madrid: Editorial MAPFRE, 1992).

11. Manuel Aguirre Elorriaga, *La Compañía de Jesús en Venezuela* (Caracas: Lucas G. Castillo, 1941), 67. See also Daniel Restrepo, *Compendio historial y galería de ilustres varones* (Bogotá: Imprenta del Corazón de Jesús, 1940). For more detailed information on Gu-

milla's Jesuit formation and work in the Orinoco missions, see works by José del Rey, especially *Una utopía sofocada: Reducciones Jesuíticas en la Orinoquia* (Caracas: Universidad Católica del Táchira / Universidad Católica Andrés Bello, 1998); and Joseph Gumilla, *P. José Gumilla: Escritos varios,* ed. José del Rey (Caracas: Fuentes para la Historia Colonial de Venezuela, 1970), an archive of letters and contemporary Jesuit correspondence.

12. My use of the term "intercultural discourse" considers not only Jesuit appropriation of indigenous knowledge, but also how its mediation and adaptation by European epistemologies affected the Hispanic Enlightenment.

13. Juan Eusebio Nieremberg's textbook, *HISTORIA NATVRAE maxime peregrinae libris XVI distincta. In quibus rarissima Naturae arcane, etiam astronomica, & ignota Indiarum animalie, quadrupedes, aues, pises, reptilia, insecta . . . nouae & curiosissimae quaestiones disputantur, ac plura sacrae Scripturae loca erudite enodantur* (Antwerp: Balthasaris Morelia, 1635) encouraged Jesuit natural historians to make a pious use of wonder as Joseph de Acosta had in his *Historia natural y moral de las Indias,* ed. Edmundo O'Gorman (1590; reprint, Mexico: Fondo de Cultura Económica, 1940/1962). See, among others, South American and Caribbean natural histories by Ovalle, Pelleprat, Labat, Charlevoix, Acuña, Rodríguez, and Peguero.

14. O'Malley, *Jesuits,* notes that "the Jesuits of this period liked to think they had a 'way of proceeding'—or better, ways of proceeding—special to themselves. The expression actually goes back to Ignatius of Loyola himself . . ." (*Jesuits,* xiv).

15. Simón Bolívar himself bestowed upon his friend Humboldt the status of " 'true discoverer of South America' . . . for captivating the world with his depictions of the region's aesthetic and scientific wonders." (Gerard Helferich, *Humboldt's Cosmos: Alexander von Humboldt and the Latin American Journey that Changed the Way We See the World* (New York: Gotham Books, 2004), 303. Mary Louise Pratt, *Imperial Eyes: Travel Writing and Transculturation* (New York: Routledge, 1992) has affirmed that Humboldt's "reinvention of America for Europe was transculturated by Euroamerican writers into a creole process of self-invention" (175). Jorge Cañizares-Esguerra, however, reminds us that this "transculturation" went both ways. Humboldt not only enlightened the locals but was enlightened by them. The Prussian scientist incurred great debts to *criollo* and Amerindian intellectuals. As Cañizares-Esguerra clarifies in "Spanish America: From Baroque to Modern Colonial Science," in *Cambridge History of Science, Volume Four: Eighteenth-Century Science,* ed. Roy Porter (Cambridge: Cambridge University Press, 2003): "Humboldt's thirty volumes should be read not only as the product of a genius working in isolation but also as a summary of the Spanish American Enlightenment" (737).

16. See Kossok, "La Ilustración en América Latina," 92.

17. In *I Have Landed: The End of a Beginning in Natural History* (New York: Harmony Books, 2002), the evolutionary theorist Stephen Jay Gould describes both Charles Darwin's and the landscape painter Frederic Edwin Church's debts to Humboldt's Enlightenment. Even Subirats grants Humboldt status as the "prototypical European Enlightenment intellectual" (*Modernidad truncada,* 21). See also Barbara Maria Stafford, *Voyage into Substance: Art, Science, Nature, and the Illustrated Travel Account, 1760–1840* (Cambridge, MA: MIT Press, 1984).

18. Riskin, *Science in the Age of Sensibility,* 283. Although focused on French Enlightenment empiricism, Jessica Riskin's critique of the long-standing stereotypes about "sober empirical observation and rational theory construction" and of any ideas about a "chilly, arid rationalism" that can be traced back to the eighteenth-century writings of encyclopedists like d'Alembert provide useful fodder for examining the culture of Hispanic Enlightenment (283–84). As Riskin's study argues: "[T]he view that sciences are cultural expressions, that intuition and emotion are crucial to understanding, and that ra-

tionalism, in its claims to absolute truth, is dogmatic, arrogant, and oppressive, all played crucial roles in Enlightenment science itself" (285).

19. The works of these pioneers have yet to be replaced. Arthur P. Whitaker's *Latin America and the Enlightenment* (New York: D. Appleton-Century Company, 1942) collected seven classic views, including Roland D. Hussey's oft-cited "Traces of French Enlightenment in Colonial Hispanic America" (Whitaker, *Latin America,* 23–51). Whitaker then contributed the opening essay to A. Owen Aldridge, *The Ibero-American Enlightenment* (Urbana: University of Illinois Press, 1971), 21–57. Jean Sarrailh's *L'Espagne éclairée de la seconde moitié du XVIIIe siècle* (Paris: Impr. nacionale, 1954) was translated from the French as *La España ilustrada de la segunda mitad del siglo XVIII* in 1957.

20. For a summary of the liberal and conservative strains of Enlightenment theory for Latin America see, for example, the editors' introduction and the contribution by Olegario Negrín Fajardo in Diana Soto Arango, Miguel Ángel Samper, and Luis Carlos Arboleda, eds., *La Ilustración en América Colonial* (Madrid: Doce Calles, 1995), 9–17, 67–89.

21. See, for example, John F. Wilhite, "The Enlightenment in Latin America: Tradition versus Change," *Dieciocho* 3, no. 1 (1980): 18–26.

22. Francisco Sánchez Blanco (1991, 1997, 1999, 2002), for example, revitalizes Mario Góngora's concept of Catholic Enlightenment and cites Vicente Rodríguez Casado, "El intento español de Ilustración cristiana," *Estudios Americanos* 9, no. 42 (1955): 141–69; and Richard Herr, *The Eighteenth-Century Revolution in Spain* (Princeton, NJ: Princeton University Press, 1958). Eduardo Subirats, who in *La ilustración insuficiente* (Madrid: Taurus, 1981) first defined Spain's Enlightenment as "insufficient" and then in 2001 described both Latin America's Enlightenment and modernity as "truncated," remains an outspoken proponent of this perception of a failed modernity. Perhaps Arthur Whitaker's 1971 summary of earlier scholars' failed attempts to reconcile "Catholic" and "Enlightenment" as "a contradiction in terms, which only proves that there was no true Enlightenment in either Spain or its colonies" has influenced this discourse of failure (Whitaker, "Changing and Unchanging Interpretations of the Enlightenment in Spanish America," in Aldridge, *Ibero-American Enlightenment,* 55).

23. Subirats *Modernidad truncada,* 17–23.

24. Although Pérez Magallón's "pluridiscursive" modernity assimilates Spain's routes to modernity within "hegemonic countries" (especially France and England) while at the same time exalting Spanish national identity, his attention to the period between 1675 and 1725 is essential for understanding Enlightenment in the colonies, which assimilated Spain's routes to modernity while underscoring regional particularities.

25. Cañizares-Esguerra, *How to Write,* 209.

26. Ruth Hill, *Sceptres and Sciences in the Spains: Four Humanists and the New Philosophy (ca. 1680–1740)* (Liverpool: Liverpool University Press, 2000), 24.

27. Joseph Gumilla, *El Orinoco ilustrado y defendido,* ed. Demetrio Ramos (Caracas: Fuentes para la Historia Colonial de Venezuela, 1963), 31. All quotes from Joseph Gumilla's *El Orinoco ilustrado* are from this edition, hereafter cited in the text. All translations are mine.

28. For example, in José Rufino Cuervo, *Diccionario de construcción y regimen de la lengua castellana* (Santa Fe de Bogotá: Instituto Caro y Cuervo, 1993), 322–328, the subcategories of definition number two reveal a supernatural Christian enlightenment alongside ordinary intellectual understanding and the accumulation of knowledge: "1. Alumbrar, iluminar. 2. a) Dar luz al entendimiento y a la voluntad. b) Teol. Alumbrar Dios interiormente a la criatura con luz sobrenatural. c) Instruir, proporcionar o adquirir conocimientos. 3. a) Adornar un impreso con láminas o grabados alusivos al texto. b) Embellecer, hermosear. c) Dicho de la lengua o del estilo, Enriquecer, embellecer. 4. a) Hacer ilustre a una persona o cosa. b) Darse a conocer (*refl.*)." The third definition demonstrates that "illus-

trated" can mean both adorned with material illustrations (drawings, etc.) or linguistically adorned with an "enriching, embellishing" style.

29. Sebastián de Covarrubias y Orozco, *Tesoro de la lengua castellana o española,* 4th ed. (1611; Barcelona: Editorial Alta Fulla, 1998), 731.

30. *Oxford English Dictionary,* 2nd ed., s.v. "illustrate." *OED Online* (Oxford University Press) 4 http://dictionary.oed.com/cgi/entry/00181778.

31. Real Academia de la Lengua, *Diccionario de la lengua castellana,* vol. 2 (1726–39; Madrid: Espasa Calpe, 1992), 212. Cuervo's *Diccionario de construcción y regimen de la lengua castellana* bases most of the definitions of *ilustrar* on dictionaries such as *Diccionario de la lengua castellana,* e.g., "Dar luz al entendimiento y a la voluntad" (323); "Embellecer, hermosear" (326); "Hacer lustre a una persona o cosa" (327).

32. Real Academia de la Lengua, *Diccionario,* 212

33. Chapter 23 is entitled "Método el más practicable para la primera entrada de un misionero en aquellas tierras de gentiles, de que trato, y en otras semejantes" [The most practical method for the first entry of a missionary in those lands of gentiles, which I deal with, and other similar ones].

34. This French edition, broken up into three small volumes but not the two parts of the Spanish version, could also be found in the personal library of the enlightened polymath and third president of the United States, Thomas Jefferson.

35. "Mucho más correcta que las anteriores, y adornada con ocho láminas finas, que manifiestan las costumbres y ritos de aquellos Americanos" [Much more correct than the previous editions, and adorned with eight fine plates, that show the customs and rites of those Americans]. Joseph Gumilla, *Historia natural, civil y geographica de las naciones situadas en las riveras del Rio Orinoco* (Barcelona: Carlos Gubert y Tutó, 1791, frontispiece.) The 1791 edition changes the title slightly, although "El Orinoco ilustrado" remains as the header on all left-hand pages. The 1745 spelling is modernized according to Royal Academy norms. Since it is a postexpulsion publication within Spain, Gumilla's "Apóstrofe a los operarios de la Compañía de Jesús" [Apostrophe to the workmen of the Company of Jesus], all Jesuit approbations, and opening letters are dropped.

36. Filippo Salvadore Gilij's well-known postexpulsion volumes treated Orinoco tribes' customs, religion, and languages. See his *Saggio di storia americana; o sia, Storia naturale, civile e sacra de'regni, e delle provincie spagnuole di Terra-Ferma nell' America Meridionale* (Rome: L. Perego erede Salvioni, 1780–84).

37. The 1741 edition's chapter 12 title, "De los mortales venenos que usan. Raro modo de fabricarlos: Maña y cautela para darlos: su efecto; y las contras que se han rascreado para evadir su eficacia" [On the mortal poisons that they use. Strange means of fabricating them: tricks and cunning to give them: their effect; and the antidotes that have been created to avoid their efficacy] becomes in the 1745 edition: "Del mortal veneno llamado curare. Raro modo de fabricarle, y de su instantánea actividad" [On the mortal poison called curare. Strange means of fabricating it, and its instantaneous activity]. To chapter 13's title, Gumilla adds the phrase "descubrió el Autor de este libro" [discovered by the Author of this book] to underscore his eyewitness authority.

38. Its inclusion in this collection suggests nineteenth-century recognition of Gumilla's importance in the development of Spain's scientific discourse and follows the 1791 edition's alteration of the original title, dropping *El Orinoco ilustrado* and emphasizing the nations along the Orinoco with this title: *Historia natural, civil y geográfica de las naciones situadas en las riveras del Rio Orinoco* [*Natural, civil and geographical history of the nations situated along the banks of the Orinoco River*].

39. Jesuit historians cite three dates for this edition (1944, 1945, and 1946). Since the copies themselves are not dated, I choose the bicentennial of the publication of the complete edition.

40. Ramos's introduction, "Gumilla y la publicación de *El Orinoco ilustrado*," constitutes the most comprehensive study of *El Orinoco ilustrado* that exists. Ramos also published separate essays on Gumilla's cartography, geography, and ethnography.

41. See this book's bibliography for a comprehensive list of all the editions of *El Orinoco ilustrado*.

42. See Stephen Greenblatt, *Marvelous Possessions: The Wonder of the New World* (Chicago: University of Chicago Press, 1991) for the history of European appropriation of wonder for both persuasive and possessive goals. For centuries after Columbus's voyages, Spanish monarchs collected and displayed New World flora, fauna, and human artifacts.

43. Ibid., 24–25.

44. For alternative "scientific revolution" narratives, see the following articles collected in Antonio Lafuente, *Mundialización de la ciencia y cultura nacional* (Madrid: Doce Calles, 1993): George Basalla, "The Spread of Western Science Revisited" (599–618); David Wade Chambers, "Locality and Science: Myths of Centre and Periphery" (605–17); Alberto Elena "La configuración de las periferias científicas: Latinoamérica y el mundo islámico" (195–218); and James E. McClellan III, "Comparative Perspectives on European Science and New-World Societies: A Comment" (454–554). See also Jorge Cañizares-Esguerra, "Iberian Science in the Renaissance: Ignored How Much Longer?" *Perspectives on Science* 12, no. 1 (2004): 86–124.

45. Francis Bacon, *The Advancement of Learning* (New York: Modern Library, 2001), 8–9.

46. See Guillermo Morón and J. A. de Armas Chitty, "Año Gumillano: Acuerdo de la Academia Nacional de la Historia," *Boletín de la Academia Nacional de la Historia,* 1987, 217–19. Gumilla's natural history has been lauded as "the genesis of national consciousness" in Venezuela, while at the same time it has earned him fame as a patriotic "son of Colombia." See, for example, Guillermo Fajardo Morón, "El escritor venezolano José Gumilla," *Boletín de la Academia Nacional de la Historia,* 1986: 1101–2. These comments point to late eighteenth-century and early nineteenth-century appropriations of Gumilla in Colombia and Venezuela as regional pride took precedence over the Spanish viceroyalty of New Granada. The "proto-criollo" and "proto-nationalistic" status afforded Gumilla provides an interesting consideration for a different study that considers a transatlantic formation of "criollo subjectivity" and points to alternate and conflicting understandings of the politics and economics of Spain and America. Today Gumilla's legacy includes Centro Gumilla, which was founded as an ecclesial base community in Venezuela in 1968 and is a supporter of Jesuit political activism and a proponent of social justice. Additionally, see this study's bibliography for several brief articles on Gumilla by the Jesuit scholar José del Rey. In a few of them he portrays *El Orinoco ilustrado* as a text that champions Venezuela and Colombia, a region that served as the cradle of Latin American independence.

Chapter 1. Cabinet of Curiosities

1. Earlier examples of textual cabinets can be found in, for example, Robert Basset's compilation, *Curiosities: or the Cabinet of Nature. Containing Phylosophical, naturall, and morall questions fully answered and resolved. Translated out of Latin, French, and Italian Authors* (London: N. and I. Okes, 1637). For further scholarship on books as cabinet displays, see Mary B. Campbell, *Wonder and Science: Imagining Worlds in Early Modern Europe* (Ithaca, NY: Cornell University Press, 1999); and Margaret Hodgen, *Early Anthropology in the Sixteenth and Seventeenth Centuries* (Philadelphia: University of Pennsylvania Press, 1971).

Stephanie Merrim, "La *Grandeza mexicana* en el contexto criollo," in *Nictimene—sacrilege: Estudios coloniales en homenaje a Georgina Sabàt-Rivers,* ed. Georgina Sabàt de Rivers, Electa Arenal, Mabel Moraña, and Yolanda Martínez-San Miguel (Mexico City: Universidad del Claustro de Sor Juana, 2003), organizes a discussion of the urban spaces evoked in two Spanish American baroque poems by Bernardo de Balbuena and Sor Juana Inés de la Cruz as exotic textual wonder cabinets. Hodgen's description of Johann Boemus's textual cabinet of ethnographic curiosities, however, has more in common with Gumilla. *The Manners, Lawes, and Customes of All Nations* (1611) used "the printed page, as others had employed the 'cabinet de curiosités,' for assembling and exhibiting the range of human custom, ritual and ceremony" (Hodgen, *Early Anthropology,* 131).

2. This phrase is from Stephen Greenblatt, *Marvelous Possessions: The Wonder of the New World* (Chicago: University of Chicago Press, 1991). As Greenblatt pointed out in "Resonance and Wonder," in *Exhibiting Cultures: The Poetics and Politics of Museum Display,* ed. Ivan Karp and Steven D. Lavine (Washington, DC: Smithsonian Institution Press, 1991), 42–56, New World chronicles penned by Spanish and American-born soldiers, scholars, and holy men continued the medieval traditions of textual descriptions that were capable of evoking so much visual wonder that the reader could almost see what was described on the page.

3. Lorraine Daston, "Attention and the Values of Nature in the Enlightenment," in *The Moral Authority of Nature,* ed. Lorraine Daston and Fernando Vidal (Chicago: University of Chicago Press, 2004) reminds us that "throughout the eighteenth century, natural theology—the worship of God through the study of His works—supplied the motivation and rationale" for close observation and natural history (105). Daston also illuminates connections between eighteenth-century natural history descriptions of God-the-artisan's workmanship and man's workmanship within a political economy. Jesuits and non-Jesuits alike "wove together theological, aesthetic, moral, and economic strands of valuation in their observations and descriptions, often deploying the single word 'utility' for this knot of the good, the beautiful, and the useful" (101).

4. These Catholic polymaths were at home in all fields of knowledge, and their early eighteenth-century writings constitute important sources for the ideas of seventeenth-century natural philosophers such as Descartes and Bacon. Feijoo's *Teatro Crítico Universal,* vol. 1 (1726) and *Cartas eruditas y curiosas,* vol. 1 (1742) have long been hailed as manifestations of the Enlightenment. Key Muratori (1672–1750) Venice publications include *Delle reflessioni sopra il buon gusto nelle Scienze e nell'Arti* (1708), *Filosofia morale esposta* (1735), *Delle forze dell' intendimento umano* (1735), and *Delle forze della fantasia* (1745). More on them in chapter 3 of the present book.

5. Joseph Gumilla, *El Orinoco ilustrado y defendido,* ed. Demetrio Ramos (Caracas: Fuentes para la Historia Colonial de Venezuela, 1963), 37. Unless otherwise indicated, all quotes from Joseph Gumilla are from this edition, hereafter cited in the text. All translations are mine.

6. See Pedro Álvarez de Miranda, *Palabras e ideas: El léxico de la Ilustración temprana en España: (1680–1760)* (Madrid: Real Academia Española, 1992) for a study of the lexicon and the ideas behind the Age of Enlightenment (*Siglo de las luces*).

7. Gumilla's authority as an eighteenth-century natural historian, enmeshed with his status as a Jesuit priest, resembles the authority of his contemporary, the widely known Spanish Benedictine Benito Jerónimo Feijoo, today praised as the Spanish father of the Enlightenment. Feijoo announced in his essay on natural history that he aimed at "el deleite de tener suspensos y admirados a sus compatriotas con la relación de cosas nunca vistas, ni oídas" [the delight of having his compatriots amazed and filled with wonder by the relation of things never seen nor heard] Benito Jerónimo Feijoo, "Historia Natural," in vol. 2 of *Teatro crítico universal* (Madrid: D. Joaquin Iberra, 1779), 61. Gumilla seeks the same.

8. Translations mine from the Oscar Rodríguez Ortiz edition of *El Orinoco ilustrado* prepared for *Los libros de "El Nacional,"* a newspaper in Caracas, Venezuela.

9. Gerard Turner, "The Cabinet of Experimental Philosophy," in Oliver Impey and Arthur MacGregor, *The Origins of Museums: The Cabinet of Curiosities in Sixteenth- and Seventeenth-Century Europe* (Oxford: Oxford University Press 1985), 241.

10. See Aguiló Alonso, "El coleccionismo de objetos procedentes de ultramar a través de los inventarios de los siglos XVI y XVII," in *Relaciones artísticas entre España y América,* ed. Enrique Arias Anglés (Madrid: Consejo Superior de Investigaciones Científicas, 1990), 107–49. See also José Miguel Morán Turino and Fernando Checa Cremades, *El coleccionismo en España: De la cámera de maravillas a la galería de pinturas* (Madrid: Cátedra, 1985).

11. See Marc-André Bernier and Réal Ouellet, "Pierre Pellepratt's Accounts of the Jesuit Missions in the Antilles and in Guyana," in *Jesuit Accounts of the Colonial Americas: Textualities, Intellectual Disputes, Intercultural Transfers,* ed. Marc-André Bernier, Clorinda Donato, and Hans-Jürgen Leusebrink (Toronto: University of Toronto Press, forthcoming) for Pierre Pellepratt's (1606–67) "amphitheater of astonishment." Bernier describes this French Jesuit's use of the spectacular to evoke moral responses as a forerunner to Noel-Antoine La Pluche's *Le spectacle de la nature; ou entretiens sur particularités de l'histoire naturelle, qui ont paru les plus propres à rendre les jeunes-gens curieux, & à former leur esprit* (Paris: Veuve-Estienne, 1732–50).

12. For the conventions of the French tradition embodied by La Pluche and "the division between a sensuous, pleasurable, or merely 'curious' *watching* and a rational, tasking, language-driven *observation* [that] arose during the eighteenth century," see Barbara Maria Stafford, "Voyeur or Observer? Enlightenment Thoughts on the Dilemmas of Display," *Configurations* 1, no. 1 (1993): 95; and Daston, "Attention and the Values of Nature," 107.

13. When articulating how early modern values were applied to acquiring and possessing nature, Stephen Greenblatt explains that "the wonder-cabinets of the Renaissance were at least as much about possession as display. The wonder derived not only from what could be seen but from the sense that the shelves and cases were filled with unseen wonders, all the prestigious property of the collector" ("Resonance and Wonder," 49–50).

14. William Eamon, *Science and the Secrets of Nature* (Princeton, NJ: Princeton University Press, 1994), 223–24.

15. See the letters to and from Gumilla in Joseph Gumilla, *P. José Gumilla: Escritos varios,* ed. José del Rey (Caracas: Fuentes para la Historia Colonial de Venezuela, 1970).

16. Dominique Deslandres, "*Exemplo aeque ut verbo,*" in *The Jesuits: Cultures, Sciences, and the Arts, 1540–1773,* ed. John W. O'Malley (Toronto: University of Toronto Press, 1999), 263.

17. Gumilla mentions speaking in front of "sujetos eruditos" [erudite subjects] (*El Orinoco ilustrado,* 305). Later, Gumilla admires Kircher's sense of wonder and describes "el nunca bastantemente alabado Padre" [the never-well-enough praised Father] Kircher's magic lights, narrating the "secret preparations" for lighting up all the candles at once. Gumilla compares the wonder produced by Kircher's "arte e ingenio" (362) [skills and genius] to the wonder evoked by nature's mysterious operations. The remains of what came to be known as "Kircher's museum" are currently on display at the Ethnographic Museum Pigorini in Rome. See Luis Millones Figueroa, "La *intelligentsia* jesuita y la naturaleza del Nuevo Mundo en el siglo XVII," in *El saber de los jesuitas, historias naturales y el Nuevo Mundo,* ed. Luis Millones Figueroa and Domingo Ledezma (Madrid: Iberoamericana, 2005), 34.

18. Eamon, *Science and the Secrets of Nature,* 225.

19. Gumilla's understanding of *elocutio* differs from modern and postmodern understandings that reduce style—and often the concept of rhetoric itself—to "window dressing." Twentieth-century critics inherited this semantic reduction from nineteenth-century romantic writers. See Luisa López Grigera, *La retórica en la España del Siglo de Oro: Teoría y práctica* (Salamanca: Ediciones Universidad de Salamanca, 1994), for an explanation of how rhetoric was "stabbed" in the back by "our Romantic great-grandparents." As she explains, "[S]till for many of our contemporaries, 'rhetoric' means Baroque ornamentation, lack of spontaneity; in a word, everything false and ugly" (133). Translation mine.

20. In addition to exposition and argumentation, classical and Renaissance oratory was based on amplification. Aristotle classified amplification as a technique of persuasion. For his part, Cicero identified *amplificatio* as a key element for the "supreme virtue" of eloquence. According to Quintilian, praise of the "beauty and utility" in demonstrative rhetoric very frequently required amplification: "But the peculiar business of panegyric is to amplify and embellish its subjects. This kind of eloquence is devoted chiefly to gods or men, though it is sometimes employed about animals and things inanimate." Quintilian, *Institutes of Oratory,* trans. John Selby Watson, 2 vols. (London: Henry G. Gohn, 1856), 3.7.6, p. 219.

21. Luis de Granada, *Los seis libros de la Retórica eclesiástica o la manera de predicar* (Madrid: Ediciones Atlas, 1945), 2.3.1, p. 508. First published in Venice in Latin 1578.

22. In addition to meaning "enlightenment," the eighteenth-century usage of the term "illustration" included the meaning of "amplification." Rhetorical treatises emphasized the ornamental aspect of illustration, defining it as the "lumen" that accentuates or amplifies the vivid description, and sacred oratory clarified ties between amplification and emotion. See Luis de Granada, *Los seis libros,* bk. 3, "En que se trata del modo de amplificar, y de los afectos" [In which the means of amplification and emotions are treated] and Quintilian *Institutes of Oratory* 7.3.61–71 for the emotive effect of detailed description and the need to study and imitate nature to achieve it. *Ratio atque Institutio Studiorum Societatis Jesu* established Ciprian Suárez's *De arte rhetorica* (1586) as the preferred text for teaching Jesuits rhetorical strategies in Spain and Spanish America. Suárez's definitions for vivid description (or as he calls it, evidentia or hypotyposis) were obviously shaped by Cicero and Quintilian: "*Hypotyposis,* which Cicero terms *descriptio,* is a representation of facts made in such vivid language that it appeals to the eye rather than to the ear. Or, it is a visual presentation of events." Suárez, *De arte rhetorica,* trans. Lawrence J. Flynn (Gainseville: University Presses of Florida, 1955), 344. Since Suárez's chapter entitled "The Rule for Amplification" clearly suggested that "we should employ the topics which can arouse expectancy, wonder, and delight" (181), it is no great surprise that these are the primary emotions Gumilla evokes through extensive amplification.

23. As Lorraine Daston and Katharine Park noted in *Wonders and the Order of Nature: 1150–1750* (New York: Zone Books, 1998): "If each object by itself elicited wonder, all of them densely arrayed floor to ceiling or drawer upon drawer could only amplify the visitor's gasp of mingled astonishment and admiration" (260).

24. If we reduced Gumilla's descriptions treating marvelous plants throughout several chapters to the essential clauses and adjectives, the list would read like an extended *incrementum,* with each plant more splendid than the last. For this climactic series of wonderful fruits of the earth, Gumilla makes constant use of figures for amplification such as *congeries* (heaping up of descriptive words), *epitheton* (where plants are qualified with appropriate adjectives), and *dirimens copulatio* (or *progressio:* a series of comparisons).

25. Throughout *El Orinoco ilustrado,* Gumilla repeatedly grants himself eyewitness authority, employing such phrases as: "Me cito a mí mismo, porque pasó delante de mis

ojos" [I cite myself, because it happened before my eyes] (126); "por mis ojos mismos" [with my own eyes] (267); and "soy testigo de vista" [I am an eyewitness] (317).

26. See Ralph Bauer, *The Cultural Geography of Colonial American Literatures: Empire, Travel, Modernity* (Cambridge: Cambridge University Press, 2003), 29.

27. According to Aristotle, that which is distant is marvelous, and he tied this to the discourse of novelties when recommending the trope of " 'I shall tell you something strange, the like of which you have never heard' or '[something] so marvelous.' " *On Rhetoric: A Theory of Civic Discourse,* ed. and trans. George A. Kennedy (New York: Oxford University Press, 1991), 263. In *European Literature and the Latin Middle Ages,* trans. Willard R. Trask (Princeton, NJ: Princeton University Press, 1953), 85, Ernst Curtius classifies this as the "I bring things never said before" topos, which belongs in the introduction of a text or in the opening section meant to capture the audience's attention, which is exactly where Gumilla employed it.

28. In an addition to the 1745 edition of *El Orinoco ilustrado,* Gumilla draws attention to "esta anticipada prevención" [this anticipated prevention] (Gumilla, *El Orinoco ilustrado y defendido . . .* (Madrid: Manuel Fernandez, 1745), 39. This strategy is highlighted in the Jesuit handbook for rhetoric, *De arte rhetorica:* "*Ante occupatio,* which Quintilian terms *praesumptio* and the Greeks, *prolepsis,* is a figure by which we forestall the other side's objections" (Ciprian Suárez, *De arte rhetorica,* 339). See also Luis de Granada, *Los seis libros,* bk. 2, chap. 14, "De la prolepsis, que se llama en latin *praesumptio* o *anticipatio*" (527).

29. In bk. 4 of *De doctrina Christiana,* St. Augustine analyzes an eloquent passage by St. Paul, thus establishing him as a model for Christian eloquence. Augustine, *De doctrine Christiana,* trans. D.W. Robertson, Jr. (New York: Macmillan, 1958), 123. Father Gumilla inserts himself within the tradition of the eyewitness testimony of the apostles by testifying to Christ's presence along the Orinoco. Like Simon Peter, Gumilla is an "eyewitness of his majesty" (2 Pet. 1:16; RSV). In the third paragraph of his prologue, Gumilla fulfills his tasks ("tareas") by order of St. Peter: "[C]omo mandó el Apóstol San Pedro, notaré [. . .]" (*El Orinoco ilustrado,* 30) ["like the Apostle Saint Peter ordered, I will note . . .].

30. Horace was one of the first to tie together words, images, and emotions with his *ut pictura poesis.* Gumilla grants authority to the great Latin poet to underscore the visual nature of wonder and to verify the trustworthy nature of an eyewitness.

31. Traditional Bible exegesis had employed rhetorical strategies for "*efficacia* and *evidentia,* or power and vividness. The latter creates emotional power by using imagery, dialogue, personification, and *hypotyposis.*" Debora Shugar, "Sacred Rhetoric in the Renaissance," in *Renaissance-Rhetorik,* ed. Heinrich F. Plett (Berlin: Walter de Gruyter, 1993), 130. Classical word portraits created by *illustratio* (illustration) traditionally employed vivid description for praising or blaming a person, place, or thing. While most of Gumilla's descriptions are laudatory (enumerating virtues), even the shortcomings of the peoples, plants, and animals in the Orinoco River region are displayed in a way that ultimately elevates New Granada. Gumilla's rhetorical painting enhances the prestige of the Orinoco River region with praise that depends heavily on amplification.

32. The strategies Gumilla employs for "vigorous ocular demonstration" (*sub oculos subiectio*) relate to those that were first attributed to Cicero. See [Cicero], *Ad C. Herennium, De ratione dicendi (Rhetorica ad Herennium),* trans. Harry Caplan (Cambridge, MA: Harvard University Press, 1954), 4.55.69. As Quintilian put it: "But as to the figure which, as Cicero says, sets things before the eyes, it is used, when a thing is not simply mentioned . . . in a general way, but in all its attendant circumstances. This figure I have noticed in the preceding book under *evidentia,* or 'illustration,' . . . by others it is called *hypotyposis,* which means a representation of things so fully expressed in words that it seems to be seen rather than heard" (*Institutes of Oratory* 9.2.40, pp. 163–64).

33. Both Jesuit and more general humanist guidebooks connect affective hypotyposis with Christian persuasion in sacred rhetorics, *ars praedicandi,* and treatises directed at Christian painters. Father Juan Interián de Ayala's late seventeenth-century treatise *El pintor cristiano y erudito ó tratado de los que suelen cometerse frecuentemente en pintar y esculpir las imágenes sagradas* (Barcelona: Imprenta de la Viuda é Hijos de J. Subirana, 1883) was one of various guidebooks for Christian painters. Another key treatise whose strategies overlap with classical rhetoric is Francesco Pacheco's guide to painting, *El arte de la pintura* (1649). Also, as Kart Josef Höltgen notes in his article in *The Jesuits: Cultures, Sciences, and the Arts, 1540–1773,* ed. John W. O'Malley (Toronto: University of Toronto Press, 1999), the French Jesuits Nicolas Caussin (*Rhétorique des peintures*) and Etienne Binet (*Essay des merveilles de la nature et des plus noble artifices (piece tres necessaire, à tous ceux qui font profession d'Eloquence)* provided a "standard encyclopedia on the marvels of nature and art for orators and preachers" (609). See also Marc Fumaroli, "The Fertility and Shortcomings of Renaissance Rhetoric: The Jesuit Case," in *The Jesuits: Cultures, Sciences, and the Arts, 1540–1773,* ed. John W. O'Malley (Toronto: University of Toronto Press, 1999), 90–106.

34. Since the mid-sixteenth century, a recurrent formula connects emotional power with the vivid representation of an excellent object. See Shugar, "Sacred Rhetoric in the Renaissance," where Shugar translates Friar Luis de Granada and notes virtually the same phraseology in other ars praedicandi by Valades, Carbo, Alsted, and Keckermann: "[E]motions are quickened (as philosophers say) both by the excellence of the objects and by placing them vividly before the eyes of the audience" (134). Shugar also quotes Keckermann's discussion of *magnitudo* (the excellence of an object) and *praesentia* (the effect of bringing something close by painting it before someone's eyes): "As philosophers say, the conjunction of *magnitudo* and *praesentia*—what the rhetorics call *hypotyposis*—elicits emotion by making the excellent but remote object present to the senses and imagination" (136). See also Perla Chinchilla Pawling, *De la compositio loci a la república de letras: Predicación jesuita en el siglo XVII novohispano* (Mexico City: Universidad Iberoamericana, 2004) for an analysis of preaching strategies and rhetorical sources for the Jesuit "rhetoric of the passions."

35. Jesuit visual strategies for persuasion are not based solely on Loyola's *Spiritual Exercises.* They are firmly rooted in a long rhetorical tradition that reaches back to the ancient Greeks and Romans. See Margaret Ewalt, "Father Gumilla, Crocodile Hunter? The Function of Wonder in *El Orinoco ilustrado,*" in *El saber de los jesuitas, historias naturales y el Nuevo Mundo,* ed. Luis Millones Figueroa and Domingo Ledezma (Madrid: Iberoamericana, 2005), 310–18, esp. the section "Classic and Jesuit Literary Strategies for Persuasion," for a comprehensive treatment of Gumilla's rhetorical strategies and sources.

36. In addition to Kart Josef Höltgen, see studies by Gauvin Alexander Bailey, by Clara Bargellini, and by Nicolas Standaert in *The Jesuits: Cultures, Sciences, and the Arts, 1540–1773,* ed. John W. O'Malley (Toronto: University of Toronto Press, 1999), 38–89, 680–98, 352–63. As we will see in later chapters of this book, Gumilla's vivid description, in addition to stirring the emotions, also enhances his scientific argumentation. For example, when reporting on investigations of natural phenomena, the Jesuit's persuasion depends as much on the emotions generated as on appeals to his readers' reason. In fact, Gumilla's rhetoric of wonder performs three main objectives through vivid description: exposition, amplification, and argumentation.

37. Joy Kenseth's comments on Pacheco's treatise on painting are revealing in light of Gumilla's purposes: "The representation of a world that in its beauty and its grandeur transcended the ordinary and that, above all, gave an exalted view of an exalted subject was singled out especially as the noblest end of art. . . . '[W]hen it is practiced as a Christian work, it acquires another more noble form and by this means advances to the high-

est order of virtue.' " *The Age of the Marvelous* (Hanover, NH: Hood Museum of Art, 1991), 50.

38. This traditional image of God as the artisan with a paintbrush served as an inevitable trope in Christian sermons. Here is an excerpt from a sermon preached in 1725 that uses similar language to Gumilla's prologue: "La Sabiduria Divina . . . Entreteniase en tirar lineas, y sombras, como el Pintor, que antes de poner la mano, y el Pincel en la perfeccion de la Imagen, dibuxa primero en el borrón sus primores. Lleva entonces hogada la mano, y el Pincel, porque aun no ha llegado á las sombras, colores, y matizes, para dar el ultimo perfil á la Pintura" [Divine Wisdom . . . amuses himself by tracing figures, and shadows, like the Painter, who before putting his hand, and the paintbrush to the perfection of the Image, first draws its exquisiteness in the first lines. Then he brings the hand, and the Paintbrush, because he still has not arrived at the shadows, colors, and shadings, to make the finishing touches on the Painting.] (Qtd. in María del Pilar Davila Fernandez, *Los sermones y el arte* (Valladolid: Publicaciones del Departamento de Historia del Arte, 1980), 231. Gumilla's promise to "add more brightness to the colors" (*El Orinoco ilustrado,* 33) echoes Friar Luis de Granada's prescriptions in his ecclesiastical rhetoric for describing objects "no sumaria y lijeramente, sino por extenso y con todos sus colores, de modo que poniéndolo delante de los ojos del que lo oye ó lo lee" [not summarily and lightly, but rather extensively and with all its colors, in a way that puts it before the eyes of the person who hears or reads it] (*Los seis libros,* 538).

39. Translation mine from the Real Academia de la Lengua, *Diccionario de la lengua castellana (1726–39)* (Madrid: Espasa Calpe, 1992), 1114: "pintar o representar un suceso histórico o fabuloso en cuadros, estampas o tapices."

40. All handbooks of rhetoric underline the importance of closing with a peroration that amplifies by returning to the *captatio benavolentiae* used in the exordium. For example, Cicero *Topica* 26:98 (published with *De inventione,* trans. H. M. Hubbell [Cambridge, MA: Harvard University Press, 1949], 459) says: "The peroration among other topics makes especial use of amplification."

41. *El Orinoco ilustrado* demonstrates Gumilla's sermonic skills. He preaches to the eyes of his readers with conversion narratives that include preaching to the eyes of Amerindians.

42. The *pluma* (pen) that Gumilla first mentions in his prologue and that he still evokes at the end of his text may also be translated as a feather, one whose plume and tip would be useful for both the broad brushstrokes and the fine-point detailing necessary in painting. In fact, the diversity of Gumilla's display of flora and fauna both mundane and marvelous underlines his belief that every detail of Creation deserves his brushstrokes. As the Roman Spaniard Quintilian affirmed in the first century after Christ's death: "Sometimes the picture, which we endeavour to exhibit, is made to consist of several particulars" (*Institutes of Oratory* 8.3.66, p. 102).

43. The primary illustrations in Gumilla's natural history are rhetorical. However, in addition to the map of the region he drew, Gumilla included a few actual illustrations in his *El Orinoco ilustrado* (namely, of a manatee, an Amerindian drum, and a trumpet).

44. See Suárez, *De arte rhetorica,* 342 for suggested uses for *sermocinatio* and prosopopoeia.

45. Translation mine from Cristóbol Colón, *Los cuatro viajes. Testamento,* ed. Consuelo Varela (Madrid: Alianza, 1986), 79. Like Colón (Columbus), Gumilla appeals to his readers' senses and offers up New World wonders to the king. According to Sebastian Münster's 1550 *Cosmography,* King Ferdinand commented that Columbus, the *Almirante* (Admiral) should be called instead the *Admirans* (Wonderer). Key studies of Columbus's wonder-provoking rhetorical strategies include Greenblatt, *Marvelous Possessions;* Mary Campbell, *The Witness and the Other World* (Ithaca, NY: Cornell University Press, 1988);

and Elvira Vilches, "Columbus's Gift: Representations of Grace and Wealth and the Enterprise of the Indies," *MLN* 119, no. 2 (2004): 201–25.

46. During his third voyage, Columbus seemed convinced that the earthly paradise could be found at the end of the Orinoco River: "Torno a mi propósito de la tierra de Gracia y río [Orinoco] y lago que allí fallé. [. . .] Mas yo muy assentado tengo el ánima que allí, adonde dixe, es el Paraíso Terrenal, y descanso sobre las razones y auctoridades sobre escriptas" (Colón, *Los cuatro viajes,* 245). [I go back to my aim of the Land of Grace and river {Orinoco} and lake that I found there. . . .But I have deeply rooted in my soul that there, where I said, is the Earthly Paradise, and I am supported by reasoning and authorities over writings.]

47. The following quotes from Columbus's first journey (October 21) to the New World clarify some of the the tropical tropes that Gumilla evokes: "[E]l arboledo en maravilla, y aquí y en toda la isla son todos verdes y las yervas como en el Abril en el Andaluzía y el cantar de los paxaritos que pareçe qu'el hombre nunca se querría partir de aquí, y las manadas de los papagayos que ascureçen el sol, y aves y paxaritos de tantas maneras y tan diversas de las nuestras que es maravilla. Y después ha árboles de mill maneras y todos [dan] de su manera fruto, y todos güelen qu'es maravilla, que yo estoy el más penado del mundo de no los cognosçer, porque soy bien cierto que todos son cosa de valía y d'ellos traigo la demuestra, y asimismo de las yervas." Colón, *Los cuatro viajes,* 77). [The woodland {is} a wonder, and here and in all the island the trees and groves of trees are all green and the grasses like April in Andalucia and the singing of the little birds such that man never would want to leave here, and the flock of parrots such that they darken the sun, and big and little birds of so many sorts and so diverse from ours that it is a wonder. And then there are trees of a thousand kinds and all {give} in their own way fruit, and they all smell such that it is a wonder, and I am the most pained man in the world not to recognize them, because I am quite certain that they are all things of great worth and from them I bring signs, and the same for the herbs.] Columbus repeats this discourse of the marvelous in his October 23 entry, which describes the land as: "muy provechosa de espeçería, mas que yo no la cognozco, que llevo la mayor pena del mundo, que veo mill maneras de árboles que tienen cada uno su manera de fruta y verde agora como en España en el mes de Mayo y Junio y mill maneras de yervas, eso mesmo con flores; y de todo no se cognosció salvo este liñáloe de que oy mandé también traer a la nao para levar a Vuestras Altezas" (*Los cuatro viajes,* 79) [very beneficial with spices, but I do not recognize them, for which I carry the greatest pain in the world, that I see a thousand kinds of trees that have each one their manner of fruit and green now like in Spain in the month of May and June and a thousand kinds of herbs, the same thing with the flowers; and about all this nothing is known except for this linoleum that today I also sent to be taken on the boat to bring to Your Highnesses]. Throughout this section of his *Diario,* Columbus increases the wonder by degrees (an incrementum to climax, or *gradation,* which is mounting by degrees, or, the ladder effect), each description more marvelous than the last, a strategy also effectively employed by Gumilla.

48. Translation mine from Colón, *Los cuatro viajes,* 77.

49. In his prologue Gumilla also comments directly and humbly on his own clear, "rustic" style, downplaying his command of rhetoric. He excuses his own "pronunciación bárbara" by noting the many years spent among men with limited abilities in Spanish: "En el estilo sólo tiraré a darme a entender con la mayor claridad que pueda; y no será poca dicha si lo consiguiere; porque acostumbrado largos años a la pronunciación bárbara, a la colocación y cláusulas de los lenguajes ásperos de aquellos indios, será casualidad si corriere mi narración sin tropiezo, ya en la frase, ya en la propiedad de las palabras" (*El Orinoco ilustrado,* 31). [As for the style I will only try to make myself understood with the greatest clarity that I can; and it will be no small thing if I achieve it; because accustomed

so many years to the barbarous pronunciation, or the positioning and clauses of the course languages of those Indians, it will be by accident if my narration flows without tripping up, either in the phrasing, or in the propriety of the words.]

50. As Debora Shugar explains, "[V]ividness . . . brings the excellent object 'near' the beholder, making it available not only to knowledge but also emotion. . . . Hence the sacred rhetorics instruct the preacher to use vivid description so that the hearer 'seems to see for himself. . . .' The Greeks call this *hypotyposis*, . . . likewise energia or vividness. . . . This has a wonderful power in moving the emotions" ("Sacred Rhetoric in the Renaissance," 136). See also Luis de Granada, *Los seis libros*, bk. 3, chap. 11, "De los afectos en particular: Del amor de Dios" (549), and his treatment of *admiración* as a figure of style used to stir the emotions (553).

51. When a topic is amplified rhetorically, this often creates a rhetorical abundance. Gumilla, apart from adhering to the typical discourse of wonderfully abundant New World natural resources (for example, in the chapter "Cosecha admirable de tortugas" [Bumper crop of tortoises], where we find the repeated insistence of the verb, noun, and adjective forms of "abundant": *abundar, abundancia,* and *abundante*), follows rhetorical prescriptions to use synonyms, comparisons, parallel constructions, parentheses, divisions that break down points into separate parts, poetic figures such as enumeration, or figures that aid enumeration, like polysyndeton. As noted earlier, the rhythm of either suspended syntax or *circuitus* (round composition), typical of the periodic sentence, also helps create abundance. Gumilla takes advantage of this throughout *El Orinoco ilustrado.*

Chapter 2. Colonization, Commerce

1. As the Bourbon monarchs saw it, "[N]atural history, experimental philosophy, astronomy, and cartography would help the crown to exploit botanical and mineral resources and to regain control over loosely controlled frontiers and borderlands." Jorge Cañizares-Esguerra, "Spanish America: From Baroque to Modern Colonial Science," in *Cambridge History of Science* vol. 4: *Eighteenth-Century Science,* ed. Roy Porter (Cambridge: Cambridge University Press, 2003), 730. For details on several key expeditions, see the classic study by Iris H. W. Engstrand, *Spanish Scientists in the New World: The Eighteenth-Century Expeditions* (Seattle: University of Washington Press, 1981).

2. A few notable exceptions include Michael T. Bravo, "Mission Gardens: Natural History and Global Expansion, 1720–1820," in *Colonial Botany: Science, Commerce, and Politics in the Early Modern World,* ed. Londa Schiebinger and Claudia Swan (Philadelphia: University of Pennsylvania Press, 2005), 49–65; Antonio Barrera, "Local Herbs, Global Medicines: Commerce, Knowledge, and Commodities in Spanish America," in *Merchants and Marvels:Commerce, Science, and Art in Early Modern Europe,* ed. Pamela Smith and Paula Findlen (New York: Routledge, 2002), 163–81; and James E. McClellan, "Missionary Naturalists," in *Colonialism and Science: Saint Domingue in the Old Regime* (Baltimore: Johns Hopkins University Press, 1992), 111–16.

3. For Humboldt's debts to Spanish Americans, see Jorge Cañizares-Esguerra, "How Derivative Was Humboldt?" in *Colonial Botany: Science, Commerce, and Politics in the Early Modern World,* ed. Londa Schiebinger and Claudia Swan (Philadelphia: University of Pennsylvania Press, 2005), 148–65. For the postponement of Spanish American Enlightenment and a summary of the liberal and conservative strains of Enlightenment theory, see the editors' introduction and the contribution by Olegario Negrín Fajardo in Diana Soto Arango, Miguel Ángel Puig Samper, and Luis Carlos Arboleda, eds., *La Ilustración en América Colonial* (Madrid: Daces Calles, 1995).

4. There is a long tradition of close cooperation between clergy and Crown in early modern Spain. As Charles C. Noel writes, "It had helped unite the Spanish kingdoms, build her American empire, dominate much of Western Europe for over a century. It had helped most Spaniards achieve their identity: To be Spanish came to mean being orthodox in religion and, most of the time, reasonably loyal to the Catholic monarch. . . . [U]ntil 1808 most clerics accepted the royal church because the crown, under the aegis of pious monarchs and religiously orthodox ministers, never came close to threatening the clergy's most vital interest—the Catholic monopolization of Spaniards' spiritual life." "Clerics and Crown in Bourbon Spain, 1700–1808," in *Religion and Politics in Enlightenment Europe,* ed. James E. Bradley and Dale K. Van Kley (Notre Dame, IN: University of Notre Dame Press, 2001): 146–47.

5. "Spanish officials also viewed the War of Jenkins' Ear, fought in part for the interests of the South Sea Company, as a ploy to increase English contraband in Spanish America." Lance Grahn, *The Political Economy of Smuggling: Regional Informal Economies in Early Bourbon New Granada* (Boulder, CO: Westview Press, 1997), 123.

6. For details on the 1732 Guayana treaty between Jesuits and Capuchins and Gumilla's direct role in this, see works by José del Rey Fajardo, including *Misiones Jesuíticas en la Orinoquia (1625–1767),* vol. 1; *Una utopía sofocada;* and Joseph Gumilla, *P. José Gumilla: Escritos varios,* ed. José del Rey Fajardo (Caracas: Fuentes para la Historia Colonial de Venezuela, 1970). The 1734 Guayana agreement included the Franciscans in this polemic. The 1736 Caracas treaty established the demarcations for these competing orders' territories that are labeled on Gumilla's map.

7. Grahn, *Political Economy of Smuggling,* 3.

8. The South Sea Company had been granted exclusive trading rights earlier in the century. This representative, James Houstoun, also "declared that illegal business in the province was commonly done with their [the Jesuits'] help" (ibid., 134).

9. Ibid., 136.

10. Ibid., 135. In fact, it wasn't until 1749 that the new king "Ferdinand VI empowered civil authorities to inspect ecclesiastical buildings suspected of containing illegal goods" (ibid., 136). When pointing out Capuchin failures, Gumilla suggests their complicity with illegal trade: "[P]or más que se han esforzado y trabajado los reverendos Padres Capuchinos de la provincia de Aragón en su ministerio apostólico, todavía hay naciones de gentiles en aquellas costas, que gustan más de la amistad y trato con los extranjeros: punto digno de la atención y reparo que requiere" (Joseph Gumilla, *El Orinoco ilustrado y defenido* ed. Demetrio Ramos [Caracas: Fuetes para la Historia Colonial de Venezuela, 1963], 47. Translation my own. ["Despite how much the reverend Capuchin fathers from the Aragón province have made an effort and worked in their apostolic ministry, there are still pagan nations on those coasts, who prefer the friendship and dealings with foreigners: a point worthy of the attention and criticism that it requires.]

11. See, for example, Demetrio Ramos' introduction to *El Orinoco ilustrado* by Joseph Gumilla, ed., Demetrio Ramos (Caracas: Fuentes para la Historia Colonial de Venezuela, 1963); idem, "La geografía de los modos de vida del valle venezolano y el jesuita valenciano P. Gumilla," *SAITABI* 6, nos. 29–30 (July–December 1948): 242–51; idem, "Las ideas geográficas del Padre Gumilla." *Estudios geográficos* 5.14 (1944): 179–199; "Un mapa inédito del Río Orinoco es el precedente del de Gumilla y el más antiguo de los conocidos," *Revista de Indias,* January–March 1944, 89–104; and idem, "Un plan de inmigración y libre comercio defendido por Gumilla para Guayana en 1739," *Anuario de Estudios Americanos* 15 (1958): 201–24. See also José del Rey Fajardo's introduction to *P. José Gumilla: Escritos varios;* and Manuel Alberto Donís Ríos, "José Gumilla S.J.: Impulsor del cambio cartográfico ocurrido en Guayana a partir de 1731;" *Boletín de la Academia Nacional de la Historia,* January–March 1986, 157–76.

12. At first glance, the cartographic map seems identical in the 1741, 1745, and 1791 editions. However, the 1791 edition does not name "Paulus Minguet" as the etcher next to "P.J.G. Delineavit" [Drawn by Father Joseph Gumilla]. Also, the differences in shading, fancier stars in the first edition to denote *presidios* (garrisons) and the additional mountains drawn into the second and third maps suggest that the three maps were printed from different copper plates.

13. "la primera aportación seria al conocimiento de aquella región mítica" (Ramos, "Un mapa inédito," 90) [the first serious contribution to knowledge of that mythical region.]

14. As the twentieth-century scholar Demetrio Ramos observed, since the early eighteenth century this valuable map and its Jesuit author have been of inestimable worth: "La explicación de la postura de Gumilla está, pues, en su base erudita, en su caudal de lecturas y conocimientos, que le convierten en uno de los hombres más meritorios de nuestro siglo XVIII, sobre todo en lo referente al Orinoco. Este prestigio fue reconocido por todos los escritores de la época, y no sólo influye poderosamente [. . .] en las ideas geográficas del Tratado de límites de 1750, sino que el mismo Humboldt, como antes Caulín, caminan sobre lo que dejó hecho el abnegado misionero" ("Las ideas geográficas," 198). ["The explanation for the position of Gumilla is, then, in his erudite foundation, in his abundance of readings and knowledge that convert him into one of the most meritorious men of our eighteenth century, above all regarding the Orinoco. This prestige was recognized by all the writers of the time, and he not only powerfully influenced . . . the geographical ideas of the Boundary Treaty of 1750, but also Humboldt himself, as previously Caulín; they walked in the footsteps of what the selfless missionary had done."]

15. Although M. Eidous corrects Gumilla's geographical mistake in a prefatory "avertissement," the translated text maintains it.

16. The discussion on Orinoco waterfalls is repeated in several of Humboldt's works, including the *Tableaux de la nature ou, considerations sur les desert, sur la physionomie des végétaux, et sur les cataracts de l'Orenoque* (Paris: F. Schoell, 1808); its translation, *Aspects of Nature, in Different Lands and Different Climates; with Scientific Elucidations* (Philadelphia: Lea and Blanchard, 1850); *Views of Nature, or Contemplation of the Sublime Phenomena of Creation* (London: H. G. Bohn, 1850); and the very famous work, *Relation historique du Voyage aux régions équinoxiales du Nouveau Continent,* 3 vols. (Stuttgart: F. A. Brockhaus, 1970), or *Personal Narrative of a Journey to the Equinoctial Regions of the New Continent,* trans. Jason Wilson (London: Penguin Books, 1995).

17. La Condamine, for his part, mentioned a conversation between his expedition's chief mathematician and Gumilla that took place in Cartagena de Indies before the first edition of *El Orinoco ilustrado* appeared. It revealed the Jesuit to be convinced of the connection between the Orinoco and Amazon rivers (Ramos, "Las ideas geográficas," 194). Román was also influenced by Gumilla to make his journey (Ramos, "Las ideas geográficas," 199).

18. Gumilla cites them all in the footnotes to *El Orinoco ilustrado.* See Gumilla, *P. José Gumilla: Escritos varios,* xciii–xciv for a discussion of Fritz's map of the Amazon (1707), which continued a long tradition of maps by highly esteemed cartographers who imagined the mountains separating the rivers including Blaeu, Laet, and de Bry. See also Ramos, "Las ideas geográficas," 186–87.

19. In most cases, the indices send readers to five chapters at the end of part 1 of *El Orinoco ilustrado* (namely chapters 20, 21, 22, 24, and 25).

20. All these marginalia pages are from the original 1745 edition of Joseph Gumilla, *El Orinoco ilustrado, y defendido.* 2nd ed. (Madrid: Manuel Fernadez, 1745).

21. As the title of this first index—"Índice de las cosas más notables que se contienen en este libro" [Index of the most noteworthy things that this book contains]—suggests, it will highlight the book's encyclopedic contents and direct readers to sections of the text

that reference geographical sites within and beyond New Granada, as well as names of European humanist–era scholars, mapmakers, Spanish conquistadors, past and present Jesuit historians, missionaries, and heroes. Finally, this index alphabetizes the names of Amerindian nations along the Orinoco and their curious customs, as well as their notable conversions and the singlularly heroic martyrdom of new Christians ("neófitos"). All quotes from the indices are from Joseph Gumilla, *El Orinoco ilustrado, y defendido.* 2nd ed. (Madrid: Manuel Fernadea, 1745).

22. For example, one index entry, "Canela silvestre en Quijos, y Macas, Orinoco, y Filipinas" [Wild cinnamon in Quijos, and Macos, Orinoco, and Philippines], sends readers to a textual claim that Orinoco cinnamon rivals the finest of the East. Here, after a description of the beauty and fragrance of his Jesuit province's wild cinnamon trees, Gumilla says enthusiastically: "[L]a tal canela no es de otra ni de inferior especie que la del Oriente, en donde también parte de los árboles aromáticos son silvestres, como dicen Guillermo y Juan Bleau" (Gumilla, *El Orinoco ilustrado,* 1963, 248) [This cinnamon is not a different, inferior species from that of the East, where also some of the aromatic trees are wild, as William and John Bleau say]. Another expensive spice Gumilla details is vanilla, which grows wild in the region but has been successfully domesticated: "entre los cuales no es de menor importancia aquella fruta o especie aromática que vulgarmente se llama vainilla; ésta, de su propia naturaleza y condición, es silvestre (si bien ya se ha hallado modo fácil y método al propósito para cultivarla)" (1963, 250) [among which is not less important that fruit or aromatic spice that is commonly called vanilla; by its own nature and condition, it is wild (though already an easy method for cultivating it has been found)].

23. As in "Corre el gran río Orinoco, como ya dije y se ve en el plan" (Gumilla, *El Orinoco ilustrado,* 247) [The great River Orinoco runs, as I already said and is seen in the map] or "como ya dije y demuestra el plan" (249) [as I already said and the map shows].

24. This second index (n.p., 1745) crafts a pharmacopoeia highlighting forty-nine botanical specimens investigated by Jesuits, such as the "Bejuca de playa, [que] es contra veneno de culebras" [Liana from the beach, which is used against snake venom], "Caña agria, útil para muchos remedios" [Sour cane, useful for many remedies], and "Piñones americanos, [que] son purga eficaz" [American pine seeds, which are an effective purgative], as well as cures that come from animal products such as the "Hueso de la cabeza de manatí, sirve contra flujo de sangre" [Bone from the manatee head, which serves against the flow of blood], "Hueso último de la cola de armadillo, es contra el dolor de oídos," [Last bone of the armadillo's tail, which works against earaches], "Piedra de la iguana, es contra mal de orina" and [Iguana stone, it works against urinary illness]. As we saw above, cornucopia index entries privilege highly valued known products that can be found in abundance in New Granada such as "Añil, nace de suyo en aquellos terrenos," [Indigo, which grows on its own in those lands], "Caña dulce abunda en aquel país" [Sweet cane abounds in that country], "Sal, hecha de raíces de polipodio" [Salt, made from the roots of polypodium], or "Vainilla, es fruta aromática silvestre" [Vanilla, which a wild aromatic fruit]. However, that longer index does include new species and remedies, as revealed in entries like "Peces muy diferentes de los de Europa" [Fish very different from those of Europe], which leads readers to chapters that boast "imponderable" fish harvests and their comestible and medicinal values alongside "innumerable" manatees, the "marine cows" first mistaken for mermaids and then much celebrated as creatures specific to the New World. See chapter 21, "Variedad de peces y singulares industrias de los indios para pescar. Piedras y huesos medicinales que se han descubierto en algunos pescados" [Variety of fish and extraordinary industriousness of the Indians for fishing. Medicinal stones and bones that have been discovered in some fish].

25. A second look at the start and close of the alphabetical pharmacopoeia suggests a variant to the Virgin's oil: a cancer-curing herb that reappears on the page referred to as a soothing balm: "Amargosa ò yerva de Santa María, es contra el cancer, pag. 296" [Tea or herb of Holy Mary, it is against cancer, page 296] and "Yerva de Santa Maria, es para llagas, p. 296" [Herb of Holy Mary, it is for sores, page 296].

26. Gumilla serves as an important precursor to the José Celestino Mutis expedition and Francisco José de Caldas's proclamations about the geography of New Granada providing the perfect location for an international trade emporium. On Mutis and Caldas spreading the news of the "fantastic economic potential of New Granada," see Cañizares-Esguerra, "How Derivative Was Humboldt?" 159–61.

27. For example, at least ten entries lead to part 2, chap. 21, "Arboles frutales que cultivan los indios. Hierbas y raíces medicinales que brota aquel terreno" [Fruit trees that the Indians cultivate. Medicinal herbs and roots that that land sprouts].

28. For more on this, see Margaret Ewalt, "Crossing Over: Nations and Naturalists in *El Orinoco ilustrado;* Reading and Writing the Book of Orinoco Secrets," *Dieciocho* 29, no. 1 (Spring 2006): 1–25.

29. Gumilla frames the defense of Jesuit missions along both the Amazon and Orinoco rivers and their tributaries as a religious and patriotic duty, thus reaffirming the ties between Spanish and Jesuit agendas. "Y como fiel y leal vasallo de nuestro invicto y católico monarca Felipe V, a quien Dios guarde y prospere para el bien de su monarquía y de la universal Iglesia Católica, debo añadir que, de no ponerse remedio, dando eficaz providencia para reprimir el empeño con que los portugueses del río Marañón hasta las riberas de Orinoco, empezaron a molestar y cautivar a los indios de ellas, desde el año de 1737, en que estaba yo en el Orinoco [. . .] del mismo modo dañarán (como se ve dañan hoy) e imposibilitarán las Misiones que mi Provincia del Nuevo Renio [mantiene] con tanto afán y costo, así de vidas de sus misioneros, como de caudales, que en tan apostólica empresa ha gastado y gasta, y quedarán frustrados los piadosos deseos de nuestro piadoso Monarca y de mi apostólica Provincia" (*El Orinoco ilustrado,* 251). [And as a faithful and loyal vassal of our unconquered and Catholic monarch Philip V, whom God protects and who prospers for the good of his monarchy and of the universal Catholic Church, I should add that, unless something is done about this, giving effective ruling to repress the incursions of the Portuguese from the Amazon River to the banks of the Orinoco, they will again start to bother and capture the Indians from these, as they have since the year of 1737, when I was in the Orinoco . . . in the same way they will harm (as they are seen harming today) and they will make impossible the missions that my province of the New Kingdom maintains with such zeal and cost, both in terms of lives of missionaries and of wealth spent in such an apostolic venture, and the pious desires of our pious monarch and of my apostolic province will remain frustrated.]

30. Gumilla blames Dutch thirst for natural remedies as one reason why they have helped the Caribs attack Jesuit missions. Here we recall his treatment of the Virgin's oil, which reveals links between political, economic, and religious agendas: "La codicia que tienen los holandeses de comprar estos aceites de mano de los caribes es la causa principal de su amistad y de los daños que han padecido y padecen nuestras Misiones. Y el anhelo con que lo buscan los extranjeros es prueba eficaz de las grandes virtudes que en dicho aceite han reconocido" (*El Orinoco ilustrado,* 219). [The extreme desire that the Dutch have to buy these oils from the Carib Indians is the primary cause of their friendship as well as the anguish that our missions have suffered and currently suffer. This thirst with which foreigners seek it out is an effective proof of the great virtues that have been recognized in this oil.] This discourse seeks protection of Jesuit missions and trade and, by extension, Spanish colonization and commerce.

31. For example, Gumilla appeals to fellow Jesuits as well as Spanish politicians to deploy missionaries and military troops while also underscoring less-cultivated products

such as Orinoco indigo, which sprouts up in the wild and is just waiting to be cultivated: "Por lo que mira al añil, brota en aquel terreno [. . .] y ya se ve cuánto diera, y con qué abundancia, sembrado y cultivado" [For what we see/one sees about indigo is it sprouts up in that land . . . and it is already seen how much it would yield, and with what abundance, planted and cultivated]. Just as the "salsafrás, tan apreciable [. . .] se halla con abundancia en los contornos de la boca del río Caura en Orinoco, donde sin buscarlo, se ha encontrado" (249) [sassafras, so precious . . . is found with abundance on the edges of the mouth of the Caura River on the Orinoco, where without looking for it, it has been found], so many wild yet valuable souls await conversion. The Jesuits have already saved countless souls, these figurative fruits of the biblical vineyards, but deploying more missionaries is essential during this period of crisis. Too many "products" deserving enlightenment and ripe for cultivation are still waiting.

32. Marginalia pagination from 1745 original, p. 368.

33. These quotes from the three subtitles to the final chapter in part 1 are found in the 1745 original *El Orinoco ilustrado, y defendido.*

34. Gumilla wrote several deliberative letters petitioning religious and political leaders during the 1730s. See Gumilla, *P. José Gumilla: Escritos varios.* In *El Orinoco ilustrado* he ties royal troop support to a peaceful future and thriving Jesuit evangelism: "Sea su divina Majestad toda la gloria. Ahora, con las especiales providencias que se ha dignado dar el católico celo de nuestro invicto monarca Felipe V, a quien Dios prospere, cometiendo sus especiales órdenes e instrucciones a don Gregorio Espinosa de los Monteros, coronel de los reales ejércitos, gobernador y capitán general de las provincias de Cumaná y la Guayana, jefe de la reputación, destreza militar y valor que sabe España, tenemos fija esperanza que amanecerá la tranquilidad en el Orinoco, y con ella los progresos en la cultura especial de aquellos retirados gentiles y la resulta de copioso fruto para el cielo" (334–35). [May His Divine Majesty be in all his glory. Now, with the special rulings that the Catholic zeal of our unconquered monarch Philip V has deigned to give, may God let him prosper, committing his special orders and instructions to Mr. Gregorio Espinosa de los Monteros, colonel of the royal armies, governor and chief of the provinces of Cumaná and Guayana, chief of the reputation, military skills, and valor that Spain knows, we have fixed hopes that peace will dawn in the Orinoco, and with it progress in the special culture of those remote pagans and the consequence of copious fruit for heaven.]

35. Gumilla underlined the obvious Dutch support of the Caribs, who fought with "tanta presteza y arte militar que causaron admiración y [. . .] después lo supimos de cierto que iban con los caribes algunos herejes embijados y disimulados" (*El Orinoco ilustrado,* 330) [such promptness and military art that they inspired wonder and . . . afterward we found out for sure that some heretics went with the Caribs painted up and concealed]. The Jesuit had already tried to move the governor to action, threatening direct complaints to the king if no troop support were offered: "[E]n el año 1733 me quejé agria, aunque modestamente, al gobernador de Esquivo, con una larga carta, en que le conté los daños que padecían nuestras Misiones, y que de no poner remedio su señoría, daría cuenta a mi católico monarca, para que su Majestad se querellase a las altipotencias de Holanda" (328) [In the year 1733 I complained bitterly, although modestly, to the governor of Esquivo, with a long letter, in which I told him the damages that our missions suffered, and that if his lordship did nothing about it, my Catholic Monarch would find out about it, so that His Majesty could complain to the higher powers of Holland].

36. When Gumilla does discuss the Jesuit missions destroyed by the Caribs in the past, however, there is a strong contrast with the fates of the Franciscans and Capuchins. Instead, he emphasizes the Jesuits' heroic rebuilding of missions, including faith-fortifying details such as the holy cross that God did not allow the "barbarous Carib Indians" to burn. Details such as this, alongside narrations of missionaries and soldiers protected

by intercession of Saint Francis Javier, whose image graced the standards raised in combat, serve as propaganda to privilege the Jesuit order and to recruit new missionaries.

37. Gumilla excuses the length and gory details of these scenes thus: "Como aún es reciente el dolor, se me fue la pluma refiriendo este trabajo [. . .] pero sirva ahora de muestra o regla para medir y entender los muchos asaltos, ardides y estratagemas con que casi siete años continuos han perseguido los caribes a sangre y fuego a aquellas Misiones y otras del mismo Orinoco, procurando de todos modos desterrar el nombre de cristiano de sus riberas [. . .]" (*El Orinoco ilustrado,* 330). [Since the pain is still recent, my pen just went on and on relating this report . . . but it now serves as proof or a ruler to measure and understand the many assaults, schemes, and tricks with which for almost seven continuous years the Caribs have persecuted with blood and fire those missions and others on the same Orinoco, trying in all ways to banish the name of Christianity from its banks. . .]

38. Gumilla supported existing Jesuit defense plans for a fortress on Fajardo Island. Careful not to upset his Jesuit superiors, he carefully proposed the Limón Island stronghold as a less expensive and perhaps more effective fort. See Ramos, "Un plan de inmigración," 201–24.

39. See Gumilla, *P. José Gumilla: Escritos varios,* 55–69 for the full text of this 1739 letter, written in Madrid: "Informe, que hace a su Majestad, en su Real, y Supremo Consejo de las Indias, el Padre Joseph Gumilla, de la Compañia de Jesús, missionero de las Missiones de Casanare, Meta, y Orinoco, Superior de dichas missiones, y Procurador General de la Provincia del Nuevo Reyno en esta Corte, sobre impedir a los indios Caribes, y a los olandeses las hostilidades, que experimentan las colonias del gran Río Orinoco, y los medios mas oportunos para este fin." [Report made to his majesty, in his Royal, and Supreme Council of the Indies, by Father Joseph Gumilla of the Company of Jesus, Missionary of the Missions of the Casanare, Meta, and Orinoco, Superior of said missions, and general solicitor of the province of New Granada in this court, on how to impede the hostilities of the Carib Indians, and the Dutch, that the colonies of the great Orinoco River are experiencing, and the most opportune means to achieve this end.]

40. For more on Lara's success and debts to Gumilla, as well as the political reasons Gumilla chose not to include the defense maps, see Donís Rios, "José Gumilla S.J.," 157–76.

41. Gumilla paints a fertile region promising prosperity to colonizing Spaniards along the Orinoco River and its branches: "[T]erminan dichas llanuras en los dilatados márgenes del río Orinoco. Las vegas de éste y de los ríos que reciben pudieran dar abrigo a muchas y grandes villas y lugares de españoles, y sus fértiles ejidos y campañas rasas dieran paso abundante a innumerables cabañas y hatos de ganado. Todo está pronto, todo convida al cultivo, y por todas partes ofrece el país larga correspondencia en ricos y abundantes frutos" (*El Orinoco ilustrado,* 249–50). [Said plains end along the extensive margins of the Orinoco River. The fertile lowlands of this and of the receiving rivers could give shelter to many and great towns and cities of Spaniards, and their fertile common lands and field patches could pave the way for abundant and countless livestock and herds of cattle. Everything is ready, everything invites cultivation, and in all parts the country offers great returns in rich and abundant fruits.]

42. For more on this plan, see Ramos, "Un plan de inmigración," 201–24.

43. As Gumilla puts it, "[E]l índice más cierto, y que más evidencia la riqueza de cualquier reino es su comercio, de modo, que por lo pingüe o débil del comercio, se conoce claramente el mayor o menor fondo de cualquier reino, sea el que se fuere" (*El Orinoco ilustrado,* 257) [The most certain index, and the one that most evidences the wealth of any kingdom, is its commerce; by the substance or weakness of its commerce is clearly known the greater or lesser wealth of any kingdom, whichever it is].

44. Gumilla quotes the Jesuit Juan de Urtassum's *Intereses de Inglaterra mal entendidos en la guerra presente con España* three times. It is unclear whether the Jesuits knew that this supposed translation from a member of Parliament was a rhetorical device first used before the Treaty of Utrecht by a French cleric, Jean Baptiste Du Bus. Du Bus invented an "anonymous" British parliament member and claimed to be merely translating his *England's interest mistaken in the present war.* In any case, all parties involved used this economic pamphlet as propaganda during more than one "present war" with England.

45. One contemporary definition that is useful for clarifying Gumilla's understanding of the term "contraband" as "all trade prohibited by law and ruled by the monarch to be contrary to the public good" was laid out by Pedro González de Salcedo in his *Tratado jurídico político del contra-bando* (1729) (qtd. in Grahn, *Political Economy of Smuggling,* 14). For Gumilla, nothing about the Jesuits' evangelical mission was contrary to the public good.

46. Also, when revising for the 1745 version this "Apostrofe," Gumilla added five key words that were not in its 1741 opening appeal (*captatio benavolentiae*) to restrict this sermon's agricultural and fishing allegory to "copiosas y abundantes cosechas *de almas para el cielo*" (*El Orinoco ilustrado,* 495; emphasis mine) [copious and abundant crops *of souls for heaven*], distancing somewhat the religious from political discourse by reinforcing the missionary's role as soul-harvester and fisher of men instead of conduit to commercial prosperity.

47. For a recent celebration of Jesuit missionary service in New Granada, see the articles and images of the exposition catalogue, *Desde Roma por Sevilla al Nuevo Reino de Granada: La Compañía de Jesús en tiempos coloniales,* ed. María Constanza Toquita Clavijo (Bogotá: Museo de Arte Colonial, 2004).

Chapter 3. An Eclectic Enlightenment

1. The diffusion-phase model constitutes only one manifestation of this time-honored story about the birth of modern science. See George Basalla, "The Spread of Western Science Revisited," in *Mundialización de la ciencia y cultura nacional,* ed. Antonio Lafuente, Alberto Elena, and María L. Ortega (Madrid: Doce Calles, 1993), 599–618; David W. Chambers, "Locality and Science: Myths of Centre and Periphery," in *Mundialización de la ciencia y cultura nacional,* ed. Antonio Lafuente, Alberto Elena, and María L. Ortega (Madrid: Des Calles, 1993), 605–19; and Roy MacLeod, "Introduction. Nature and Empire: Science and the Colonial Enterprise," *Osiris* 15 (2001): 1–13.

2. Peter Dear, "The Cultural History of Science: An Overview with Reflections," *Science, Technology, and Human Values* 20 (1995): 150–70; idem, *Revolutionizing the Sciences: European Knowledge and Its Ambitions: 1500–1800* (Princeton, NJ: Princeton University Press, 2001); Bruno Latour, *Laboratory Life: The Social Construction of Scientific Facts* (Beverly Hills, CA: Sage Publications, 1979); idem, "Give Me a Laboratory and I Will Raise the World," in *The Science Studies Reader,* ed. Mario Biagioli (New York: Routledge, 1999), 258–75; Simon Schaffer, "Natural Philosophy," in *The Ferment of Knowledge: Studies in the Historiography of Eighteenth-Century Science,* ed. Roy Porter and George Sebastian Rousseau (Cambridge: Cambridge University Press, 1980), 55–91; and idem, "Natural Philosophy and Public Spectacle in the Eighteenth Century," *History of Science* 21 (1983): 1–43. See also Steven Shapin, *The Scientific Revolution* (Chicago: University of Chicago Press, 1996); and idem, "Social Uses of Science," in *The Ferment of Knowledge: Studies in the Historiography of Eighteenth-Century Science,* ed. Roy Porter and George Sebastian Rousseau (Cambridge: Cambridge University Press, 1980), 93–139.

3. Lorraine Daston, "Afterword: The Ethos of Enlightenment," in *The Sciences in Enlightened Europe,* ed. William Clark, Jan Golinski, and Simon Schaffer (Chicago: University of Chicago Press, 1999), 495–504; idem, "Enlightenment Fears, Fears of Enlightenment," in *What's Left of Enlightenment? A Postmodern Question,* ed. Keith Michael Baker and Peter Hanns Reill (Stanford, CA: Stanford Univesity Press, 2001), 115–28; idem, "The Ideal and Reality of the Republic of Letters in the Enlightenment," *Science in Context* 4 (1991): 367–86; Dorinda Outram, *The Enlightenment;* (Cambridge: Cambridge University Press, 1995), idem, "The Enlightenment Our Contemporary," in *The Sciences in Enlightened Europe,* ed. William Clark, Jan Golinski, and Simon Schaffer (Chicago: University of Chicago Press, 1999), 32–40; and Roy Porter, introduction to *The Ferment of Knowledge: Studies in the Historiography of Eighteenth-Century Science,* ed. Roy Porter and George Sebastian Rousseau (Cambridge: Cambridge University Press, 1980), 1–10. See also Barbara Maria Stafford, *Artful Science: Enlightenment Entertainment and the Eclipse of Visual Education* (Cambridge, MA: MIT Press, 1994).

4. See Walter Mignolo, *Local Histories/Global Designs: Coloniality, Subaltern Knowledges, and Border Thinking* (Princeton, NJ: Princeton University Press, 2000) for his discussion of two phases of modernity (this one would be the transition to the second phase). See also Walter Mignolo, introduction to *Natural and Moral History of the Indies,* ed. Jane E. Mangan, trans. Frances M. López-Morillas (Durham, NC: Duke University Press, 2002), xvii–xxviii, idem, "Commentary: José de Acosta's Historia natural y moral de las Indias: Occidentalism, The Modern/Colonial World, and The Colonial Difference," Mangan, *Natural and Moral History,* 451–518. Both phases of modernity were propelled by the exploration of the New World. The first phase is the humanistic Renaissance period when Spain and Portugal dominated. During the second phase England and France defeat Spain and Germany gains prominence. In Mignolo's narrative, the new philosophies of science break with, instead of mediate, Christian paradigms and culminate in the displacement of God from the Enlightenment.

5. Juan Pimental, "The Iberian Vision: Science and Empire in the Framework of a Universal Monarchy, 1500–1800," *Osiris* 15 (2001): 18.

6. "Not surprisingly, all meta-narratives of modernity and progress that came of age in the eighteenth century have found no place for the technological and philosophical contributions of the Iberians in the early modern period." (Jorge Cañizares-Esguerra, "Iberian Science in the Renasiannce: Ignored How Much Longer?" *Perspectives on Science* 12, no. 1 (2004): 96. Here he refers to the North Atlantic narratives of modernity (109). See also David Freedburg, *The Eye of the Lynx: Galileo, His Friends and the Beginnings of Modern Natural History* (Chicago: University of Chicago Press, 2002); and Latour, "Give Me a Laboratory."

7. Mignolo, introduction, 497.

8. On this see David T. Gies, "Dos preguntas regeneracionistas: '¿Qué se debe a España?' y '¿Qué es España,' Identidad nacional en Forner, Moratín, Jovellanos y la generación de 1898," *Dieciocho* 22, no. 2 (1999): 307–30.

9. Pimental, "Iberian Vision," 18.

10. Mignolo discusses the "overarching imaginary of the modern/colonial world system" and locates the production of knowledge in Europe. This Eurocentrism, which he describes as Occidentalism, "is the visible face in the building of the modern world" (*Local Histories/Global Designs,* 20). He also notes how an "imperial epistemic difference" generating "internal borders" within the "modern world system" allowed for the "absence of Spain from the scientific revolution of the seventeenth century," the period historians have traditionally marked as when southern Europe fell behind the scientific and philosophical production of northern Europe (Mignolo, "Commentary," 480).

11. Ralph Bauer, *The Cultural Geography of Colonial American Literatures,* (Cambridge: Cambridge University Press, 2003), 2. Bauer discusses the East/West, center/periphery,

and the North Atlantic / South American contrasts in his comparison of Spanish and British texts from the sixteenth through eighteenth centuries. Bauer's reminder that knowledge is never unilateral and that diverse cultures and locations affect each other underscores Spain and Spanish America's "important role in the making of the very culture of modernity by which it has been subsequently marginalized" (3). My exploration of how Jesuit intercultural mediation affected the Hispanic Enlightenment is in line with his thesis about "how various places and histories are connected and act upon each other in new cultural formations" (3).

12. See Jean Sarraihl, *L'Espagne éclairée de la seconde moitié du XVIII siècle* (Paris: Impr. Nacional, 1954); and Jonathan Israel, *Radical Enlightenment: Philosophy and the Making of Modernity, 1650–1750* (New York: Oxford University Press, 2001). See also G. H. W. Vanpaemal, "Jesuit Science in the Spanish Netherlands," in *Jesuit Science and the Republic of Letters*, ed. Mordechai Feingold (Cambridge, MA: MIT Press, 2003), 389–432, who reminds us that "it is now recognized that within the Catholic Church the Jesuits formed an 'intellectual and religious elite' " (390).

13. Vanpaemal, "Jesuit Science," 389. See also Peter Dear, "Jesuit Mathematical Science and the Reconstitution of Experience in the Early Seventeenth Century," *Studies in the History and Philosophy of Science* 18 (1987): 133–75. The European Enlightenment owes much to the Jesuits, whose reconciliations of the ancients and the moderns encouraged scientific advances that might otherwise never have occurred. As Francisco Sánchez-Blanco affirmed in *La mentalidad ilustrada* (Madrid: Taurus, 1999), Jesuit expulsion from Spanish territories did slow scientific progress. See also Sánchez-Blanco, El *absolutismo y las luces en el reinado de Carlos III* (Madrid: M. Pons, 2002). Still, the Society of Jesus constitutes a prime example of how, despite "religious impediments," neither Spain nor Spanish America was cut off from modern European scientific ideas.

14. Most scholarship on Jesuits continues to focus on scientific contributions after their 1767 expulsion from all Spanish territories, when a surge of natural histories detailing Spanish America were published from exile. For this reason, Jesuits have been granted a perhaps disproportionate role in the late eighteenth-century and early nineteenth-century formation of "proto-national" identities and discourses in Spanish America.

15. This became the slogan most often printed on Jesuit standards/ensigns. See Luis Millones Figueroa and Domingo Ledezma, "Introducción: Los jesuitas y el conocimiento de la naturaleza americana," in *El saber de los jesuitas, historias naturales y el Nuevo Mundo*, ed. Milliones Figueroa and Ledesma (Madrid: Iberoamericana, 2005), 9; and Luis Millones Figueroa, "La *intelligentsia* Jesuita," in *El saber de los jesuitas, historias naturales y el el Nuevo Mundo*, ed. Milliones Figueroa and Ledezma (Madrid: Iberoamericana, 2005), 36.

16. Harris, "Jesuit Scientific Activity in the Overseas Missions, 1570–1773," *ISIS* 96 (2005): 76. Harris surveys the Society of Jesus's global reach but focuses mostly on Spain and her possessions: "At its peak around 1750, the Society operated more than 500 colleges and universities in Europe, a hundred more in overseas colonies (mostly in Spanish America), and roughly 270 mission stations scattered around the globe" (72). See this entire issue of *ISIS*, the premiere interdisciplinary journal published by the History of Science Society. See also Mordechai Feingold, ed., *Jesuit Science and the Republic of Letters* (Cambridge, MA: MIT Press, 2003); and John W. O'Malley, *The Jesuits II: Cultures Sciences and the Arts, 1540–1773* (Toronto: University of Toronto Press, 2005).

17. Harris, "Jesuit Scientific Activity," 79.

18. See, for example, recent prominent journal issues dedicated to nature, science, and empire, such as *Osiris* 15 (2001), *ISIS* 96 (March 2005), and *CLAR* (June 2006), and anthologies: *Merchants and Marvels: Commerce, Science, and Art in Early Modern Europe*, ed. Pamela Smith and Paula Findlen (New York: Routledge, 2002); *Colonial Botany: Science, Commerce, and Politics in the Early Modern World*, ed. Londa Schiebinger and Claudia Swan

(Philadelphia: University of Pennsylvania Press, 2005); and the *Cambridge History of Science, vol. 4: Eighteenth-Century Science,* ed. Roy Porter (Cambridge: Cambridge University Press, 2003).

19. See Israel, *Radical Enlightenment,* 528–40 for an excellent study of Spanish Baconianism based on analysis of several primary texts. As Israel concludes, "[I]f there was one part of continental Europe of which it can be justly said that English empiricist ideas almost completely ousted every other competing variety of Enlightenment, that part was the Iberian Peninsula" (528).

20. As Rebecca Haidt has noted, "[S]keptics such as Feijoo held that human understanding is indeed limited, but inquiry and experimentation are licit and useful." Haidt, *Embodying Enlightenment: Knowing the Body in Eighteenth-Century Spanish Literature and Culture* (New York: St. Martin's Press, 1998), 44. Jeremy Robbins succinctly summarizes the renovated conceptions of *utilitas* in the early eighteenth century and Feijoo's role in linking it with skepticism and supporting "the twin Baconian methods of experimentation and empirical observation" laid out in Bacon's *Novum Organum.* (Robbins, "From Baroque to Pre-Enlightenment: Resolving the Epistemological Crisis," *Bulletin of Spanish Studies* 87, no. 8 (2005): 232. In contrast to baroque notions of skepticism, which warned of the deceits of the senses and deemed reliance on them problematic to the Catholic faith, Feijoo's eighteenth-century skepticism accepted sense-based data "as a means of gaining knowledge about reality" (Robbins, "From Baroque," 247). See also Robbins, "From Baroque," 223–27 and 250–53.

21. Feijoo's articles in his *Teatro crítico* and *Cartas eruditas y curiosas* frequently cite and praise the Jesuits. Bacon openly acknowledged his debt to them, viewing Jesuits as models for the advancement of knowledge. As he wrote about Jesuit influences in his *Advancement of Learning,* "[P]artly in themselves and partly by the emulation and the provocation of their example, they have much quickened and strengthened the state of learning." Qtd. in J. L. Heilbron, "The Physicists: Jesuits," in *Elements of Early Modern Physics* (Berkeley and Los Angeles: University of California Press, 1982), 99. And he envied Jesuit education: "[C]onsult the schools of the Jesuits, nothing in use is better. . . . If only they were ours" (qtd. in Heilbron, "Physicists," 94).

22. Román de la Campa describes the persistent conception of modernity in Spanish America as a "failed, expendable tradition." De la Campa, *Latin Americanism* (Minneapolis: University of Minnesota Press, 1999), ix. Through Enrique Dussel's concept of "transmodernity," however, we might incorporate solidarity between center and periphery, Western culture and Third World cultures, and so on. See Enrique Dussel, "Eurocentrism and Modernity (Introduction to the Frankfort Lectures)," *Boundary 2* 20, no. 3 (Fall 1993): 65–76. See also idem, *The Invention of the Americas: Eclipse of "the Other" and the Myth of Modernity,* trans. Michael D. Barber (New York: Continuum, 1995), cited extensively in Mignolo's *Local Histories/Global Designs* to support the view of Spanish America as the "first periphery" of modern Europe as well as the post-Enlightenment concept of modernity occluding Europe's own Iberian periphery (especially Spain). Mignolo also discusses ways Homi Bhaba's "Countermodernity" takes into account diverse sites of knowledge production "from the margins."

23. Amos Funkenstein, *Theology and the Scientific Imagination from the Middle Ages to the Seventeenth Century* (Princeton, NJ: Princeton University Press, 1986), 49. See also Ruth Hill, *Sceptres and Sciences in the Spains: Four Humanists and the New Philosophy (ca. 1680–1740)* (Liverpool: Liverpool Unviersity Press, 2000), 1–7. For the roots of Spanish empiricism and eclectic philosophy in the Middle Ages, see José Antonio Maravall, "Empirismo y pensamiento político (Una cuestión de origenes)," in *Estudios de historia del pensamiento español* (Madrid: Ediciones cultura hispánica, 1984), 15–38.

24. The expansion of New World exploration played a key role in these challenges to available knowledge. New knowledge about previously unknown flora, fauna, and nat-

ural phenomena as well as Amerindian heuristics opened Europe to new paradigms. The intersection of European and American knowledge encouraged reciprocal exchanges, as with the Amerindian "border epistemology" that adapted Renaissance principles to their epistemic frame. See Mignolo, "Commentary"; and Denise Albanese, *New Science, New World* (Durham, NC: Duke University Press, 1996). See also Millones Figueroa and Ledezma, "Introducción," on the "necesidad de crear o replantear nuevos sistemas de comprensión" (17) [necessity of creating or reopening new systems of comprehension] brought on by New World geographies (such as the Andes). As they point out, Nieremberg saw the New World as offering the "missing piece" for symbolic cosmovision.

25. Christian philosophers have always employed eclectic solutions, as when they first mediated pagan and Christian ideas. See Maravall, "Empirismo y pensamiento político"; and Funkenstein, *Theology and the Scientific Imagination.*

26. The Jesuits were not only participants, but also the secondary school and university educators of the elite. As J. L. Heilbron notes: "The powerful Jesuit educational system, its celebrated pedagogues and part-time researchers, made the Society the leading patron of physical and mathematical sciences during the seventeenth century" ("Physicists," 99).

27. Víctor Navarro, "Tradition and Scientific Change in Early Modern Spain: The Role of the Jesuits," in *Jesuit Science and the Republic of Letters,* ed. Mordechai Feingold (Cambridge, MA: MIT Press, 2003), 332.

28. These "modern" ideas even showed up in poetry. For example, see chapter 2 of Hill, *Sceptres and Sciences in the Spains.* On the topic of academic societies, see also Israel, *Radical Enlightenment,* 529; and Jesús Pérez-Magallón, *Construyendo la modernidad: La cultura española en el tiempe de los novatares* (Madrid: Consejo Superior de Investigaciones Cienificas, 2002), 69–77.

29. See, for example, José Luis Abellán-García, "El triunfo de la renovación científica: Andrés Piquer," in *Historia crítica del pensamiento español* (Madrid: Espasa-Calpe, 1981), 449–61.

30. See Francisco Puy Múñoz, "El problema del conocimiento en el pensamiento del siglo XVIII (1700–1760)," *Anales de la Cátedra Francisco Suárez* 1, no. 2 (1961): 191–226; and Francisco Puy Múñoz, "La comprensión de la moralidad en el pensamiento español del siglo XVIII (1700–1760)," *Anales de la Cátedra Francisco Suárez* 2, no. 1 (1962): 87–118. Valencia, Seville, and Madrid are cited as the key sites of "the first stirrings of the Spanish Enlightenment, that is, the initial assault on scholasticism and Galenist medicine, [which] began only in the 1680s and 1690s" (Israel, *Radical Enlightenment,* 528).

31. Andrés Piquer, *Fisica moderna, racional, y experimental,* 2nd ed. (Madrid: Joachin Ibarra, 1780 [1745]), 2.

32. Ibid.

33. Andrés Piquer, *Discurso sobre la aplicacion de la Philosophia a los assuntos de religion para la juventud española* (Madrid: Joachin Ibarra, 1757), 30.

34. Piquer, *Fisica moderna,* 8.

35. "En estos ultimos siglos se han descubierto algunas operaciones maravillosas de la Naturaleza, ocultas à los tiempos passados; y tal vez con la diligencia y aplicacion à examinar, y descubrir sus obras por el camino de la observacion, y experienca, se llegaràn en lo venidero à entender muchas cosas, que hoy no alcanzamos" (Piquer, *Discurso,* 121). [In these recent centuries some marvelous operations of nature have been discovered, hidden to earlier times; and perhaps with the diligence and application to examine and discover its works through the pathway of observation and experiments, many things that today we do not understand will come to be understood in the future.]

36. "Para el jesuita francés Noël Regnault no hay nada nuevo, ni hay ruptura: todo está de alguna manera en Aristóteles o en los antiguos. Titula una de sus obras *L'origine ancienne de la physique nouvelle* [For the French Jesuit Noël Regnault there is nothing new,

nor any rupture: all is in some way in Aristotle or in the ancients. He entitles one of his works *The ancient origin of the new physics* (Paris 1734), Francisco Sánchez-Blanco, *Europa y el pensamiento español del siglo XVIII* (Madrid: Alianza, 1991), 103. Two histories of science that did emphasize the modern character of eclecticism are Olga Victoria Quiroz-Martínez, *La introducción de la filosofía moderna en España (El eclecticismo español de los siglos XVII y XVIII)* (Mexico City: Fondo de Cultura Económica, 1949); and María del Carmen Rovira, *Eclécticos portugueses del siglo XVIII y algunas de sus influencias en América* (Mexico City: El Colegio de México, 1958).

37. As Sánchez-Blanco affirmed: "Los eclécticos se contentan con escoger entre las opiniones aceptadas en la Antigüedad y con introducir nuevos conocimientos, también de orden experimental, allí donde no provoquen conflictos" [The eclectics contented themselves with choosing between the accepted opinions in Antiquity and with introducing new knowledge, also of an experimental order, only when it did not provoke conflicts] (*Europa y el pensamiento español,* 108).

38. See Maravall, "Empirismo y pensamiento político," 15–38.

39. "Aprobacion del M.R.P. Carlos de la Reguera, de la Compañia de Jesus, de la Academìa Española, y Cathedratico de Mathematicas en el Colegio Imperial de esta Corte." Antonio María Herrero, *Physica moderna, experimental, systematica donde se contiene lo mas curioso, y util de quanto se ha descubierto en la Naturaleza* (Madrid: 1738), n.p.

40. Piquer, *Discurso,* 30.

41. This skeptical line of reasoning was also put forth in Diego Matheo Diego Zapata's *Ocaso de las formas aristotelicas* (1745). Zapata's idea that the church fathers were divinely inspired only in sacred matters, and that thus "there is no 'blind deference' to their views beyond theological issues," coexisted in Spanish America (quoted in Israel, *Radical Enlightenment,* 533). Caracciolo Parra cites contemporary natural philosophers across the Atantic in his *La instrucción en Caracas (1567–1725). Filosofía universitaria venezolana. Cronistas de Venezuela* (Madrid: Editorial J.B., 1954) as affirming: "En lo natural no ha de seguirse la autoridad, sino la experiencia y la razón" and "la autoridad de los Santos no constituye argumento cierto en las ciencias naturales: tanto vale cuanto persuada la razón" [In the natural it is not necessary to follow authority, but rather experience and reason" and "the authority of the church fathers does not constitute a certain argument in the natural sciences: only that which reason persuades is worthwhile.] (328).

42. Piquer, *Física moderna,* 26.

43. See Piquer, *Discurso,* 14–15. As we will see in the second part of this chapter, Gumilla follows an identical path of reasoning.

44. This rhetorical strategy did not, however, prevent criticism from his mentor, Mayans, as well as from philosophers who still preferred Aristotelian philosophy. See Vicente Catalayud, *Cartas eruditas por la preferencia de la Philosophia aristotélica para los estudios de religión* (Valencia, 1758–60).

45. Piquer, *Discurso,* 121.

46. "[N]o hemos visto hasta ahora la Física experimental de los Modernos escrita en lengua comun, ni con la extension necesaria para instruirse en ella" (Piquer, *Física moderna,* n.p.). [We have not seen up to now the experimental physics of the moderns written in common language, nor with the necessary length or range to be able to train oneself in it.] Here the Valencian doctor categorizes the "moderns" as either systematic (Descartes, Gassendi, and Newton) or experimental (Bacon, Boyle, and Boerhaave). Piquer's survey of "the moderns" also mentions Tosca, Feijoo, Malebranche, Muratori, and Regnault.

47. Herrero, *Physica moderna,* 3. Here the rejection of Aristotle is explicit: "los sentimientos de Aristoteles, que en esta Obra se desprecian [. . .]" [the opinions of Aristotle, which are rejected in this Work . . .].

48. Piquer, *Discurso,* 56

49. "[N]o solo deba ser Theologo, sino tambien Physico. [. . .] La Physica Experimental puede aprovechar mucho en este caso, porque por ella se puede averiguar quáles sean las leyes que guardan los cuerpos dentro de la natural en sus movimientos" (Piquer, *Discurso,* 121). [Not only should one be a theologian, but also a physicist. . . . Experimental physics can benefit a lot in this case, because by it one can investigate which are the laws that bodies observe within their natural movements.]

50. Piquer, *Física moderna,* 4.

51. Piquer, *Discurso,* 51. On this page Piquer also cites Muratori's "defense of the great St. Augustine."

52. Piquer, *Fisica moderna,* 4, 48.

53. Cabriada's *Carta philosophica* (1686), cited in Israel, *Radical Enlightenment,* 530.

54. Israel, *Radical Englightenment,* 536. "As the Spanish and Portuguese Enlightenment matured in the 1730s, 1740s, and 1750s, no leading figure contested or doubted that Bacon, Boyle, Locke, and Newton provided the best intellectual basis for a viable moderate Enlightenment" (Israel, *Radical Enlightenment,* 537). For more on Spanish Baconianism, see Bauer, *Cultural Geography.*

55. See Clelia Pighetti, *Atomi e Lumi nel Mondo Spagnolo* (Milan: Franco Angeli, 1993), 62–63. Qtd. in Hill, *Sceptres and Sciences,* 6.

56. Translation mine from Ludovico Antonio Muratori, *Delle reflessioni sopra il buon gusto nelle Scienze e nell'Arti* (Venezia: Niccolò Pezzana, 1766 [1708]), 31: "[E] dell'accrescimento delle Scienze quel chiarissimo Filosofo Francesco Bacone da Verulameio, le cui Opere sono state, e saran sempre un Seminario di ottime Leggi per raggiugnere l'ottimo Gusto."

57. This phrase is from Muratori, who enumerates all the major scientific societies of the time (London, Paris, St. Petersburg, Rome, Florence, Bologna) in his *Fuerza de la humana fantasia,* trans. Vicente María de Tercilla (Madrid: D. Manuel Martin, 1777), which was originally published as *Della forza della fantasia umana* [*Force of Human Imagination*] in 1740. As Muratori puts it: "Todos se persuaden que con su imaginario systema han dado con la verdad. Mas que esto no es asi, puede inferirse de tantas guerras literarias como permanecen en las Escuelas. [. . .] No obstante, concluimos, que ningun systema, ninguna opinion puede conducirnos á la certeza de la verdad" [Everyone persuades themselves that with their imaginative system they have come upon the truth. But this is not so, as can be inferred by the many literary wars that remain in the schools. However, we conclude that no system, no opinion can direct us to the certainty of truth]. (*Fuerza,* 283–85). See also 286–87.

58. Piquer, *Fisica moderna,* 3. The entire quote reads: "Así estuvo la Física mal cultivada por muchos siglos, hasta que Francisco Bacon de Verulamio, Gran Canciller de Inglaterra, ácia el fin del siglo décimosexto empezó á renovarla, librándola de la superfluidad de los razonamientos, y manifestando, que el verdadero modo de adelantarla era por el camino de la experiencia" (Piquer, *Fisica moderna,* 3) [So Physics was poorly cultivated for many centuries, until Francis Bacon Verulam, Great Chancellor of England, toward the end of the sixteenth century started to renovate it, freeing it of the superfluousness of reasoning and showing that the true means for advancing it was by the pathway of experience].

59. Feijoo's influence reached across the ocean to Spanish America and even as far as Spanish territories in the Philippines. Israel cites a contemporary of Feijoo in the viceroyalty of New Spain as observing already in 1734 "that Feijóo's volumes had fundamentally transformed thinking not only in Spain itself but in the viceroyalties [and] . . . the distant Philippines" (Arturo Ardao, *La filosofía polémica de Feijóo* (1962): 18, qtd. in Israel, *Radical Enlightenment,* 534). By the mid-eighteenth century, notes Israel, Feijoo's brand

of Enlightenment was officially endorsed by the state and Catholic Church: "On 23 June 1750 the new King of Spain, Ferdinand VI (ruled 1746–59) issued an edict proclaiming that the works of Feijóo, and those containing similar concepts and views, had the full approval of the Crown and that all attacks on them in print were to cease forthwith" (Israel, *Radical Enlightenment,* 540).

60. Piquer, *Física moderna,* 5.

61. Benito Jerónimo Feijoo, *Cartas eruditas y curiosas* (Madrid: Imprenta real de la Gazetta, 1773), Letter 23 (1745), vol. 2, p. 286.

62. Feijoo, *Teatro crítico,* 2:15: "[C]uanto de un siglo a esta parte se adelantó en la física, todo se debe al canciller Bacon. Éste rompió las estrechas márgenes en que hasta su tiempo estuvo aprisionada la filosofía. Éste derribó las columnas que con la inscripción *Non plus ultra* habían fijado tantos siglos ha la cienca de las cosas naturales" [However much physics has advanced in a century to this moment, all is because of Chancellor Bacon. This man broke the tight magins within which philosophy was imprisoned up to his day. This man knocked down the columns that with the inscription *non plus ultra* had determined for so many centuries the knowledge of natural things.] (Qtd. in Russell P. Sebold, "Colón, Bacon y la metáfora heroica de Feijoo," in *Homenaje a don Agapito Rey,* ed. Joseph Roca-Pons (Bloomington: Indiana University Press, 1980), 349. For a discussion of this quotation and the "heroic metaphor" for Bacon's "discoveries" see Sebold, "Colón, Bacon," 333–54.

63. Feijoo expresses "el desengaño de Sistema" [the disenchantment with systems] with the rhetorical figure *contentio* (a series of antitheses) culminating in an exaggerated *exclamatio:* "[L]os Sistemas . . . no hicieron otra cosa, que tomar sueños por realidades, sombras por luces, ilusiones por aciertos, parhelias por Soles. Si lo que dieron a especulaciones vagas, dieran a observaciones experimentales, ¡Oh! ¡qué Gazofilacio tan opulento de Física hubieran dejado a la posteridad, en vez de los inútiles harapos que hemos heredado de ellos!" [The systems did nothing other than take dreams as realities, shadows as lights, illusions as wise ideas, mock suns as real suns. If only the systems granted to experimental observations that which they grant to unreliable speculations, Oh! Such an opulent gazophylacium {treasure chest} of physics would have been left for posterity, instead of the useless rags that we have inherited from the systems.] Feijoo, *Cartas eruditas y curiosas,* Letter 23, (1745), vol. 2, 284.

64. Ibid., 285. According to Feijoo, Bacon's "new method" was revealing the flaws in imaginative systems like Descartes': "del nuevo método, iban al mismo paso haciendo perder el gusto de los Sistemas, ayudando a ello no poco el descubrimiento de algunos considerables defectos en el más plausible de todos; esto es en el Cartesiano [. . .]" [From the new method they are going along the same path to losing their taste for the systems, this helped in no small degree by the discovery of some considerable defects in the most plausible of all the systems; that is in Cartesianism.] (ibid., 287).

65. "fijando la atención en los efectos, para colegir de ellos, en cuanto se pudiese, las causas inmediatas. Este proyecto, formado entre varios Sabios de una, y otra Nación, ocasionó el origen de las dos célebres Academias, la Real de las Ciencias en París, y la Sociedad Regia en Londres" [focusing attention on the effects in order to infer from them in as much as possible the immediate causes. This project formed among various wisemen of one and another nation caused the origin of the two celebrated academies, the Royal Academy of Sciences in Paris and the Royal Society in London] (ibid., 286).

66. Feijoo, the "new Spanish Bacon," also extended his praise to two other British Christian natural philosophers—Boyle and, eventually, Newton. See translations from *Teatro crítico universal,* 7:332–33 in Israel, *Radical Enlightenment,* 535.

67. Ramon Ceñal, "Fuentes jesuíticas francesas en la erudición filosófica del padre Feijoo," in *El Padre Feijoo y su siglo* (Oviedo: Universidad de Oviedo, 1966), 285–314. See

also idem, "Feijoo y la filosofía de su tiempo," *Pensamiento: Revista trimestral de investigación e información filosófica* 21 (1965): 251–72.

68. See Ivy Lillian McClelland, "The Significance of Feijoo's Regard for Francis Bacon," *Studium ovetense* 4 (1976): 250. McClelland compares Feijoo's *Teatro crítico* not only with the Jesuits' *Mémoires de Trévoux* but also with the premiere British periodical of the times, Addison's *Spectator.*

69. Joseph Gumilla, *El Orinoco ilustrado y defendido,* ed. Demetrio Ramos (Caracas: Fuentes para la Historia Colonial de Venezuela, 1963), 381. Unless otherwise indicated, all references in the text and notes to *El Orinoco ilustrado,* are to this edition. All translations are my own.

70. See José del Rey Fajardo, "Denis Mesland, introductor del cartesianismo en América?" *Latinoamérica; Revista mensual de cultura* 10 (1958): 102–4 for a case study of Jesuit mediation of Cartesian philosophy. Gumilla attended the Javeriana University in Bogotá well after the establishment of "la flexibilidad ecléctica que imprime el sistema jesuítico" (104) [the eclectic flexibility that imprints the jesuit system].

71. With *Historia natural y moral de las Indias,* which in Acosta's words "se podrá tener esta Historia por nueva, por ser juntamente historia y en parte filosofía," [this History will be taken as new, for being both history and philosophy] Acosta invented a new Jesuit model for combining "los hechos y costumbres de hombres" [the deeds and customs of man] with "las obras de naturaleza" [the works of nature] for the glory of God and the advancement of evangelism. Joseph de Acosta, *Historia natural y moral de las Indias,* ed. Edmundo O'Gorman (Mexico City: Fondo de Cultura Economíca, 1940/1962), 13–14.

72. Gregorio Mayans y Siscar, *Pensamientos literarios,* 248–49, quoted in Hill, *Sceptres and Sciences,* 29–30 n. 35.

73. Piquer, *Física moderna,* 53.

74. Piquer, *Discurso,* 28–29.

75. "For by the grace given to me I say to everyone among you not to think of yourself more highly than you ought to think, but to think with sober judgment, each according to the measure of faith that God has assigned" (Rom. 12:3 [Revised Standard Version]).

76. Ludovico Antonio Muratori, *Il cristianesimo felice nelle missioni de' padri della Compagnia di Gesù nel Paraguai,* (Venice: Giambatista Pasquali, 1743) united various letters from Jesuit missionaries and travelers, furnished to Muratori by an Italian Jesuit priest. It was soon translated into Spanish, French, German, and English.

77. Muratori, *Fuerza,* 288. Piquer cited Muratori's defense of St. Augustine: "Muratori en defensa del Grande Augustino" (*Discurso,* 51).

78. See, for example Gumilla, *El Orinoco ilustrado,* 388. For my understanding of Nieremberg, I am greatly indebted to Domingo Ledezma, "Una legitimación imaginativa del Nuevo Mundo: La *Historia Naturae, Maxime Peregrinae* de Juan Eusebio Nieremberg," in *El saber de los jesuitas, historias naturales y el Nuevo Mundo,* ed. Luis Millones Figueroa and Domingo Ledezma (Madrid: Iberoamericana, 2005), 53–83.

79. For example, in book. 4, chap. 1 of *Los seis libros,* Luis de Granada suggests the use of allegory to declare "la admirable fuerza y eficacia de la divina gracia que por él se nos concede. Porque estas cosas exactamente expuestas y amplificadas, arrebatan maravillosamente los entendimientos humanos á la admiracion de cosas tan grandes, é inflaman poderosamente el amor a la divina bondad, benignidad, caridad y misericordia" [the wonderful force and effectiveness of divine grace which by {allegory} is conceded to us. Because these things exactly set forth and amplified marvelously capture human understanding to wonder at great things, and they powerfully inflame love for divine goodness, benevolence, charity, and forgiveness] (555). Luis de Granada, *Los seis libros de la Retórica eclesiástica o la manera de Predicar* (Madrid: Ediciones Atlas, 1945).

80. Mimosa pudica, or "the sensitive plant." See Michael Allaby, *The Concise Oxford Dictionary of Botany* (Oxford: Oxford University Press, 1992), 258.

81. This quote is reminiscent of an eclectic handbook of natural philosophy published in 1738 by Antonio María Herrero and mentioned earlier in these notes. Herrero also invites his readers to look at the soil and observe natural spectacles too numerous, too diverse, and too marvelous to paint in words: "Si bolvemos los ojos á la Tierra, no hallará nuestra curiosidad menor delicia [. . .] en fin, son tantos, tan diversos, y tan prodigios los expectaculos, que nuestra atencion advierte en la naturaleza, que faltan las expresiones para pintarlos, quando sobra el assombro para advertirlos" (*Physica moderna,* 4) ["If we return our eyes to the earth, our curiosity will not find less delight . . . in short, the spectacles that our attention notices in nature, are so many, so diverse, and so wonderful, but we lack expressions to paint them, when the astonishment abounds for noticing them].

82. Gumilla first suggests that "gentle plants" are actually young women when he introduces the "Gran lección para el recato en todas las mujeres, especialmente para las tiernas plantas" (*El Orinoco ilustrado,* 444) [Great lesson for modesty in all women, especially for the gentle plants]. This metaphor is made more explicit in the exclamation with which he closes the section: "¡Oh, y qué enseñanza para las tiernas bellezas, que salen a ser vistas, a que las miren y remiren!" (445) [Oh, and what a lesson for the gentle beauties, who go out to be seen, so that they look at them and look again!]. For an excellent discussion of the Renaissance tradition of allegorical, emblematic nature, see Katherine Park, "Nature in Person," in *The Moral Authority of Nature,* ed. Lorraine Daston and Fernando Vidal (Chicago: University of Chicago Press, 2004), 50–73.

83. Gumilla seems to be referencing Noël Regnault, *Entriens physiques d'Ariste et d'Eudoxe, ou, Physique nouvelle en dialogues.* I consulted the *Philosophical Conversations, or a New System of Physics by way of Dialogue,* trans. Thomas Dale (London: W. Innys, C. Davis, and N. Prevost, 1731).

84. The *Oxford English Dictionary* dates use of the idea of effluvia as an "outflow of material particles too subtle to be perceived by touch or sight" back to the mid-seventeenth century. Effluvia can serve as either an attracting or repelling agent. See http://dictionary.oed.com/cgi/entry/00181778.

85. Caroline Walker-Bynum, "Wonder," *American Historical Review* 102 (1997): 10. As she explains, Aquinas Christianizes Aristotle and connects wonder with the desire for knowledge that culminates in an encounter with Majesty.

86. In John W. O'Malley, *The Jesuits: Cultures, Sciences, and the Arts, 1540–1773* (Toronto: University of Toronto Press, 1999), O'Malley notes that "the Jesuits of this period [1540–1773] liked to think they had a 'way of proceeding'—or better, ways of proceeding— special to themselves. The expression actually goes back to Ignatius of Loyola himself . . ." (xiv). Dominique Deslandres describes some "methods used to attain the missionary's goals": "On both sides of the Atlantic, the Jesuit method consisted principally of attracting people and moving them sufficiently to induce them to convert. . . . The missionaries knew how to move people, how to alternate hot and cold 'showers' of emotion with perfect precision. . . . The missionaries would bring their audiences to a climax and then, having profoundly moved them, change the tone of the discourse to one of reassurance" (*"Exemplo aeque ut verbo,"* in O'Malley, *Jesuits,* 263).

87. Gumilla (*El Orinoco ilustrado,* 462) holds up the Jesuit relation of Nicolás Trigault as a model for dispelling Chinese superstitions and enlightening the "doctos astrólogos del imperio de la China" [learned astronomers of the empire of China] who, despite their "nobles y muy cultivados ingenios, especialmente en orden a la contemplación de los astros y planetas" [noble and well-cultivated intellects, especially with regard to the contemplation of the stars and planets], formerly held the same mistaken beliefs as the

Orinoco tribes: "tan neciamente como las gentes bárbaras del Orinoco" [just as foolishly as the barbarous peoples of the Orinoco].

88. Quoted in Lorraine Daston and Katherine Park, *Wonders and the Order of Nature: 1150–1750* (New York: Zone Books, 1998), 227.

89. To my best knowledge, Russell Sebold was the first to compare Bacon's heroic metaphor, Feijoo's framing of his own efforts to share new "países intelectuales," and the efforts of the early discoverers Columbus and Vasco de Gama. See Sebold, "Colón, Bacon y la metáfora heroica de Feijoo," in *Homenaje a don Agapito Rey,* ed. Joseph Roca-Pons (Bloomington: Indiana University Press, 1980), 348–51.

90. Francis Bacon, book 1, aphorism 92, and book 2, aphorism 36, in *Novum Organum: with Other Parts of the Great Instauration,* ed. Peter Urbach and John Gibson (Chicago: Open Court, 1994), 103, 211.

91. William Eamon, *Science and the Secrets of Nature* (Princeton, NJ: Princeton University Press, 1994), 283. See also Robbins for Juan de Cabriada's use of the Bacon's heroic metaphor to warn Spaniards against falling behind in discoveries of the new sciences, "by showing that it is other Europeans who are not the intellectual conquistadores" (Robbins, "From Baroque," 227).

92. Bacon's own plans for galleries to display collections of the world's natural and man-made objects were dubbed "Solomon's House" for the biblical wisdom of King Solomon.

93. On this topic, see Paula Findlen, *Possessing Nature: Museums, Collecting, and Scientific Culture in Early Modern Italy* (Berkeley and Los Angeles: University of California Press, 1994), especially part 1 ("Locating the Museum") and part 2 ("Laboratories of Nature").

94. Ibid., 4–5.

95. Eamon, *Science and the Secrets of Nature,* 299.

96. As Paula Findlen observes, "the first science museums appeared—repositories of technology, ethnographic curiosities, and natural wonders. They emerged at a time when all of Europe seemed to be collecting" (*Possessing Nature,* 2). For more on European cabinets and collecting, see Oliver Impey and Arthur MacGregor, *The Origins of Museums* (Oxford: Oxford University Press, 1985); Adalgisa Lugli, *Naturalia et mirabilia: Il collezionismo enciclopedico nelle Wunderkammern d'Europa* (Milan: G. Mazzotta, 1983); Krzysztof Pomian, *Collectors and Curiosities: Paris and Venice, 1500–1800* (Cambridge: Polity Press, 1990); and Julius von Schlosser, *Die Kunst- und Wunderkammern der Spätrenaissance* (Leipzig: Klinkhardt & Biermann, 1908).

97. El Inca Garcilaso de la Vega's three stages were articulated in the first book of his *Comentarios reales de los incas* (1609). The first pre-Incan "barbaric" stage, second Incan "civilizing" stage, and third stage of Christianity became a template for discussing the stages of Amerindian men in America.

98. As Gumilla explains, "En fin, quede por fijo que por los mismos grados por donde blanquea la mestiza, blanquea también la mulata a la cuarta generación" (*El Orinoco ilustrado,* 86) [In short, it remains certain that the same stages through which the *mestiza* whitens, the mulatto also whitens at the fourth generation].

99. Despite this language of "superior" or "inferior," "advancement" or "regression," Gumilla actually expresses a highly optimistic view of interracial harmony. He concludes this section on racial classification with a command that celebrates the power of color-blind, Christian love: "En fin, el amor es ciego, y el ciego en punto de colores ni distingue ni tiene voto; y, caso que lo tuviera, es nulo. Miremos, pues, los colores con la indiferencia que ya dije" (*El Orinoco ilustrado,* 89). [In short, love is blind, and the blind man with regard to colors neither distinguishes them nor has a say in the matter; and, even if he

did, it is null and void. Let us examine, then, colors with the indifference that I spoke of.] As long as one's partner is a Christian, skin color should not be a factor when deciding whom to marry. His missionary experience has made him realistic about human nature. At the same time he cites Pope Clement's decree allowing missionaries to conduct "licit, valid, and equal" marriages among new Christians and accelerate their whitening process —"de modo que los cuarterones y ochavones se reputan y se deben tener por blancos" [so that the quadroons and octoroons are deemed and should be taken for whites]—he also reports his New Granada encounters with happily married women who both "advance" and "descend" along the whitening steps: "[H]oy en día, en Cartagena de Indias, en Mompox y en otras partes se hallan españoles honrados casados (por su elección libre) con negras, muy contentos y concordes con sus mujeres; y al contrario, vi en la Guayana una mulata blanca casada con un negro atezado y en los Llanos de Santiago [. . .] una mestiza blanca casada con otro negro" (89). [Nowadays, in Cartagena de Indies, in Mompox, and in other parts are found honorable Spaniards married (by their own free choice) to black women, very content and in agreement with their wives; and on the contrary, I saw in Guayana a white mulatto married to a tan black and in the Plains of Santiago . . . a white *mestiza* married with another black man.]

100. Still not understood by doctors today, this pigment disorder is referred to as vitiligo and not as nature's "plaything," "toy," or "perversity."

101. Gumilla writes of hiding in a Jesuit-run hacienda this mother and daughter, by now objects of intense curiosity with many wanting to buy the baby at any cost: "[Q]uerían comprarla sin reparar en costo" (*El Orinoco ilustrado,* 101) [They wanted to buy her without considering the cost]. One of the portraits Gumilla mentions—"[S]i bien su copia corrió por todo el Nuevo Reino y Provincia de Caracas, y aun me aseguraron que los cónsules de la Factoría inglesa habían enviado a Londres una copia muy individual de ella" (101) [Her portrait ran all around the New Kingdom and province of Caracas, and they even assured me that the consuls of the English Trading Post had sent to London a very singular portrait of her]—made its way into the 1777 volume of Buffon's *Natural History.* I thank John Wood Sweet for pointing out that the famous case of Maria Sabine, the "marvelously spotted" baby born in Cartagena whose portrait was published by Buffon, is the same *niña* (child) of the *negra casada* (married slave woman) in *El Orinoco ilustrado.* See his *Bodies Politic: Negotiating Race in the American North, 1730–1830* (Baltimore: Johns Hopkins University Press, 2003), 276.

102. Gumilla's elevation of Augustine as the "gran Padre de la Iglesia" (*El Orinoco ilustrado,* 95) [great Father of the Church] and the "Sol de la Iglesia San Agustín" (97) [Sun of the Church], coupled with his reference to Aquinas as "Doctor Santo Tomás," participates in the relegation of Aquinas to "church doctor" instead of church father by Spanish eclectics like Piquer.

103. The value of his ethnographic and cultural data is still cited in the twenty-first century. See, for example, Rodrigo Navarrete, "Behind the Palisades: Sociopolitical Recomposition in the Unare Depression, the Eastern Venezuelan Llanos (Sixteenth to Eighteenth Centuries)," *Ethnohistory* 47, nos. 3–4 (2000): 535–59.

104. See Feijoo, *Cartas eruditas y curiosas* vol. 3, Letters 15 and 17, and vol. 4, Letters 6 and 9.

105. Feijoo directs readers to a specific page, "el Tom. 1. del Orinoco Ilustrado, pág 311," in "Pidió un amigo al Autor su dictamen en orden a los Polvos Purgantes del Doctor Ailhaud, Médico de Aix en la Provenza; y fue respondido en ésta" (ibid., Letter 9 (1753), vol. 4, p. 117.

106. Luis Millones Figueroa and Ledezma, "Introducción"; Pimental, "Iberian Vision"; and Cañizares-Esguerra, "Iberian Science" for Bacon's intellectual and iconographic debts to Acosta as well as the influence of Andrés García de Céspedes' etching

representing Bacon's hunt for knowledge and the route "beyond the pillars of Hercules" in the frontispiece to *The Great Instauration* (1620). On this topic, see also Bauer, *Cultural Geography.*

Chapter 4. ¡Oh Monstruo, Oh Bestia!

1. This approach to Gumilla's "theory of knowledge" joins Walter Mignolo's discussion of the possibilities of a Jesuit hierarchy of knowledge "beyond epistemic hegemony." Walter Mignolo, "Commentary" to *Natural and Moral History of the Indies,* ed. Jane E. Mangan (Durham, NC: Duke University Press 2002), 495. While my notion of the intercultural is informed by Mignolo's incorporation of recent criticism and discussions of European cultural intersections with subaltern perspectives in *Local Histories/Global Designs,* there is, of course, much more to be said about European and Amerindian knowledge systems and their influence on each other during the eighteenth century beyond the realm of what the current book attempts.

2. As we saw in chapter 3, this pious wonder was articulated in contemporary Spanish treatises on experimental physics. Works like Herrero's *Physica moderna, experimental, systematic* (Madrid, 1738) reveal the simultaneously religious and scientific value of wonder during the Hispanic Enlightenment; the primary goal of "physics" was still to heighten wonder at the works of God. Herrero's treatise ties the curious and useful value of investigating "hidden causes" of "nature's marvels"—"han de hallar los curiosos ingenios la mas gustosas, y util ocupacion de sus meditaciones" (6) [curious wits are bound to find the most pleasurable and useful occupation for their meditations]—to increasing wonder. If the external parts of nature stir such wonder and amazement, asks Herrero, how much more amazing would it be to understand its internal parts?

3. Joseph Gumilla, *El Orinoco ilustrado y defendido,* ed. Demetrio Ramos (Caracas: Fuentes para la Historia Colonial de Venezuela, 1963), 313. Unless otherwise indicated, all references to *El Orinoco ilustrado* are to this edition. All translations are my own.

4. Aristotle, *A New Aristotle Reader,* ed. J. L. Ackrill (Princeton, NJ: Princeton University Press, 1987), 255, 258. Aquinas articulated the Christian *utilitas* of wonder as "the best way to grab the attention of the soul" and Christianized Aristotle's treatise on *Rhetoric* "to connect wonder with pleasure and draw on the *Metaphysics* to associate it with a desire that culminates not so much in knowledge as in encounter with majesty [i.e., closer to God]." Qtd. by Caroline Walker Bynum, "Wonder," *American Historical Review* 102 (1997): 10. For more on early modern Christian connections between wonder and knowledge, see Mary B. Campbell, *Wonder and Science* (Ithaca: NY: Cornell University Press, 1999); Lorraine Daston and Katharine Park. *Wonders and the Order of Nature: 1150–1750* (New York: Zone Books, 1998); and Stephen Greenblatt, *Marvelous Possessions* (Chicago: University of Chicago Press, 1991).

5. "Quiero concluir y roborar este punto con la autoridad del venerable Padre Juan Eusebio Niremberg [*sic*] [. . .] un jesuita a cuyo cargo está toda la maniobra de la botica del Colegio Imperial, el cual alega a favor de esta mi opinión [. . .]" (Gumilla, *El Orinoco ilustrado,* 388) [I want to conclude and support this point with the authority of the venerable father Juan Eusebio Nieremberg . . . a Jesuit in whose charge is all the operations of the pharmacy of the Imperial College who approves of my opinion . . ."]. See also "De Serpentibus Bambae," chapter 38 of Juan Eusebio Nieremberg, *Historia Natvrae* (Antwerp: Balthasaris Marelia, 1635).

6. Nieremberg laid out this aspect of the Jesuit theory of knowledge when stating, "Admiratio non debet obstare veritatii in effectibus naturae" (*Historia Natvrae,* qtd. in

Ledezma, "Una legitimación imaginativa del Nuevo Mundo" 66) [Wonders should not impede the quest for truth]. Although he did not investigate in the field and instead collected his natural history information from various Jesuit accounts, this scholar remained an essential footnote to eighteenth-century natural histories. On Nieremberg, see also William B. Ashworth, Jr., "Natural History and the Emblematic World View," in *Reappraisals of the Scientific Revolution,* ed. David C. Lindberg and Robert S. Westman (Cambridge: Cambridge University Press 1990).

7. Gumilla opens his discussion of El Dorado with a stirring rhetorical amplification that imagines a series of questions and answers and culminates in a long exclamation about the real El Dorado: "Preguntémosle a Keymisco, ingles, y a otros jefes sus paisanos: 'Amigos. ¿qué viajes son éstos? ¿Para qué tanta repetición de peligrosas navegaciones?' Preguntemos en el Perú y en Quito a uno y otro Pizarro; en Santa Fe de Bogotá, a uno y otro Quesada; en el Marañón, a Orellana; y en Meta, a Berrío y otros muchos famosos capitanes: '¿Para que os afanáis? ¿A qué fin tantas leves, marcha y viajes arduos, difíciles e intolerables?' 'Buscamos (dicen) el famoso y riquísimo Dorado. [. . .]' ¡Notable asunto el ir aquellos jefes españoles tropezando a cada paso en un Dorado de tesoro inagotable, cual realmente es todo el Nuevo Reino de Granada y Tierra Firme, tan lleno de fecundas minas de oro, plata y esmeraldas [. . .]!" (*El Orinoco ilustrado,* 253). [Let us ask Keymis, British, and other commanders that are his countrymen: 'Friends, what kind of expeditions are these? Why so much repetition of dangerous navigations?' Let us ask in Peru and in Quito one or another Pizarro; in Santa Fe de Bogotá, one or another Quesada; in the Amazon, Orellana; and in the Meta, Berrío and many other famous captains: 'Why do you all work so hard?' What is the purpose of so many triflings, departures, and arduous trips, both difficult and intolerable?' 'We are searching for (they say) the famous and very rich El Dorado. . . .' Noteworthy matter the going of those Spanish commanders stumbling at each step on an El Dorado of inexhaustible treasure, which is actually all of the New Kingdom of Granada and its Tierra Firme provinces, so full of fecund mines of gold, silver, and emeralds . . . !]

8. As we will see in the conclusion, Gumilla's negative views of curare contrasted with those of later European scientific travelers who foresaw its value as a general anesthesia and who valued it as much as they valued quinine.

9. "¿Qué misionero, qué español, qué soldado pudiera vivir entre ellos, si despreciada por los mismos la silenciosa furia de su saeta y curare, no se aturdieran al estrépito contingente del fusil?" (Gumilla, *El Orinoco ilustrdo,* 364). [What missionary, what Spaniard, what soldier could live among them, if these same tribes did not underestimate the silent fury of their arrows and curare, and were not stunned by the contingent clatter of the gun?]

10. "[P]ara esconderse más buscó o le señaló el Autor de la Naturaleza, no la tierra común al resto de las plantas, sino el cieno podrido y corrupto [. . .]" (ibid., 364). [To better hide itself it sought, or the Author of all Nature showed it, not the common llanoslands of the rest of the plants, but rather the festering and corrupt swamp. . . .]

11. Walter Mignolo, *Local Histories/Global Designs* (Princeton, NJ: Princeton University Press, 2000) 60. See also Anibal Quijano, "Colonialidad y modernidad-racionalidad," in *Los Conquistados: 1492 y la población indígena de las Américas,* ed. Robin Blackburn and Heraclio Bonilla (Bogotá: Tercer Mundo Editores, 1992), 437–47 for "modernity's" suppression of group production of knowledge in favor of individual, rational, empirical methodologies.

12. Mignolo, *Local Histories/Global Designs,* 187. This "intersubjective" or "enactive" production of knowledge is, in the case of the Caverre tribe, both production of knowledge about and production of curare itself.

13. Here we should recall Michel Foucault's discussion of popular "subjugated" knowledge in "Lecture One: 7 January 1976," in *Power/Knowledge: Selected Interviews and Other Writings, 1972–1977,* ed. C. Gordon (New York: Pantheon Books, 1980), 82. Fou-

cault wrote of the *savoir des gens* as "naïve knowledges, located low down on the hierarchy, beneath the required level of cognition and scientificity" (82).

14. These samples are from the first alphabetical index appended to *El Orinoco ilustrado,* entitled "Índice de raíces, frutas, yerbas, aceites, resinas y otras cosas medicinales, que se han descubierto en el Río Orinoco y sus vertientes" [Index of the roots, fruits, herbs, oils, resins, and other medicinal things that have been discovered along the Orinoco River and its slopes].

15. This popular wisdom contained a germ of truth. Despite eighteenth-century developments in germ theory (by Leeuwenhoek, Joblot, Bradley, and Vallisnieri, among others), Gumilla appears unaware of the bacterial causes of intestinal distress. Even though his Jesuit forefather Athanasius Kircher had postulated invisible living bodies as sources of contagion nearly one hundred years earlier, Gumilla cites lack of proof or personal experience, rejecting what might have been Amerindian awareness of bacteria: "[A]firma el vulgo y común de aquellas gentes, y muchos que no son parte del vulgo lo creen, que un mal muy común [. . .] que se llama *bicho,* es un animalejo vivo, nacido en los intestinos o entremetido en ellos. [. . .] A mí no me han dado prueba ni razón que me haya inclinado a creer que este tal bicho sea animalejo viviente" (Gumilla, *El Orinoco ilustrado,* 411). [The common people among them affirm, and many who are not part of the masses believe it, that a very common sickness . . . that is called *bicho,* is a tiny live aninal, born in the intestines or introduced into them. . . . They have not given me proof or reason that makes me believe that this said *bicho* is a living tiny animal.] Still, he draws the attention of medical doctors to this particular mediation of Amerindian beliefs with his own observations: "Pero éste y otros puntos sólo los apunto para que los doctos tengan este campo más para sus discursos, propios de los profesores de física" (411). [But these and other points I am only making a note of so that the learned will devote more attention to these in the discourses appropriate for professors of physics.]

16. "Ya dije arriba el modo bárbaro, cruel y necio, con que los indios, en su ciega gentilidad, curaban; erré, no curaban a los mordidos de culebra. Ahora será muy del caso, porque este libro también se ordena al bien de aquellas pobres gentes, apuntar aquí brevemente los remedios usuales que los Padres misioneros tienen prontos [. . .] para bien de aquellos pobres ignorantes indios, a cuya noticia no había llegado especie de tales antídotos" (Gumilla, *El Orinoco ilustrdo,* 400). [I already spoke above of the barbarous, cruel, and foolish way with which the Indians, in their blindness as pagans, cure; I was wrong, they did not cure the snakebites. Now it will be very relevant, since this book also is directed at the well-being of those poor peoples, to briefly point out here the typical remedies that the missionary fathers have at their disposal . . . for the good of those poor ignorant Indians, to whom no type of news had arrived about such antidotes.]

17. Gumilla's experiments here bring to mind the procedures discussed in William Harvey's *Anatomical Exercises on the Motion of the Heart and Blood in Animals* (1628), which are also mentioned in Federico Bottoni, *Evidencia de la Circulación de la Sangre* (Lima: Ignacio de Luna, 1723), n.p.: "Gulilemo Harvéo, célebre Medico Ynglés, fué el primero, que claramente habló, y escribió de este Movimiento, en el Año 1628" [William Harvey, famous British doctor, was the first who clearly spoke and wrote about this movement, in the year 1628].

18. "Cuatro cosas debemos por ahora considerar en el sonido y en la voz: Primera, la producción; segunda, la propagación; tercera, la reflexión, y cuarta, el aumento" (Gumilla, *El Orinoco ilustrado,* 348). [We should for now consider four things in sound and in voice: First, the production; second, the dissemination; third, the reflection, and fourth the amplification.] To help his readers visualize what he has observed firsthand, Gumilla provides one of only three illustrations he drew for *El Orinoco ilustrado* apart from his detailed map: a log-shaped war drum about 2 1/2 yards long.

19. "[L]os Padres misioneros recién llegados al río Orinoco y a otros muchos pasajeros, que se aturdían y llenaban de pavor" (ibid., 354) [The missionary fathers recently arrived at the Orinoco River along with many other passengers, who were stunned and full of fear].

20. "[E]n algunas iglesias los ecos del predicador le atormentan y confunden, y aturden y exasperan a los oyentes" (ibid., 351) [In some churches the echoes of the preacher torment and confuse him, and stun and exasperate the congregation].

21. The seventeenth-century Jesuit Francesco Grimaldi studied both sound and light waves. Father Marin Mersenne, "whose monastic cell acted as a central clearing house for the European scientists of the period" (Allen G. Debus, *Man and Nature in the Renaissance* [Cambridge: Cambridge University Press, 1978], 106), published *Harmonie universelle* in 1636. Gumilla cites: "In *Arm. Univ.*, lib. III, pág. 214" (*El Orinoco ilustrado,* 349) [In *Harmonie universelle,* book 3, page 214] and "*Diario de los sabios parisienses,* Día 16 de agosto de 1677" (ibid.) and [*Journal of the Wise Parisians,* August 16, 1677].

22. Of course, twenty-first-century physics understands sound waves to be longitudinal. Its particles oscillate around a center; their movement is back and forth, but in a line with direction. Gumilla's experimental philosophy did, however, confirm that the compression and expansion of air create amplitude, decibel level, intensity, and pitch in both the human voice and musical instruments.

23. "Buío" (or "guío") is a regional name for the anaconda still used in Colombia.

24. For example, in his *Natural History* (AD 77), Pliny the Elder compared the Bagradas River's large snake with a Roman baby-eating boa in a chapter entitled, "Of monstrous great Serpents, and namely of those called Boa." Both the Bagradas River incident in North Africa (256 BC) and Alexander the Great's expedition to India (327–325 BC) contributed to the portrayal of giant boas as man-eating monsters. See Richard B. Stothers's fascinating article for a catalogue of both ancient and current sources detailing giant boas and their "poisonous breath." Stothers, "Ancient Scientific Basis of the 'Giant Serpent' from Historical Evidence," *ISIS* 95, no. 2 (2004): 220–38. Stothers cites everything from Polybius's *Histories* (ca. 150 BC) to John C. Murphy and Robert W. Henderson's *Tales of Giant Snakes: A Historical Natural History of Anacondas and Pythons* (Malabar, Fla.: Krieger, 1997).

25. For additional discussion of New World snakes, see chapter 4 of Tzvetan Todorov, *The Conquest of America: The Question of the Other* (New York: HarperPerennial, [1984]). Todovov includes an image from the *Florentine Codex* labeled "The fabulous serpent." More recently, Anita Been, *Animals & Authors in the Eighteenth-Century Americas* (Providence, RI: John Carter Brown Library, 2004) features two striking non-Jesuit images from the Atlantic coastline between the Orinoco and Amazon rivers: an enormous close-up of a caiman wrestling with an anaconda and Stedman's captured Guiana anaconda.

26. There are far too many "unexpected encounters" to discuss in such limited space. A brief yet representative selection includes Gumilla's own encounters: "las muchas y repetidas veces que en veintidós años de continuos viajes [. . .] me encontré repentinamente con los buíos y siempre con sobresalto y horror" (*El Orinoco ilustrado,* 381) [the many and repeated times that in twenty-two years of continuous expeditions . . . I suddenly met up with the buíos and always with a start and horror]; other Jesuits' face-to-face experiences: "con mucho susto un Padre a quien yo traté [. . .] pasando de Caracas a las Misiones de Orinoco, se halló repentinamente con el espectáculo más horrendo que se puede pensar, y era un tremendo buío" (379) [with much fear a father with whom I deal . . . passing from Caracas to the Orinoco missions, suddenly found himself with the most horrendous spectacle that one can imagine, and it was a tremendous buío]; as well as those of Amerindian natives: "[N]o hay año en que no se desaparezcan hombres

campesinos, de los que salen o a pescar, o a cazar" (379), [A year does not pass without peasants disappearing, of those who go out to fish, or to hunt].

27. "Yo sé y todos pueden ver, y saber por experiencia, que los efluvios del imán, incorporados en el hierro y en el acero, le atraen y tenazmente resisten; nadie habrá que no halle la misma virtud atractiva en los efluvios que el azabache imprime en las pajas, si quiere hacer el experimento" (ibid., 386–87). [I know and everyone can see, and know by experiments, that the effluvia of the magnet, incorporated in the iron and steel, attract and tenaciously resist; there is noone who will not find the same attractive virtud in the effluvia that the lodestone fixes on the straw, if he wants to do the experiment.] Here Gumilla footnotes book 21 of Augustine's *City of God.*

28. Refuting Aristotle, Gumilla affirms the existence of "el vacuo, que tanto aborrece la Naturaleza" (*El Orinoco ilustrado,* 392) [the vacuum, which nature so abhors]. For Gumilla, the vacuum does not constitute an "imaginary space." He compares the attractive force of the anaconda's breath to an air pump, and to matter's contraction to avoid the vacuum. "[P]uede el curioso filosofar acerca de la virtud atrayente del buío [. . .] que de las fauces del culebrón sale un turbillón de efluvios malignos [. . .] de lo que se infiere (aunque no se vea) que en dicho aire está el turbillón o remolino de efluvios venenosos y en su centro la virtud atrayente de este venenoso turbillón del buío con la similitud de la bomba aspirante y atrayente con cuyo movimiento se extrae el agua de la sentina y fondo de los navíos, arrebatada contra todo su peso e inclinación natural hacia lo alto del navío, sin que hallemos otra razón que dar en esta maniobra sino decir que sube el agua y deja violentamente su centro para evitar el vacuo que (por más experimentos que se añadan) lo tiene la Naturaleza desterrado a los espacios imaginarios" (392–93). [The curious can philosophize about the attractive virtue of the buío . . . that a spinning stream of malignant effluvia goes out of the jaws of the snake . . . from this it can be inferred (even though it is not seen) that in said air is a stream or whirlpool of poisonous effluvia and that in its center the attractive virtue of this poisonous whirlpool from the buío works much like the attractive air pump by whose movement water is extracted from the bilge and bottom of ships. Water is snatched despite all its weight and natural inclination toward the height of the ship, without us finding any other reason for this motion except that the water rises and violently contracts from its center in order to avoid the vacuum that (no matter how many experiments are added) nature supposedly has exiled to imaginary spaces.]

29. The *Oxford English Dictionary* dates use of the idea of effluvia as an "outflow of material particles too subtle to be perceived by touch or sight" to the 1640s. Today's definition is "A stream of minute particles, formerly supposed to be emitted by a magnet, electrified body, or other attracting or repelling agent, and to be the means by which it produces its effects. Chiefly *pl.* (Now only *Hist.;* but it probably survived the theory which it strictly implies.) Also *fig.*" See http://dictionary.oed.com/cgi/entry/00181778. See also the definition for "magnetism." Gumilla most likely consulted the Jesuit Athanasius Kircher's *Magneticum naturae regnum.* Since Gumilla wrote a good forty years before the Austrian doctor Mesmer (1734–1815) scandalized the Paris Academy of Sciences, we cannot think in terms of Gumilla "mesmerizing" readers or of the "animal magnetism" of anaconda effluvia, for neither phrase had yet been introduced.

30. The comparison with the magnet supports Gumilla's inductive process: "[E]s también preciso que del estrago lastimoso que causa el vaho del buío, monstruo corpulento, se infiera y reconozca una actividad atrayante; y sea enhorabuena tan oculta y difícil de averiguar, como lo es la que confesamos en la piedra imán" (*El Orinoco ilustrado,* 387) [It is also clear that from the pitiful ruin caused by the vapor of the buío, corpulent monster, can be inferred and recognized an attractive activity; and that it is, thank God, as hidden and difficult to investigate as that which we admit in the magnetic stone]. Gumilla viewed

comparisons by analogies, similtudes, and contrasts as "good philosophy": "Supuesto que se procede bien arguyendo *asimili,* inquiriendo unos efectos a vista de otros [. . .] guiándonos por la similtud de ellos, no debe despreciarse en la Filosofía natural la argumentación a *contrariis*" (390) [I suppose that if one proceeds well by arguing *asimili,* investigating some effects in light of others . . . guiding ourselves by the similarity of them, natural philosophy should not reject *a contrariis* argumentation].

31. This cite underscores Gumilla's commitment to the Baconian methodology outlined in chapter 3. By way of a contrariis argumentation, Gumilla details differences from Sloan's rattlesnake described in the *Philosophical Transactions* and bases his argument on his Orinoco River region gathering of natural knowledge and induction as well as on a contrariis argumentation: "De lo dicho se ve que el culebrón de que habla el caballero Esloane en las *Memorias Filosóficas de la Regia Sociedad de Londres* es de especie diversa, porque el buío no tiene colmillos ni dientes, y por eso no come, sino que engulle la presa que atrajo. Y al contrario, Mr. Esloane supone que su culebrón primero hiere, y luego sigue con la vista la presa, que por instinto sabe morirá luego que el veneno, que lleva consigo, difunda toda su actividad; no así el buío, que, como dije, primero ve, v. gr., al venado, luego abre la boca, le arroja el vaho, e inficionado y aturdido, lo atrae y se lo engulle" (*El Orinoco ilustrado,* 378). [From the aforementioned it can be seen that the snake about which Mr. Sloan speaks in the *Philosophical Transactions of the Royal Society of London* is of another species, because the buío does not have fangs or teeth, and for this reason it does not eat, but rather swallows, the prey that it has attracted. And to the contrary, Mr. Sloan supposes that his snake first wounds and then follows with its eyes its prey, which knows by instinct that it will die as soon as the poison that it carries inside spreads all its activity; this is unlike the buío, which, as I said, first sees, for example, a stag, then it opens its mouth, flings out its vapor, and with the stag infected and stunned, attracts it and swallows it up.] Gumilla's footnote cites "Tom. XXXVIII, en cuarto, del año de 1738" (378) [Volume 38 in the fourth, from the year of 1738], but this article by Sloan actually appears in 1733–34. See Hans Sloan, "Conjectures on the Charming or Fascinating Power Attributed to the Rattle-Snake: Grounded on Credible Accounts, Experiments and Observations," *Philosophical Transactions* 38 (1733–734): 321–31. Gumilla's anaconda discourse constitutes a significant departure from Jesuit natural histories preceding *El Orinoco ilustrado* and is more daring than posterior American and European accounts integrating native popular wisdom about the "fascinating faculty" of rattlesnakes. On this, see the facsimiles of philosophical journal entries from naturalists such as Kalm, Bartram, Barton, and Audobon in Kraig Adler, *Early Herpetological Studies and Surveys in the Eastern United States* (New York: Arno Press, 1978).

32. In the section entitled "On the poisonous snakes of these lands," the subsections are: "I. Del culebrón espantoso llamado buío. [. . .] II. Reflexión sobre el párrafo antecedente y confirmación de lo que él contiene. [. . .] III. Trata de la acción y fatal atractivo del buío. [. . .] IV. De la acción o vibración de los efluvios. [. . .] V. De la fuerza atractiva del vaho del buío. [. . .] VI. De algunas señas para filosofar sobre la dicha virtud atrayente. [. . .] VII. De otras culebras malignas y de algunos remedios contra sus venenos" (*El Orinoco ilustrado,* 375–402). [I. On the frightening snake called buío. . . . II. Reflection on the preceding paragraph and confirmation of what it contains. . . . III. Treats the action and fatal attraction of the buío. . . . IV. On the action or vibration of the effluvia. . . . V. On the attractive force of the vapor of the buío. . . . VI. On some signs for philosophizing about said attractive virtue. . . . VII. On other malicious snakes and some remedies against their poisons.]

33. One translation of verse 180 of Horace's *Ars poetica* is: "Actions that have been admitted to our consciousness through our having heard them have less of an impact on our minds that those that have been brought to our attention by our trusty vision and for

which the spectator himself is an eyewitness." (*Horace for Students of Literature,* ed. Leon Golden and O. B. Hardison, Jr. Gainesville: University Presses of Florida, 1995).

34. Pliny the Elder, *The Historie of the World,* (London: Adam Islip, 1634), 199.

35. "Verdad es que aunque los científicos de éste y del Mundo Nuevo confiesen uniformes la atracción cuestionada, siempre quedarán suspensos, con anhelo y ansia por descubrir la raíz de ella, que es la virtud activa atrayente" (Gumilla, *El Orinoco ilustrado,* 389). [It is true that even if the scientists of this and the New World uniformly admit the attraction in question, they will always remain amazed, with longing and eagerness to discover the root of it, which is the active attractive virtue.]

36. Several French reviews discussed Gumilla's treatment of the existence of the anaconda's magnetic effluvia. For example, the January 1748 edition of the Jesuit journal *Mémoires pour l'Histoire des Sciences & Beaux-Arts* [*Mémoires de Trévoux*] warns of this "monstre, appellé *Buio* . . . il répand un soufflé empoisonné qui engourdit tout ce qu'il atteint. [. . .] On ne peut s'en dégager; les efforts que fait un homme pour s'en retirer n'aboutissent qu'à le précipiter plus surement dans la gueule de monstre" [Monster called *Buio* . . . spreads a poisonous breath that numbs all who await it . . . One cannot disengage from it; the efforts that a man makes to withdraw from it do nothing but precipitate his certain end in the mouth of the monster] (33–34). Ten years later a review article for the French translation in *L'Année Littéraire* reaffirmed that "il pousse un soufflé empesté qui étourdit la personne ou l'animal qui passé par l'endroit où il le dirige, & le force meme de s'avancer ver lui, & de venire se presenter à sa gueule" [It pushes out a pestilent breath that stuns the person or the animal that passes to the right of where it is directed, and it forces this same being to advance toward it, and to come and present itself to its mouth] (Elie Frèron, *L'Année Littéraire* [1758; Geneva: Slatkin Reprints, 1966], 79–80). Reviewers of Gumilla's natural history in the *Journal Encyclopedique* in 1759 acknowledged that while what he said may seem incredible, they remained persuaded that Gumilla only reported what he saw. Cépède's *Histoire naturelle des quadrupèdes ovipares et des serpens* (1788–89) also believed in the "miraculous power" of snakes' "infectious breath." Cépède was cited by Benjamin Smith Barton, who did doubt this power. See Barton, "A Memoir Concerning the Fascinating Faculty which has been Ascribed to the Rattlesnake, and Other American Serpents," *Transactions of the American Philosophical Society* 4 (1799): 74–113.

37. Benito Jerónimo Feijoo, "Descubrimiento de una nueva Facultad, o Potencia Sensitiva en el hombre a un Filósofo," *Cartas eruditas y curiosas* (Madrid: Imprenta real de la Gazeta, 1773), Letter 6 (1756), vol. 4, pp. 76–77. Here Feijoo's treatment of the anaconda validates Gumilla's firsthand report and proof through various testimonies of the irresistible attraction of its breath (the emanating vapor that draws in its prey).

38. The renowned herpetologist John Thorbjarnarson notes that "more than any other snake, stories about anacondas have been subject to wild exaggeration and hyperbole. Padre Gumilla, a Jesuit priest . . . wrote that the anaconda hypnotized and captured its prey using invisible poisonous vapors that issued from its mouth." Thorbjarnarson, ("Trailing the Mythical Anaconda," *Americas* 47, no. 4 (1995): 38. Even though Thorbjarnarson cites Gumilla in a hyperbolic context, he does concede that "it is not hard to understand why some encounters with free-living anacondas in the remote backwaters of South American have led to stories of animals of mythic proportions" (38).

39. Gumilla's description of the "dragon with four horrible feet, frightening on land and dreadful in water" recalls book 8, chap. 25 of Pliny's *Natural History,* minus the "dragon" name: "The river Nilus nourisheth the Crocodile; a venomous creature, foure footed, as dangerous on water as land." (Pliny the Elder, *Historie of the World,* 208). In fact, this was standard phrasing for introducing the crocodile in natural histories. Luis Joseph Peguero, who quotes both Pliny and Gumilla in his *Historia de la Conquista de la Isla Española de Santo Domingo,* (1762; reprint, Santo Domingo: Museo de las Casas Reales, 1975), 260, opens

his discussion thus: "[E]l Cocodrilo es bestia de cuatro pies; en tierra o agua es nocivo" [The Crocodile is a beast with four feet; on land or in water it is harmful].

40. "El caimán es pescado: al pescado ha dado Dios toda la agilidad que ha menester para nadar, subir y bajar en el agua; luego el caimán no necesita de piedras para sumirse en el río" (Gumilla, *El Orinoco ilustrado,* 420) [The caiman is a fish: to fish God has given all the agility that is necessary to swim, rise, and fall in the water; therefore the caiman does not need stones to submerge itself in the river].

41. As noted before, educated readers would have recognized Gumilla's evocation of Thomas Salmón's *Historia universal.* In a footnote (*El Orinoco ilustrado,* 423n) Gumilla reproduces a quote from Salmón, *Lo stato presente di tutti I paesi e popoli del mondo naturale, político, e morale, con nuove osservazioni e correzioni degli antichi, e moderni viaggiatori* (Venice: Presso giambatista Albrizzi Q. Gir, 1736): "Aperti alcuni di essi coccdrilli, si sono trovate, nel loro ventre, ossa de huomini, e di animali; como ancor pietre, che inghiottono, peremplersi lo stomaco. Tom. II, cap. IX, pág. 225." [Once some of those crocodiles were opened up, there were found in their bellies bones of humans and animals that they swallow like stone ballast to fill up their stomach].

Conclusion

1. Juan Pimentel, "The Iberian Vision" *Osiris* 15 (2001):24.

2. See ibid., 27, and Steven Harris, "Jesuit Scientific Activity in the Overseas Missions, 1550–1773," *ISIS* 96 (2005), for recent details about "Jesuit visions" that incorporated local knowledge in missions with modern science.

3. On this late eighteenth-century paradigm shift, which discredited many missionary and traveler tales, see Jorge Cañizares-Esguerra, "Spanish America in Eighteenth-Century European Travel Compilations: A New 'Art of Reading' and the Transition to Modernity," *Journal of Early Modern History* 2, no. 4 (1998): 329–49; idem, *How to Write the History of the New World* (Stanford, CA: Stanford University Press, 2001), an essential study for the examination of Spain and Spanish America's "making of modernity." An excellent synthesis of Cañizares-Esguerra's contributions can be found in Ralph Bauer, "The Postcolonial Origins of Modernity," *William and Mary Quarterly* 59, no. 4 (October 2002): 975–81.

4. Bauer, "Postcolonial Origins of Modernity," 976–77. Here Bauer challenges the tradition of writing off Catholic cultures as "pre-"or "unmodern." See also Cañizares-Esguerra, "Iberian Science in the Renaissance," *Perspectives on Science* 12, no. 1 (2004), where, among other topics, he discusses Anglo-American scholarship's continued exclusion of Iberia from modernity.

5. "Lettre XIV," in *L'Année Littéraire, Tome VII* (1758; reprint, Geneva: Slatkine Reprints, 1966), 350. Gumilla's French edition significantly widened his reading public to include philosophers such as the abbé Raynal, who borrowed some passages from this work for his *Histoire Philosophique et politique, des etablissemens & du commerce des Europeéns dans les deux Indes* (Amsterdam: 1770); and Adam Smith, who cited Gumilla in a footnote about El Dorado in book 4, chapter 6 of *An Inquiry into the Nature and Causes of the Wealth of the Nations* (London: W. Strahan and T. Cadell, 1776). As for "nonphilosophers," Jules Vernes must have consulted Joseph Gumilla, *Histoire naturelle, civile et geographique de l'Orenoque* (Avignon: Chez la Veuve de F. Girard, 1758) in addition to Humboldt's *Relation historique du voyage aux regions équinoxiales du nouveau continent* (1805–38) when composing his novel, *Le superbe Orénoque* (Paris: J. Hetzel, 1898).

6. See, for example, the entry "Orinoco" in Antonio de Alcedo, *Diccionario geográfico histórico de las Indias Occidentales o América. Es a saber: de los Reynos del Perú, Nueva España,*

Tierra Firme, Chile, y Nuevo Reyno de Granada (Madrid: B. Cano, 1786–89). After listing many of the Orinoco tribes the Jesuits worked with along the banks of the Orinoco, this renowned encyclopedia laments that the Society of Jesus's flourishing missions passed to the Capuchin order after the Jesuit expulsion. A quick check of the most recent editions of classic encyclopedias (for example, the Grolier online encyclopedia) reveals the Jesuits Gumilla and Gilij as sources of eighteenth-century knowledge of the Orinoco and Humboldt as the nineteenth-century source.

7. For placement of Spain and Spanish America within the centers and peripheries of modernity, see Pimental, "Iberian Vision." For the theorization of the "peripheries" see George Basalla, "The Spread of Western Science Revisited in *Mundiclización de la ciencia y cultura nationel,* ed. Antonio Lafuente, Alberto Elena, and María L. Ortega (Madrid: Doce Calles, 1993); David W. Chambers, "Locality and Science: Myths of Centre and Periphery," and Enrique Dussel, "Eurocentrism and Modernity (Introduction to the Frankfort Lectures)," *Boundary 2* 20, no. 3 (Fall 1993).

8. For more on readers' passion for natural histories, see David Freedburg, *The Eye of the Lynx: Galileo, His Friends and the Beginnings of Modern Natural History* (Chicago: University of Chicago Press, 2002).

9. For details on how narratives of colonial-era real-life experiences repeatedly found their way into nineteenth- and twentieth-century fiction, see, for example, David Bost, "Historiography and the Contemporary Narrative: Dialogue and Methodology," *Latin American Literary Review* 16, no. 31 (1988): 34–44; and idem, "Historians of the Colonial Period: 1620–1700," in *The Cambridge History of Latin American Literature,* ed. Roberto González Echevarría and Enrique Pupo-Walker, vol. 1 (Cambridge: Cambridge University Press, 1996), 143–90.

10. In " 'Semejante a la noche' de Alejo Carpentier: Historia/Ficción," *MLN* 87, no. 2 (1972): 280, Roberto González Echevarría notes that entire pages of *Los pasos perdidos* were lifted from *El Orinoco ilustrado.*

11. Alejo Carpentier, *Los pasos perdidos,* ed. Roberto González Echevarría (Madrid: Cátedra, 1985), 174. All translations mine. Despite this indirect textual acknowledgment, however, Carpentier does not have his musicologist-narrator enumerate actual chronicles. Instead, he fabricates sources such as a seventeenth-century missionary whom readers might not recognize as an invention without González Echevarría's annotations. See Raul Silva Cáceres, "Una novela de Carpentier," *Mundo nuevo* 17 (1967): 33–37, for the first hypothesis that this invented Friar Servando de Castillejos was actually Joseph Gumilla. He writes, "[A]demás de describir íntegramente el río, su flora, su fauna y muchos pueblos indígenas que viven en sus orillas, incluye un diseño preciso de los instrumentos musicales buscados por el personaje principal" (30) [As well as entirely describing the river, its flora, its fauna and many indigenous peoples that live along its banks, Gumilla includes a precise design of the musical instruments sought by the main character]. Silva Cáceres compares two very similar descriptive passages of the Orinoco River taken from Gumilla and then Carpentier.

12. Carpentier, *Los pasos perdidos,* 89. In "Semejante a la noche" and in his introduction and notes to *Los pasos perdidos,* Roberto González Echevarría writes about general connections between Gumilla's facts and Carpentier's fiction. In *Myth and Archive* he assigns to twentieth-century Latin American novelists a self-consciously anthropological perspective and discusses the authority they sought "by mimicking the texts that constitute anthropological discourse." *Myth and Archive: A Theory of Latin American Narrative* (Cambridge: Cambridge University Press, 1990), 144. He includes Carpentier in this appropriation of a legitimizing "hegemonic discourse" backed by and embodying the system of anthropology: "The critic that the novelist becomes is essentially an anthropologist, because anthropology furnishes the only discourse capable of authoritatively analyzing and

narrating the autochthonous, hence the fable of legitimation" (156). In "Hearts of Darkness: The celebration of Otherness in the Latin American *Novela de la selva*," *Romance Studies* 23, no. 2 (2005): 105–16, Lesley Wylie takes this one step further by applying the postcolonial theories of Homi Bhabha, Jean-François Lyotard, and Edward Said and asserting that Carpentier's "colonial mimicry" of anthropological discourse constitutes a parodic appropriation of "imperial tropes" of travel writing in order to reclaim and then reinscribe the South American rain forest in a manner demonstrating the limits of Western epistemology. While this reading is compelling in the context of recent trends in literary criticism and, in fact, the line between imitation and parody can be nebulous, I still prefer to read Carpentier's metatextual moments as resisting postmodernism and instead as appropriating and contributing to the historical, scientific, and literary clout accorded chronicles well into the twentieth century.

13. "Los hombres llamados salvajes" and "El mundo del tiempo detenido," the two 1952 articles from *El Nacional* where Carpentier mentions Gumilla, *El Orinoco ilustrado*, and the Saliva tribe's musical instrument with two mouthpieces, are cited by González Echevarría in a footnote (see Carpentier, *Los pasos perdidos*, 90). While positing the importance of anthropological discourse on twentieth-century novelists in *Myth and Archive*, González Echevarría includes among Carpentier's influences texts ranging from those by twentieth-century French anthropologists (Claude Lévi-Strauss and Michel Leiris) and Orinoco travelers (Alain Gheerbrant) to early modern chronicles by scientific travelers and missionaries.

14. In *El Orinoco ilustrado y defendido*, ed. Demetrios Ramos (Caracas: Fuentes para la Historia Colonial de Venezuela, 1963), Gumilla even compares his accumulation of man-eating beasts to a "vortex": "todo este torbellino de especies funestas" (394) [this whole vortex of fatal species]. Unless otherwise indicated, all references to Gumilla in the text and notes are to this edition. Translations are mine.

15. This progression is exemplified in brief when the protagonist (Cova) and his lover (Alicia) first flee to the Casanare region, where they are overcome by the beauty of the sun rising on the llanos. As the Orinoco region grasslands turn to swamplands, however, both the narrator's foreboding and the actual dangers increase. See José Eustasio Rivera, *La vorágine*, ed. Montserrat Ordóñez (Madrid: Cátedra, 1998), 91–96.

16. Rivera died at forty after working with a commission to mark the Colombian, Venezuelan, and Peruvian borders in the Orinoco and Amazon River regions. The Mexican poet Juan José Tablada called Rivera's death a "revenge of the jungle," which first shot a metaphorical curare-poisoned dart across the continent and then killed him with malaria. Cited in the introduction to Rivera, *La vorágine*, 14–16.

17. In a footnote, Montserrat Ordóñez points out the similarities between Rivera's passage, " 'Tápelo con el pañuelo pa que le sirva de cedazo.' Así lo hice varias veces, sacudiendo los animalillos. . . ." ['Cover it with a handkerchief so that it serves as a sieve.' I did this various times, removing the little animals] and what "ya el padre Gumilla registra así, en el siglo XVIII, esta manera de colar el agua" [father Gumilla already similarly notes, in the eighteenth century, this means of straining the water]—that is, with a "pañuelo doblado" [folded handkerchief] that strains "innumerables animalejos, casi imperceptibles a la vista, que transferidos al estómago, se aferran de él" [countless little animals, almost undiscernible to the eye, that if transferred to the stomach, anchor into it] (Rivera, *La vorágine*, 166). On pages 107 and 174 Montserrat Ordóñez provides two other footnotes that refer readers to Gumilla. See also *El Orinoco ilustrado*, 414.

18. Rivera, *La vorágine*, 94.

19. See González Echevarría, *Myth and Archive*, 103. He discusses it within the context of "conventional literary history, which focuses on works that fall within the sphere of influence of European literature such as Jorge Isaacs' *María*" (103). González Echevar-

ría is in fact warning against traditional approaches "that hardly take into account the powerful influence of scientific travel books on those very novels and on Latin American narrative of the nineteenth century in general" (103). See also *Myth and Archive,* 40. González Echevarría's comments on *María* were published a year before Doris Sommer's *Foundational Fictions: The National Romances of Latin America* (Berkeley and Los Angeles: University of California Press, 1991).

20. Jorge Isaacs, *María,* ed. Donald McGrady (Madrid: Cátedra, 1991), 300–306.

21. Ibid., 303.

22. Here I echo a persuasion technique for creating rhetorical eyewitness status that reaches far back into classical rhetorical treatises but was more recently appropriated into modern anthropological writing. As Clifford Geertz explains, "Ethnographers need to convince us . . . not merely that they themselves have truly 'been there,' but that had we been there we should have seen what they saw, felt what they felt, concluded what they concluded." *Works and Lives: The Anthropologist as Author* (Stanford, CA: Stanford University Press, 1988), 16.

23. In addition to Spanish-language articles such as María Matilde Suárez, "El contenido etnográfico del *Orinoco Ilustrado,*" *Montalban* 3 (1974): 309–35, Gumilla is cited as a reliable source in various articles from journals like Eduard Conzemius, "Ethnographical Notes on the Black Carib," *American Anthropologist* 30, no. 2 (1928): 183–205; John Gillin, "Crime and Punishment among the Barama River Carib of British Guiana," *American Anthropologist* 36, no. 3 (1934): 331–44; and Rodrigo Navarrete, "Behind the Palisades," *Ethnohistory* 47, nos. 3–4 (2000).

24. Naturally, Gumilla had an influence on exiled Jesuits such as Filippo Salvatore Gilij, Mario Cicala, and Antonio Julián. For over two centuries Jesuits have continued to mine *El Orinoco ilustrado* for facts and have taken Gumilla's missionary propaganda and religious enlightenment of the Orinoco River region as a source of pride, crafting hagiographic accounts of Gumilla's legacy.

25. The Paris Royal Academy of Sciences managed to convince the Spanish government to allow La Condamine's expedition to try to resolve competing theories about the size and shape of the world proposed by Newtonians and Cartesians. Juan and Ulloa were sent along as Spanish emissaries and were supposed to guard against the theft of secrets. However, La Condamine managed to spirit away the seedlings of two extremely valuable trees: the *Cinchona officinalis* (a source of quinine) and the *caoutchouc* (rubber).

26. Charles Marie de La Condamine, *Relation abrégée d'un voyage fait dans l'interieur de l'Amérique Méridionale. Depuis la côte de la Mer du Sud, jusqu'aux côtes du Brésil & de la Guiane, en descendant la riviere des Amazones* (Paris: Veuve Pissot, 1745).

27. Jorge Juan and Antonio de Ulloa, *Relacion historica del viage a la America Meridional Hecho de orden de S. Mag. para medir algunos grados de Meridiano Terrestre, y venir por ellos en conocimiento de la verdadera Figura, y Magnitud de la Tierra, con otras varias Observaciones Astronomicas, y Phisicas: Por Don Jorge Juan, Comendador de Aliaga, en el Orden de San Juan, Socio correspondiente de la Real Academia de Ciencias de Paris, y Don Antonio de Ulloa, de la Real Sociedad de Londres; ambos Capitanes de Fragata de la Real Armada* (Madrid: Antonio Marin, 1748).

28. This disdain for the *savoir de gens* is evidenced in statements like the following conclusion to the section on the anaconda's breath: "Lo mas de esto fue con vulgaridad supuesto de aquellas incultas Gentes, y creído de los otros con buena fé; porque ninguno por satifacer la curiosidad se havrá arrojado al peligro del examen" (ibid., 540). [The majority of this was assumed with commonness by those ignorant people, and believed with good faith by others; because nobody would have flung himself into the danger of testing it to satisfy their curiosity.] Such disregard for regional "common sense"—or, as Michel Foucault put it in his lectures on the relationship between power and knowledge,

disregard for "a particular, local, regional knowledge"—participates in an authority-building discourse that subjugates "naïve knowledges, located low down on the hierarchy, beneath the required level of cognition and scientificity." Michel Focault, "Lecture One: 7 January 1976." In *Power/Knowledge: Selected Interviews and Other Writings, 1972–1977* ed. C. Gordon (New York: Pantheon Boks, 1980), 82.

29. Juan and Ulloa, *Relacion historica,* 539. To compare with Gumilla, see *El Orinoco ilustrado,* 376–95.

30. Juan and Ulloa, *Relacion historica,* 540.

31. Ibid., 538.

32. Cañizares-Esguerra, *How to Write the History,* 28. See also Cañizares-Esguerra, "Spanish America in Eighteenth-Century European Travel Compilations," 340–41.

33. La Condamine famously refuted Gumilla's initial denial of a connection between the Orinoco and Amazon rivers in a communication to the Paris Royal Academy of Sciences that was excerpted in the translator's preface to the French edition of *El Orinoco ilustrado* (1758). However, as we know from volume 1, book 1, chapter 4 of Filippo Salvadore the Jesuit Gilij's *Saggio di storia Americana,* Gumilla had revised this geographical error based on Manuel Román's daring 1744 voyage, but not quite in time for the 1745 printing of his revised and expanded second edition. According to Gilij, in 1749 Gumilla had read aloud these additions to him. See Felipe Salvador Gilij, *Ensayo de Historia Americana,* vol. 1 (Caracas: Academia Nacional de Historia, 1987), 53. Humboldt repeated this information.

34. Instead of reaffirming its fabulous geography, Gumilla's discussion of El Dorado actually included a realistic analysis that connected evangelical and economic goals and was later praised by Antonio Julián and Adam Smith.

35. Juan and Ulloa, *Relacion historica,* 70–71.

36. While Gumilla, Juan, and Ulloa take the rhetorical strategy of prosopopoeia much further, I would be remiss if I did not note here that Gumilla himself modeled his plant personification after his Jesuit forefather's lengthy description, which attributed rational faculties to the virgin plant. See Manuel Rodríguez, *El Marañon y Amazonas. Historia de los descubrimientos, entradas, y reducción de naciones. Trabajos malogrados de algunos conquistadores, y dichosos de otros, así temporales, como espirituales, en las dilatadas montañas, y mayores rios de la America* (Madrid: Antonio Gonzalez de Reyes, 1684), 376. See also a Dominican friar's treatment of the "Sensitive plante," which shares much detail with Gumilla and Rodríguez: Jean-Baptiste Labat, *Nouveau voyages aux isles de l'Amerique. Contentant L'histoire naturelle des ces pays, l'Origine, les moeurs, la Religion & le Gouvernement des Habitans anciens & modernes . . . Avec une Description exacte & curieuse de toutes ces Isles* (Paris: Chez Pierre-Francois Giffart, 1722), 200.

37. See Alexander von Humboldt, *Personal Narrative of a Journey to the Equinoctial Regions of the New Continent,* trans. Jason Wilson (London: Penguin Books, 1995), xxxix. Humboldt's writings famously detail the natural and civil (or political) histories of Mexico, Cuba, Peru, and Venezuela, and their importance to emerging American republics has been discussed since the early nineteenth century. As Mary Louise Pratt famously affirms, "Humboldt's writings . . . became essential raw material for American and Americanist ideologies forged by creole intellectuals in the 1820s, 1830s, and 1840s. His writings were a touchstone for the civic literature that claimed Spanish America's literary independence. . . . Over and over in the founding texts of Spanish American literature, Humboldt's estheticized primal América provided a point of departure for moral and civic prescriptions for the new republics. His reinvention of América for Europe was transculturated by Euroamerican writers into a creole process of self-invention." *Imperial Eyes: Travel Writings and Transcultivation* (New York: Routledge, 1992), 175.

38. The opening lines of Humboldt's *Personal Narrative* articulate an Aristotelian trope: "[W]hatever is far off and suggestive excites our imagination; such pleasures tempt us

far more than anything" (*Personel Narrative,* trans. Wilson, 15). Implicit in this is Aristotle's dictum that from wonder grows knowledge, as articulated in the classic paradigm of *admiratio—scientia.* See chapter 4 of this book, "¡Oh Monstruo, Oh Bestia! Pathways to Knowledge in *El Orinoco ilustrado,*" for more on this wonder-to-knowledge paradigm as manifest in Gumilla's natural history.

39. Humboldt's writings include the single-volume *Ansichten der Natur* (1808), simultaneously published as *Tableaux de la nature ou, considerations sur les desert, sur la physionomie des végétaux, et sur les cataracts de l'Orenoque* (1808) and later translated as *Aspects of Nature, in different lands and different climates; with Scientific Elucidations,* trans Mrs. Sabine (Philadelphia: Lea and Blanchard, 1850) and *Views of Nature, or contemplation of the Sublime Phenomena of Creation* (London: H. G. Bohn, 1850). His most famous work is perhaps the *Relation historique du Voyage aux régions équinoxiales de Humboldt et Bonpland,* or *Personal Narrative of a Journey to the Equinoctial Regions of the New Continent* (1805–38). The last part of the five volumes of *Kosmos: Entwurf einer physischen Weltbeschreibung* (1845–62), translated first in 1849 as *Cosmos: A Sketch of the Physical Description of the Universe,* was published posthumously. Although explicitly tied to his Catholic agenda, Gumilla was already striving for the kind of useful and pleasing text described by Humboldt in his *Ansichten der Natur:* "I venture to hope that these descriptions of the varied Aspects which Nature assumes in distant lands may impart to the reader a portion of that enjoyment which is derived from their immediate contemplation by a mind susceptible of such impressions. As this enjoyment is enhanced by insight into the more hidden connection of the different powers and forces of nature, I have subjoined to each treatise scientific elucidations and additions" (Humboldt, *Aspects of Nature,* vi).

40. Cañizares-Esguerra, "Spanish America," 737. Cañizares-Esguerra joins other scholars who are challenging traditionally hagiographic conclusions about Humboldt's role in South American identity formation and instead exploring the Prussian's dependence on *criollo* and Amerindian knowledge for his compilation of natural knowledge. For example, see Cañizares-Esguerra, "How Derivative was Humboldt?" in *Colonial Botany: Science, Commerce, and Politics in the Early Modern World,* ed. Londa Schiebingerand Claudia Swan (Philadelphia: University of Pennsylvania Press, 2005), for an example of such revisionist accounts' contributions to the history of science.

41. As Roberto González Echevarría summarizes it, nineteenth-century "scientific exploration brought about the second European discovery of America, and the traveling naturalists were the new chroniclers" (*Myth and Archive,* 11). For Humboldt's leading role in this rediscovery, see Pratt, *Imperial Eyes,* 111–43 and Gerard Helferich for Humboldt's fame in his time "for exploring more of the continent's six million square miles than anyone before." Helferich, *Humboldt's Cosmos: Alexander von Humboldt and the Latin Americna Journey that Changed the Way We See the World* (New York: Gotham Books, 2004), 303.

42. Especially in volume 2 of the *Relation historique du Voyage aux régions équinoxiales du Nouveau Continent,* where in addition to indirectly alluding to Gumilla, Humboldt directly references the Spanish original of *El Orinoco ilustrado* twenty-seven times as a source. See Alexander von Humboldt, *Relation historique du voyage aux régions équinoxiales du nouveau continent,* 3 vols. (Stuttgart: F. A. Brockhaus, 1970), 661. Some of the facts Humboldt borrowed from Gumilla included information on the Orinoco tortoise, its egg harvest, and the Otomac tribe's niopo powder snuff and earth-eating. For the latter, Humboldt passes judgment on Gumilla: "It is curious that the usually credulous and uncritical Father Gumilla positively denies the earth-eating as such (Historia del Rio Orinoco, nueva impr. 1791, t. 1. p. 179.). He affirms that the balls of clay had maize-meal and crocodile fat mixed with them" (*Aspects of Nature,* 157). A few lines later, however, he excuses the Jesuit: "May Gumilla, by a confusion of things wholly distinct, have been alluding to the preparation of bread . . . which is previously buried in the earth in order to hasten the

commencement of the first stage of decay?" (*Aspects of Nature,* 158). For slightly different versions of these same passages, see also chapter 2.24 of Alexander von Humboldt, *Personal Narrative of Travels to the Equinoctial Regions of America, during the years 1799–1804,* trans. and ed. Thomasina Ross (London: Henry G. Bohn, 1852–53).

43. Humboldt, *Personal Narrative,* trans. Wilson, 177.

44. While Gumilla remained loyal to the king of Spain, Humboldt, as a foreign "outside observer" who wrote much of his *Personal Narrative* during the wars for South American independence, looked ahead to Spanish American economies. In fact, as the introduction to his *Personal Narrative* reveals, Humboldt envisioned a future political and even foundational role for his own writings: "I also venture to hope, once peace has been established, that this work may contribute to a new social order. If some of these pages are rescued from oblivion, those who live on the banks of the Orinoco or Atabapo may see cities enriched by commerce and fertile fields cultivated by free men on the very spot where during my travels I saw impenetrable jungle and flooded lands" (*Personal Narrative,* trans. Wilson, 13).

45. See Nancy Leys Stepan, "Going to the Tropics," in *Picturing Tropical Nature* (Ithaca, NY: Cornell University Press, 2001), 31–56 for an account of the fascinating power of Humboldt's tropical tropes, literary results of his own youthful "urge to travel to distant lands seldom visited by Europeans," a desire spurred by the "special charm" of what Humboldt himself was able to "glean from travellers' vivid descriptions" (Humboldt, *Personal Narrative,* trans. Wilson, 15).

46. As Humboldt explained early in his multiyear publishing process, "[T]hroughout the entire work I have sought to indicate the unfailing influence of external nature on the feelings" (*Aspects of Nature,* vi). Later, in a chapter on plant physiognomy, Humboldt famously encouraged mining these feelings with an alliance between studying plants and painting them into landscapes. With a wonder-provoking blend of the visual and tactile senses, he said that "it is under the burning rays of a tropical sun that vegetation displays its most majestic forms" (ibid., 244–45).

47. Ibid., vi. This vivid description of tropical majesty so integral to Humboldt's cosmos was already an integral part of Gumilla's.

48. Ibid., vii.

49. Ibid., v.

50. "Y al modo que (si falta la luz) en la más curiosa galería [. . .] no de otra manera la más curiosa Historia . . ." (Gumilla, *El Orinoco ilustrado,* 37). [And just as within the most curious gallery (if lacking light) . . . so the most curious history]. See chapter 1 of this book, "Father Gumilla's Textual Cabinet of Curiosities, Missionary Authority, and Rhetoric of Wonder."

51. Humboldt, *Personal Narrative,* trans. Wilson, 187.

52. Ibid., 196.

53. Gumilla illustrates this marvel twice: first in part 1, chapter 9, and again among a more general description of trees in part 2, chapter 20.

54. Humboldt, *Aspects of Nature,* 238–39. By the eighteenth century, praise of the palm as manna sent to the Amerindians from God had appeared repeatedly in New World chronicles, letters, accounts, and natural histories. But Humboldt clearly read Gumilla's chapter on the Orinoco palm, entitled, "Genios y vida rara de la nación guaraúna. Palma singular de que se visten, comen, beben y tienen todo cuanto han menester" [Temperament and strange life of the Guaraúna nation. The singular palm from which they dress, eat, drink and have everything they need]. As Pratt and others have pointed out, the friendship between Humboldt and Andrés Bello, who took some field trips with the Prussian, resulted in Bello translating excerpts of Humboldt's writings into Spanish for the London-based *Repertorio Americano.* This suggests an important indirect legacy of Gu-

milla. Bello takes the first part of his article "Palmas americanas" directly from Humboldt, and states that his "Descripción del Orinoco" is from book 7, chapters 23–24 of Humboldt. However, when he mentions Gumilla by name it is not exactly what Humboldt had written. See Andrés Bello, *Obras completas de don Andrés Bello, vol. 14: Opúsculos Científicos* (Santiago de Chile: Imprenta Cervantes, 1872), 177–185, 241–62.

55. Humboldt, *Aspects of Nature,* 148–49.

56. The first time Humboldt mentions Gumilla by name, he writes, "Gumilla terms the Mauritia flexuosa of the Guaranis the tree of life, arbol de la vida" (*Aspects of Nature,* 149–50), echoing the Jesuit's description of the palm as a new tree of life: "esta admiración [. . .] un nuevo árbol de la Vida, que así se debe llamar" (Gumilla, *El Orinoco ilustrado,* 135). [this wonder . . . a New tree of Life, which is what it should be called]. Later in *El Orinoco ilustrado* Gumilla praises the banana's utility for Amerindian nations: "[H]an hallado [. . .] en cierto modo su árbol de la vida en solo el plátano" (437) [They have found . . . in a certain manner their tree of life in only the banana].

57. Humboldt, *Aspects of Nature,* 239.

58. The lengthy section on curare from *El Orinoco ilustrado* was also translated word-for-word into the *Mémoires pour l'Histoire des Sciences & Beaux-Arts,* ed. François Catrou et al. (1747; reprint, Geneva: Slatkine Reprints); 27–29, and this suspense-filled account enlightened many. Gumilla's details about curare fabrication also made their way into several other reviews of both the Spanish original and the French translation (1758) of *El Orinoco ilustrado* in Switzerland, Holland, and Belgium. See *Journal Étranger* (Geneva: 1756), *L'Année Littéraire* (Amsterdam, 1758), *Journal des savants combine avec les Mémoires de Trévoux* (Amsterdam, 1758), and *Journal Encyclopédique* (Liège, 1759).

59. Humboldt, *Personal Narrative,* trans. Ross, 438–39. La Condamine gets credit for removing the first curare poison from South America, but Humboldt bragged that he was "first to bring a considerable quantity to Europe" (*Aspects of Nature,* 165). For their part, Juan and Ulloa's expressions of dread at the instantly lethal effects of curare echo passages where Gumilla marvels at how man can eat the animals killed by curare-tipped arrows without being poisoned himself.

60. Humboldt, *Personal Narrative of Travels,* trans. Ross, 439–40.

61. Ibid., 439.

62. Ibid., 441–47.

63. Ibid., 445–46.

64. See Gumilla, *El Orinoco ilustrado,* 384. My perception of Gumilla's amateur scientist status owes much to sociological trends in the history of science. See, for example Bruno Latour, *Laboratory Life: The Social Construction of Laboratory Life;* (Beverly Hills, CA: Sage Publications, 1979); and Roy Porter and George Sebastian Rousseau, *The Ferment of Knowledge: Studies in the Historiography of Eighteenth-Century Science* (Cambridge: Cambridge University Press, 1980).

65. For example, see John Thorbjarnarson, "Trailing the Mythical Anaconda," *Americas* 47, no. 4 (1995): 38–46; N. G. Bisset, "War and Hunting Poisons of the New World," *Journal of Ethnopharmacology* 30, no. 1 (1992): 1–26; and M. R. Lee, "Curare: The South American Arrow Poison," *Journal of College Physicians of Edinburgh* 35, no. 1 (February 2005): 83–92.

Bibliography

Abellán-García, José Luis. *Historia crítica del pensamiento español.* Madrid: Espasa-Calpe, 1981.

Acosta, Joseph de. *Historia natural y moral de las Indias.* 1590. Edited by Edmundo O'-Gorman. Mexico City: Fondo de Cultura Económica, 1940.

———. *Natural and Moral History of the Indies.* Edited by Jane E. Mangan. Translated by Frances M. López-Morillas. Durham, NC: Duke University Press, 2002.

Adas, Michael. *Machines as the Measure of Men: Science, Technology and Ideologies of Western Dominance.* Ithaca, NY: Cornell University Press, 1989.

Adler, Kraig. *Early Herpetological Studies and Surveys in the Eastern United States.* New York: Arno Press, 1978.

Aguiló Alonso, María Paz. "El coleccionismo de objetos procedentes de ultramar a través de los inventarios de los siglos XVI y XVII." In *Relaciones artísticas entre España y América,* edited by Enrique Arias Anglés, 107–49. Madrid: Consejo Superior de Investigaciones Científicas, 1990.

Aguirre Elorriaga, Manuel. *La Compañía de Jesús en Venezuela.* Caracas: Lucas G. Castillo, 1941.

Albanese, Denise. *New Science, New World.* Durham, NC: Duke University Press, 1996.

Alcedo, Antonio de. *Diccionario geográfico histórico de las Indias Occidentales o América. Es a saber: De los Reynos del Perú, Nueva España, Tierra Firme, Chile, y Nuevo Reyno de Granada (1786–1789).* Biblioteca de Autores Españoles. Madrid: Ediciones Atlas, 1967.

Aldridge, A. Owen. *The Ibero-American Enlightenment.* Urbana: University of Illinois Press, 1971.

Allaby, Michael. *The Concise Oxford Dictionary of Botany.* Oxford: Oxford University Press, 1992.

Álvarez de Miranda, Pedro. *Palabras e ideas: El léxico de la Ilustración temprana en España: 1680–1760.* Madrid: Real Academia Española, 1992.

Aristotle. *A New Aristotle Reader.* Edited by J. L. Ackrill. Princeton, NJ: Princeton University Press, 1987.

———. *On Rhetoric: A Theory of Civic Discourse.* Edited and translated by George A. Kennedy. New York: Oxford University Press, 1991.

Ashworth William B, Jr. "Natural History and the Emblematic World View." In *Reappraisals of the Scientific Revolution,* edited by David C. Lindberg and Robert S. Westman, 303–32. Cambridge: Cambridge University Press, 1990.

Augustine. *De doctrina Christiana (On Christian Doctrine).* Translated by D. W. Robertson, Jr. New York: Macmillan, 1958.

Bacon, Francis. *The Advancement of Learning.* New York: Modern Library, 2001.

———. *Novum Organum: With Other Parts of the Great Instauration.* Edited by Peter Urbach and John Gibson. Chicago: Open Court, 1994.

Barrera, Antonio. "Local Herbs, Global Medicines: Commerce, Knowledge, and Commodities in Spanish America." In *Merchants and Marvels: Commerce, Science, and Art in Early Modern Europe,* edited by Pamela Smith and Paula Findlen, 163–81. New York: Routledge, 2002.

Barton, Benjamin Smith. "A Memoir Concerning the Fascinating Faculty which has been Ascribed to the Rattlesnake, and Other American Serpents." *Transactions of the American Philosophical Society* 4 (1799): 74–113.

Basalla, George. "The Spread of Western Science Revisited." In *Mundialización de la ciencia y cultura nacional,* edited by Antonio Lafuente, Alberto Elena, and María L. Ortega, 599–604. Madrid: Doce Calles, 1993.

Basset, Robert. *Curiosities: or the Cabinet of Nature. Containing Phylosophical, naturall, and morall questions fully answered and resolved. Translated out of Latin, French, and Italian Authors.* London: N. and I. Okes, 1637.

Batllori, Miguel. *La cultura hispano-italiana de los jesuitas expulsos: Españoles, hispano-americanos, filipinos, 1767–1814.* Madrid: Gredos, 1966.

Bauer, Ralph. *The Cultural Geography of Colonial American Literatures: Empire, Travel, Modernity.* Cambridge: Cambridge University Press, 2003.

———. "The Postcolonial Origins of Modernity." *William and Mary Quarterly* 59, no. 4 (October 2002): 975–81.

Been, Anita. *Animals & Authors in the Eighteenth-Century Americas.* Providence, RI: John Carter Brown Library, 2004.

Bello, Andrés. *Obras completas de don Andrés Bello. Vol. 14, Opúsculos Científicos.* Santiago de Chile: Imprenta Cervantes, 1872.

Bernier, Marc-André, and Réal Ouellet. "Pierre Pellepratt's Accounts of the Jesuit Missions in the Antilles and in Guyana." In *Jesuit Accounts of the Colonial Americas: Textualities, Intellectual Disputes, Intercultural Transfers,* edited by Marc-André Bernier, Clorinda Donato, and Hans-Jürgen Leusebrink. Toronto: Toronto University Press, forthcoming.

Bisset, N. G. "War and Hunting Poisons of the New World." *Journal of Ethnopharmacology* 30, no. 1 (1992): 1–26.

Bost, David. "Historians of the Colonial Period: 1620–1700." In vol. 1 of *The Cambridge History of Latin American Literature,* edited by Roberto González Echevarría and Enrique Pupo-Walker, 143–90. Cambridge: Cambridge University Press, 1996.

———. "Historiography and the Contemporary Narrative: Dialogue and Methodology." *Latin American Literary Review* 16, no. 31 (1988): 34–44.

Bottoni, Federico. *Evidencia de la circulación de la sangre.* Lima: Ignacio de Luna, 1723.

Bravo, Michael T. "Mission Gardens: Natural History and Global Expansion, 1720–1820." In *Colonial Botany: Science, Commerce, and Politics in the Early Modern World,* edited by Londa Schiebinger and Claudia Swan, 49–65. Philadelphia: University of Pennsylvania Press, 2005.

Campbell, Mary B. *The Witness and the Other World.* Ithaca, NY: Cornell University Press, 1988.

———. *Wonder and Science: Imagining Worlds in Early Modern Europe.* Ithaca, NY: Cornell University Press, 1999.

Cañizares-Esguerra, Jorge. "How Derivative Was Humboldt? Microcosmic Nature Narratives in Early Modern Spanish America and the (Other) Origins of Humboldt's

Ecological Sensibilities." In *Colonial Botany: Science, Commerce, and Politics in the Early Modern World,* edited by Londa Schiebinger and Claudia Swan, 148–68. Philadelphia: University of Pennsylvania Press, 2005.

———. *How to Write the History of the New World.* Stanford, CA: Stanford University Press, 2001.

———. "Iberian Science in the Renaissance: Ignored How Much Longer?" *Perspectives on Science* 12, no. 1 (2004): 86–124.

———. "Nation and Nature: Natural History and the Fashioning of Creole National Identity in Late Colonial Spanish America." In *Cultural Encounters in Atlantic Societies, 1500–1800, International Seminar on the History of the Atlantic World.* Working Paper Series Cambridge, MA: The Charles Warren Center for Studies in American History, 1998.

———. "Spanish America: From Baroque to Modern Colonial Science." In *Cambridge History of Science, vol. 4: Eighteenth-Century Science,* edited by Roy Porter, 718–38. Cambridge: Cambridge University Press, 2003.

———. "Spanish America in Eighteenth-Century European Travel Compilations: A New 'Art of Reading' and the Transition to Modernity." *Journal of Early Modern History* 2, no. 4 (1998): 329–49.

Carlyon, Jonathan. *Andrés González de Barcia and the Creation of the Colonial Spanish American Library.* Toronto: University of Toronto Press, 2005.

Carpentier, Alejo. *Los pasos perdidos.* Edited by Roberto González Echevarría. Madrid: Cátedra, 1985.

Cassirer, Ernst. *The Philosophy of the Enlightenment.* Translated by Fritz C. A. Koelln and James P. Pettegrove. Princeton, NJ: Princeton University Press, 1951. Translation of *Die Philosophie der Aufklarung,* 1932.

Castro, Américo. *El pensamiento de Cervantes.* Edited by Julio Rodríguez-Puértolas. 1925. Barcelona: Noguer, 1972.

Catalayud, Vicente. *Cartas eruditas por la preferencia de la Philosophia aristotélica para los estudios de religión.* Valencia, 1758–60.

Catrou, François, et al. *Mémoires pour l'Histoire des Sciences & des beaux-arts [Mémoires de Trévoux].* 1701–67. Geneva: Slatkine Reprints, 1747.

Ceñal, Ramon. "Feijoo y la filosofía de su tiempo." *Pensamiento: Revista trimestral de investigación e información filosófica* 21 (1965): 251–72.

———. "Fuentes jesuíticas francesas de la erudición filosófica de Feijoo." In *El padre Feijoo y su siglo,* edited by Jean-Louis Flecniakoska, 285–314. Oviedo: University of Oviedo, 1966.

Chambers, David W. "Locality and Science: Myths of Centre and Periphery." In *Mundialización de la ciencia y cultura nacional,* edited by Antonio Lafuente, Alberto Elena, and María L. Ortega, 605–17. Madrid: Doce Calles, 1993.

Chinchilla Pawling, Perla. *De la compositio loci a la república de letras: Predicación jesuita en el siglo XVII novohispano.* Mexico City: Universidad Iberoamericana, 2004.

Cicala, Mario. *Descripción histórico-topográfica de la Provincia de Quito de la Compañía de Jesús.* Quito: Biblioteca Ecuatoriana "Aurelio Espinosa Polit," 1994.

[Cicero]. *Ad C. Herennium, De ratione dicendi (Rhetorica ad Herennium).* Translated by Harry Caplan. Cambridge, MA: Harvard University Press, 1954.

Cicero. *De inventione.* Translated by H. M. Hubbell. Cambridge, MA: Harvard University Press, 1949.

Clifford, James, and George E. Marcus. *Writing Culture: The Poetics and Politics of Ethnography.* Berkeley and Los Angeles: University of California Press, 1986.

Colón, Cristóbol. *Los cuatro viajes. Testamento.* Edited by Consuelo Varela. Madrid: Alianza, 1986.

Conzemius, Eduard. "Ethnographical Notes on the Black Carib." *American Anthropologist* 30, no. 2 (1928): 183–205.

Covarrubias y Orozco, Sebastián de. *Tesoro de la lengua castellana o española.* 4th ed. 1611. Barcelona: Editorial Alta Fulla, 1998.

Cuervo, José Rufino. *Diccionario de construcción y regimen de la lengua castellana.* Santa Fe de Bogotá: Instituto Caro y Cuervo, 1993.

Curtius, Ernst Robert. *European Literature and the Latin Middle Ages.* Translated by Willard R. Trask. Princeton, NJ: Princeton University Press, 1953.

Daston, Lorraine. "Afterword: The Ethos of Enlightenment." In *The Sciences in Enlightened Europe,* edited by William Clark, Jan Golinski, and Simon Schaffer, 495–504. Chicago: University of Chicago Press, 1999.

———. "Attention and the Values of Nature in the Enlightenment." In *The Moral Authority of Nature,* edited by Lorraine Daston and Fernando Vidal, 50–73. Chicago: Chicago University Press, 2004.

———. "Curiosity in Early Modern Science." *Word and Image* 11, no. 4 (October–December 1995): 391–404.

———. "Enlightenment Fears, Fears of Enlightenment." In *What's Left of Enlightenment? A Postmodern Question,* edited by Keith Michael Baker and Peter Hanns Reill, 115–28. Stanford, CA: Stanford Univesity Press, 2001.

———. "The Ideal and Reality of the Republic of Letters in the Enlightenment." *Science in Context* 4 (1991): 367–86.

Daston, Lorraine, and Katharine Park. *Wonders and the Order of Nature: 1150–1750.* New York: Zone Books, 1998.

Dávila Fernández, María del Pilar. *Los sermones y el arte.* Valladolid: Publicaciones del Departamento de Historia del Arte, 1980.

Dear, Peter. "The Cultural History of Science: An Overview with Reflections." *Science, Technology, and Human Values* 20 (1995): 150–70.

———. "Jesuit Mathematical Science and the Reconstitution of Experience in the Early Seventeenth Century." *Studies in the History and Philosophy of Science* 18 (1987): 133–75.

———. *Revolutionizing the Sciences: European Knowledge and Its Ambitions: 1500–1800.* Princeton, NJ: Princeton University Press, 2001.

Debus, Allen G. *Man and Nature in the Renaissance.* Cambridge: Cambridge University Press, 1978.

de la Campa, Román. *Latin Americanism.* Minneapolis: University of Minnesota Press, 1999.

del Rey Fajardo, José. *Bio-bibliografía de los Jesuitas en la Venezuela Colonial.* Caracas: Universidad Católica "Andrés Bello" Instituto de Investigaciones Históricas, 1974.

———. "Denis Mesland, introductor del cartesianismo en América?" *Latinoamérica: Revista mensual de cultura* 10 (1958): 102–4.

———. "Gumilla y su obra literaria." *SIC* 257 (1963): 323–24.

———. Introduction to *P. José Gumilla: Escritos varios,* edited by Jose del Rey Fajardo. Carcas: Fuentes para la Historia Colonial de Venezuela, 1970.

———. *Misiones Jesuíticas en la Orinoquia (1625–1767).* Vol. 1. San Cristóbal: Universidad Católica del Tachira, 1992.

———. "El Orinoco y un clásico colonial." *SIC* 196 (1957): 258–60.

———. "El Padre José Gumilla, S.J.: Un sociólogo audaz y un americanista olvidado." *Revista Javeriana* (1958): 5–22.

———. *Una utopía sofocada: Reducciones Jesuíticas en la Orinoquia.* Caracas: Universidad Católica del Táchira / Universidad Católica Andrés Bello, 1998.

———. "Venezuela y la ideología gumillana." *SIC* 262 (1964): 74–76.

Deslandres, Dominique. "*Exemplo aeque ut verbo.*" In *The Jesuits: Cultures, Sciences, and the Arts, 1540–1773,* edited by John W. O'Malley, 258–73. Toronto: University of Toronto Press, 1999.

Diccionario de la lengua castellana de la Real Academia Española [DRAE]. 1726–39. Madrid: Espasa Calpe, 1992.

Donís Ríos, Manuel Alberto. "José Gumilla S.J.: Impulsor del cambio cartográfico ocurrido en Guayana a partir de 1731." *Boletín de la Academia Nacional de la Historia* (January–March 1986): 157–76.

Dussel, Enrique. "Eurocentrism and Modernity (Introduction to the Frankfort Lectures)." *Boundary 2* 20, no. 3 (Fall 1993): 65–76.

———. *The Invention of the Americas: Eclipse of "the Other" and the Myth of Modernity.* Translated by Michael D. Barber. New York: Continuum, 1995.

Eamon, William. *Science and the Secrets of Nature.* Princeton, NJ: Princeton University Press, 1994.

Engstrand, Iris H. W. *Spanish Scientists in the New World: The Eighteenth-Century Expeditions.* Seattle: University of Washington Press, 1981.

Ewalt, Margaret. "Crossing Over: Nations and Naturalists in *El Orinoco ilustrado:* Reading and Writing the Book of Orinoco Secrets." *Dieciocho* 29, no. 1 (Spring 2006): 1–25.

———. "Father Gumilla, Crocodile Hunter? The Function of Wonder in *El Orinoco ilustrado.*" In *El saber de los jesuitas, historias naturales y el Nuevo Mundo,* edited by Luis Millones Figueroa and Domingo Ledezma, 303–33. Madrid: Iberoamericana, 2005.

———. "Frontier Encounters and Pathways to Knowledge in the New Kingdom of Granada." *The Colorado Review of Hispanic Studies* 3 (Fall 2005): 41–55.

Fajardo Morón, Guillermo. "El escritor venezolano José Gumilla." *Boletín de la Academia Nacional de la Historia,* 1986, 1101–2.

Fajardo Morón, Guillermo, and J. A. de Armas Chitty. "Año Gumillano: Acuerdo de la Academia Nacional de la Historia." *Boletín de la Academia Nacional de la Historia,* 1987, 217–19.

Feijoo, Benito Jerónimo. *Cartas eruditas, y curiosas en que, por la mayor parte, se continúa el designio del Teatro Crítico Universal, impugnando, o reduciendo a dudosas, varias opiniones communes.* Madrid: Imprenta real de la Gazeta, 1773.

———. "Historia Natural." In vol. 2 of *Teatro crítico universal,* 27–70. 1728. Madrid: D. Joaquín Ibarra, 1779.

———. *Teatro crítico universal: Discursos varios en todo género de materias, para desengaño de errores communes.* Facsimile of the 1726 edition. http://www.filosofia.org/bjf/.

Feingold, Mordechai. ed. *Jesuit Science and the Republic of Letters.* Cambridge, MA: MIT Press, 2003.

Findlen, Paula. *Possessing Nature: Museums, Collecting, and Scientific Culture in Early Modern Italy.* Berkeley and Los Angeles: University of California Press, 1994.

Fortique, José Rafael. *Aspectos médicos en la obra de Gumilla.* Caracas: Italgráfica, 1971.

Foucault, Michel. "Lecture One: 7 January 1976." In *Power/Knowledge: Selected Interviews and Other Writings, 1972–1977,* edited by C. Gordon, 78–92. New York: Pantheon Books, 1980.

Freedburg, David. *The Eye of the Lynx: Galileo, His Friends and the Beginnings of Modern Natural History.* Chicago: University of Chicago Press, 2002.

Frèron, Elie. *L'Année Littéraire.* 1758. Geneva: Slatkine Reprints, 1966.

———. *Journal Étranger ou notice exacte et détaillée des ouvrages des toutes les nations étrangéres, en fait d'arts, des sciences, de littérature.* 1756. Geneva: Slatkine Reprints, 1968.

Fumaroli, Marc. "The Fertility and Shortcomings of Renaissance Rhetoric: The Jesuit Case." In *The Jesuits: Cultures, Sciences, and the Arts, 1540–1773,* edited by John W. O'Malley, 90–106. Toronto: University of Toronto Press, 1999.

Funkenstein, Amos. *Theology and the Scientific Imagination from the Middle Ages to the Seventeenth Century.* Princeton, NJ: Princeton University Press, 1986.

Gay, Peter. *The Enlightenment: An Interpretation.* 2 vols. New York: Knopf, 1966–69.

Geertz, Clifford. *Works and Lives: The Anthropologist as Author.* Stanford, CA: Stanford University Press, 1988.

Gies, David T. "Dos preguntas regeneracionistas: '¿Qué se debe a España?' y '¿Qué es España,' Identidad nacional en Forner, Moratín, Jovellanos y la generación de 1898." *Dieciocho* 22, no. 2 (1999): 307–30.

Gilij [Gilii], Filippo Salvadore. *Ensayo de Historia Americana.* Vol. 1. Caracas: Academia Nacional de Historia, 1987.

———. *Saggio di storia americana; o sia, Storia naturale, civile e sacra de'regni, e delle provincie spagnuole di Terra-Ferma nell' America Meridionale.* Rome: L. Perego erede Salvioni, 1780–84.

Gillin, John. "Crime and Punishment among the Barama River Carib of British Guiana." *American Anthropologist* 36, no. 3 (1934): 331–44.

Góngora, Mario. *Studies in the Colonial History of Spanish America.* Cambridge: Cambridge University Press, 1975.

González Echevarría, Roberto. *Myth and Archive: A Theory of Latin American Narrative.* Cambridge: Cambridge University Press, 1990.

———. "'Semejante a la noche' de Alejo Carpentier: *Historia/Ficción.*" *MLN* 87.2 (1972): 272–285.

Gould, Stephen Jay. *I Have landed: The End of a Beginning in Natural History.* New York: Harmony Books, 2002.

Grahn, Lance. *The Political Economy of Smuggling: Regional Informal Economies in Early Bourbon New Granada.* Boulder, CO: Westview Press, 1997.

Granada, Luis de. *Los seis libros de la Retórica eclesiástica o la manera de predicar.* 1578. Madrid: Ediciones Atlas, 1945.

Greenblatt, Stephen. *Marvelous Possessions: The Wonder of the New World.* Chicago: University of Chicago Press, 1991.

———. "Resonance and Wonder." In *Exhibiting Cultures: The Poetics and Politics of Museum Display,* edited by Ivan Karp and Steven D. Lavine, 42–56. Washington, DC: Smithsonian Institution Press, 1991.

Gumilla, Joseph. *Histoire naturelle, civile et geographique de l'Orenoque, et des principales rivieres que s'y jettent. . . .* Avignon: Chez la Veuve de F. Girard, 1758.

———. *Historia natural, civil y geographica de las naciones situadas en las riveras del Rio Orinoco.* Barcelona: Carlos Gubert y Tutó, 1791.

———. *Historia natural, civil y geographica de las naciones situadas en las riberas* [*sic*] *del Rio Orinoco.* Barcelona: Imprenta de la Viuda e Hijos de J. Subirana, 1882.

———. *Historia natural civil y geográfica de las naciones situadas en las riveras del Río Orinoco.* Calí: Carvajal, 1984.

———. *El Orinoco ilustrado, historia natural, civil y geographica de este gran río, y de sus caudalosas vertientes; govierno, usos, y costumbres de los Indios sus habitadores, con nuevas, y utiles noticias de animales, arboles frutos, aceytes, resinas, yervas, y raíces medicinales.* Madrid: Manuel Fernandez, 1741.

———. *El Orinoco ilustrado, y defendido, historia natural, civil y geographica de este gran río, y de sus caudalosas vertientes; govierno, usos, y costumbres de los Indios sus habitadores, con nuevas, y utiles noticias de animales, arboles frutos, aceytes, resinas, yervas, y raíces medicinales,* 2nd ed. Madrid: Manuel Fernandez, 1745.

———. *El Orinoco ilustrado.* Edited by Constantino Bayle. Madrid: M. Aguilar, [1945].

———. *El Orinoco ilustrado: Historia natural, civil y geográfica de este gran río.* Edited by José Rafael Arboleda. Bogotá: Editorial ABC, 1955.

———. *El Orinoco ilustrado y defendido.* Edited by Demetrio Ramos. Caracas: Fuentes para la Historia Colonial de Venezuela, 1963.

———. *El Orinoco ilustrado.* Edited by Marc-Aureli Vila. Valencia: Generalitat Valenciana, Comissió per al V Centenaria del Descobriment d'America, 1988.

———. *El Orinoco ilustrado: Historia natural, civil y geographica de este gran río.* Bogotá: Imagen Editores, 1994.

———. *P. José Gumilla: Escritos varios.* Edited by José del Rey. Caracas: Fuentes para la Historia Colonial de Venezuela, 1970.

———. *Tríbus indígenas del Orinoco.* Caracas: Instituto Nacional de Cooperación, 1973.

Haidt, Rebecca. *Embodying Enlightenment: Knowing the Body in Eighteenth-Century Spanish Literature and Culture.* New York: St. Martin's Press, 1998.

Harris, Steven. "Jesuit Scientific Activity in the Overseas Missions, 1570–1773." *ISIS* 96 (2005): 71–79.

Hazard, Paul. *European Thought in the Eighteenth Century, from Montesquieu to Lessing.* 1935. New Haven, CT: Yale University Press, 1954.

Heilbron, J. L. "The Physicists: Jesuits." In *Elements of Early Modern Physics,* 93–106. Berkeley and Los Angeles: University of California Press, 1982.

Helferich, Gerard. *Humboldt's Cosmos: Alexander von Humboldt and the Latin American Journey that Changed the Way We See the World.* New York: Gotham Books, 2004.

Hellyer, Marcus. *Catholic Physics: Jesuit Natural Philosophy in Early Modern Germany.* Notre Dame, IN: University of Notre Dame Press, 2005.

Herr, Richard. *The Eighteenth-Century Revolution in Spain.* Princeton, NJ: Princeton University Press, 1958.

Herrero, Antonio María. *Physica moderna, experimental, systematica donde se contiene lo mas curioso, y util de quanto se ha descubierto en la Naturaleza.* Madrid, 1738.

Hill, Ruth. *Sceptres and Sciences in the Spains: Four Humanists and the New Philosophy (ca. 1680–1740).* Liverpool: Liverpool University Press, 2000.

Hodgen, Margaret T. *Early Anthropology in the Sixteenth and Seventeenth Centuries.* Philadelphia: University of Pennsylvania Press, 1971.

Höltgen, Kart Josef. "Henry Hawkins: A Jesuit Writer and Emblematist in Stuart England," In *The Jesuits: Cultures, Sciences, and the Arts, 1540–1773,* edited by John W. O'-Malley, 600–626. Toronto: University of Toronto Press, 1999.

Horace. "Ars Poetica." In *Horace for Students of Literature,* edited by Leon Golden and O. B. Hardison, Jr., Gainesville: University Press of Florida, 1995.

Humboldt, Alexander von. *Aspects of Nature, in Different Lands and Different Climates; with Scientific Elucidations.* Translated by Mrs. Sabine. Philadelphia: Lea and Blanchard, 1850.

———. *Personal Narrative of a Journey to the Equinoctial Regions of the New Continent.* Translated by Jason Wilson. London: Penguin Books, 1995.

———. *Personal Narrative of Travels to the Equinoctial Regions of the New Continent, during the Years 1799–1804. Written in French by Alexander de Humboldt, and Translated into English by Helen Maria Williams.* Philadelphia: M. Carey, 1815.

———. *Personal Narrative of Travels to the Equinoctial Regions of America, during the years 1799–1804.* Translated and edited by Thomasina Ross. London: Henry G. Bohn, 1852–53.

———. *Relation historique du voyage aux régions équinoxiales du nouveau continent.* 3 vols. Stuttgart: F. A. Brockhaus, 1970.

———. *Tableaux de la nature, ou Considérations sur les déserts, sur la physionomie des végétaux, et sur les cataractes de l'Orénoque.* Paris: F. Schoell, 1808.

———. *Views of Nature, or contemplation of the Sublime Phenomena of Creation.* Edited by Henry George Bohn. London: H. G. Bohn, 1850.

Impey, Oliver, and Arthur MacGregor. *The Origins of Museums: The Cabinet of Curiosities in Sixteenth- and Seventeenth-Century Europe.* Oxford: Oxford University Press, 1985.

Interián de Ayala, Juan. *El pintor cristiano y erudito ó tratado de los errores que suelen cometerse frecuentemente en pintar y esculpir las imágenes sagradas.* Barcelona: Imprenta de la Viuda é Hijos de J. Subirana, 1883.

Isaacs, Jorge. *María.* Edited by Donald McGrady. Madrid: Cátedra, 1991.

Israel, Jonathan. *Radical Enlightenment: Philosophy and the Making of Modernity, 1650–1750.* New York: Oxford University Press, 2001.

Jacob, Margaret. *The Enlightenment: A Brief History with Documents.* New York: Bedford/St. Martin's, 2001.

Jardine, Nicholas, James A. Secord, and Emma C. Spary. *Cultures of Natural History.* Cambridge: Cambridge University Press, 1996.

Journal Encyclopédique. Liège: Bouillin, 1759.

Journal des savants combiné avec les Mémoires de Trévoux. Amsterdam: Rey, 1758.

Juan, Jorge, and Antonio de Ulloa. *Relacion historica del viage a la America Meridional hecho de orden de S. Mag. para medir algunos grados de Meridiano Terrestre e, y venir por ellos en conocimiento de la verdadera Figura, y Magnitud de la Tierra, con otras varias Observaciones Astronomicas, y Phisicas.* Madrid: Antonio Marin, 1748.

Julián, Antonio. *La perla de la América, provincia de Santa Marta, reconocida, observada y expuesta en discursos históricos, por el sacerdote don Antonio Julian, á mayor bien de la Católica monarquía, fomento del comercio de España, y de todo el Nuevo Reyno de Granada, é incremento de la Christiana religión entre las naciones bárbaras, que subsisten todavía rebeldes en la provincia.* 1787. Facsimile Edition. Bogotá: Academia Colombiana de Historia, 1980.

Kenseth, Joy, ed. *The Age of the Marvelous.* Hanover, NH: Hood Museum of Art, 1991.

Kossok, Manfred. "La Ilustración en América Latina. ¿Mito o Realidad?" *Ibero-Americana Pragensia* 12 (1973): 89–100.

Labat, Jean-Baptiste. *Nouveau voyages aux isles de l'Amerique. Contentant L'histoire naturelle de ces pays, l'Origine, les moeurs, la Religion & le Gouvernement des Habitans anciens & modernes . . . Avec une Description exacte & curieuse de toutes ces Isles.* Paris: Chez Pierre-Francois Giffart, 1722.

La Condamine, Charles Marie de. *Relation abrégée d'un voyage fait dans l'interieur de l'Amérique Méridionale. Depuis la côte de la Mer du Sud, jusqu'aux côtes du Brésil & de la Guiane, en descendant la riviere des Amazones.* Paris: Veuve Pissot, 1745.

Lafuente, Antonio, Alberto Elena, and M. L. Ortega. *Mundialización de la ciencia y cultura nacional.* Madrid: Doce Calles, 1993.

La Pluche, Noel-Antoine. *Le spectacle de la nature; ou entretiens sur particularités de l'histoire naturelle, qui ont paru les plus propres à rendre les jeunes-gens curieux, & à former leur esprit.* Paris: Veuve Estienne, 1732–50.

Latour, Bruno. *Laboratory Life: The Social Construction of Scientific Facts.* Beverly Hills, CA: Sage Publications, 1979.

———. "Give Me a Laboratory and I Will Raise the World." In *The Science Studies Reader,* edited by Mario Biagioli, 258–75. New York: Routledge, 1999.

Lee, M. R. "Curare: the South American Arrow Poison." *Journal of College Physicians of Edinburgh* 35, no. 1 (February 2005): 83–92.

Ledezma, Domingo. "Una legitimación imaginativa del Nuevo Mundo." In *El Saber de los jesuitas, historias naturales yel Nuevo Mundo,* edited by Luis Millones Figueroa and Domingo Ledezma, 53–83. Madrid: Iberoamericana, 2005.

López Grigera, Luisa. *La Retórica en la España del Siglo de Oro: Teoría y práctica.* Salamanca: Ediciones Universidad de Salamanca, 1994.

Loyola, Ignacio. *Los ejercicios espirituales de San Ignacio de Loyola.* Edited by Luis María Badiola. Mexico City: Librería Parroquial de Clavería, 1990.

Lucretius. *De rerum natura. On the Nature of the Universe.* Translated by R. E. Latham and edited by John Godwin. London: Penguin Books, 1951.

Lugli, Adalgisa. *Naturalia et mirabilia: Il collezionismo enciclopedico nelle Wunderkammern d'Europa.* Milan: G. Mazzotta, 1983.

MacLeod, Roy. "Introduction. Nature and Empire: Science and the Colonial Enterprise." *Osiris* 15 (2001): 1–13.

Manuel, Frank. *The Age of Reason.* Ithaca, NY: Cornell University Press, 1951.

Maravall, José Antonio. "Empirismo y pensamiento político (Una cuestión de origenes)." In *Estudios de historia del pensamiento español,* 15–38. Madrid: Ediciones cultura hispánica, 1984.

McClellan, James E. "Missionary Naturalists." In *Colonialism and Science: Saint Domingue in the Old Regime,* 111–16. Baltimore: Johns Hopkins University Press, 1992.

McClelland, Ivy Lillian. "The Significance of Feijoo's Regard for Francis Bacon." *Studium ovetense* 4 (1976): 249–74.

Menéndez y Pelayo, Marcelino. *Historia de los heterodoxos españoles.* Madrid: Librería católica de San José, gerente V. Sancho-Tello, 1880–81.

Merrim, Stephanie. "La *Grandeza mexicana* en el contexto criollo." In *Nictimene—sacrilege: Estudios coloniales en homenaje a Georgina Sabat-Rivers,* edited by Georgina Sabàt de Rivers, Electa Arenal, Mabel Moraña, and Yolanda Martínez-San Miguel. Mexico City: Universidad del Claustro de Sor Juana, 2003.

Mignolo, Walter. "Introduction" and "Commentary" to *Natural and Moral History of the Indies,* edited by Jane E. Mangan and translated by Frances M. López-Morillas, xvii–xxviii, 451–518. Durham, NC: Duke University Press, 2002.

———. *Local Histories / Global Designs: Coloniality, Subaltern Knowledges, and Border Thinking.* Princeton, NJ: Princeton University Press, 2000.

Millones Figueroa, Luis. "La *intelligentsia* jesuita y la naturaleza del Nuevo Mundo en el siglo XVII." In *El saber de los jesuitas, historias naturales y el Nuevo Mundo,* edited by Luis Millones Figueroa and Domingo Ledezma, 27–46. Madrid: Iberoamericana, 2005.

Millones Figueroa, Luis, and Domingo Ledezma. "Introducción: Los jesuitas y el conocimiento de la naturaleza americana." In *El saber de los jesuitas, historias naturales y el Nuevo Mundo,* edited by Luis Millones Figueroa and Domingo Ledezma, 9–24. Madrid: Iberoamericana, 2005.

Morán Turino, José Miguel, and Fernando Checa Cremades. *El coleccionismo en España: De la cámera de maravillas a la galería de pinturas.* Madrid: Cátedra, 1985.

Muratori, Ludovico Antonio. *Il cristianesimo felice nelle missioni de' padri della Compagnia di Gesù nel Paraguai.* Venice: Giambatista Pasquali, 1743.

———. *Delle reflessioni sopra il buon gusto: Nelle Scienze e nell'Arti.* Venice: Niccolò Pezzana, 1708.

———. *Fuerza de la humana fantasia.* Translated by Vicente María de Tercilla. Madrid: D. Manuel Martin, 1777.

Murphy, John C., and Robert W. Henderson. *Tales of Giant Snakes: A Natural History of Anacondas and Pythons.* Malabar, Fla.: Krieger Publishing Company, 1997.

Navarro, Víctor. "Tradition and Scientific Change in Early Modern Spain: The Role of the Jesuits." In *Jesuit Science and the Republic of Letters,* edited by Mordechai Feingold, 331–87. Cambridge, MA: MIT Press, 2003.

Navarrete, Rodrigo. "Behind the Palisades: Sociopolitical Recomposition of Native Societies in the Unare Depression, the Eastern Venezuelan Llanos (Sixteenth to Eighteenth Centuries)." *Ethnohistory* 47, nos. 3–4 (2000): 535–59.

The New Oxford Annotated Bible. New York: Oxford University Press, 1994.

Nieremberg, Juan Eusebio. *Curiosa y oculta filosofía: Primera y segunda parte de las maravillas de la naturaleza, examinadas en varias questiones naturales . . .* Madrid: Imprenta Real, 1643.

———. *Historia Natvrae maxime peregrinae libris XVI distincta. In quibus rarissima Naturae arcane, etiam astronomica, & ignota Indiarum animalie, quadrupedes, aues, pises, reptilia, insecta . . . nouae & curiosissimae quaestiones disputantur, ac plura sacrae Scripturae loca erudite enodantur.* Antwerp: Balthasaris Morelia, 1635.

Noel, Charles C. "Clerics and Crown in Bourbon Spain, 1700–1808." In *Religion and Politics in Enlightenment Europe,* edited by James E. Bradley and Dale K. Van Kley, 119–53. Notre Dame, IN: University of Notre Dame Press, 2001.

O'Malley, John W. *The First Jesuits.* Cambridge, MA: Harvard University Press, 1993.

———. *The Jesuits: Cultures, Sciences, and the Arts, 1540–1773.* Toronto: University of Toronto Press, 1999.

———. *The Jesuits II: Cultures Sciences and the Arts, 1540–1773.* Toronto: Toronto University Press, 2005.

Ortega y Gasset, José. *Meditaciones del Quijote: Meditación preliminar, Meditación primera.* Madrid: Residencia de Estudiantes, 1914.

Outram, Dorinda. *The Enlightenment.* Cambridge: Cambridge University Press, 1995.

———. "The Enlightenment Our Contemporary." In *The Sciences in Enlightened Europe,* edited by William Clark, Jan Golinski, and Simon Schaffer, 32–40. Chicago: University of Chicago Press, 1999.

The Oxford English Dictionary. 2nd ed. 1989. *OED Online.* Oxford University Press. http://dictionary.oed.com/cgi/entry/00181778.

Park, Katherine. "Nature in Person." In *The Moral Authority of Nature,* edited by Lorraine Daston and Fernando Vidal, 50–73. Chicago: University Press of Chicago, 2004.

Parra, Caracciolo. *La instrucción en Caracas (1567–1725). Filosofía universitaria venezolana. Cronistas de Venezuela.* Madrid: Editorial J.B., 1954.

Peguero, Luis Joseph. *Historia de la Conquista de la Isla Espanola de Santo Domingo.* San Domingo: Museo de las Casas Reales, 1975.

Pérez Magallón, Jesús. *Construyendo la modernidad: La cultura española en el tiempo de los novatores.* Madrid: Consejo Superior de Investigaciones Científicas, 2002.

Pighetti, Clelia. *Atomi e Lumi nel Mondo Spagnolo.* Milan: Franco Angeli, 1993.

Pimental, Juan. "The Iberian Vision: Science and Empire in the Framework of a Universal Monarchy, 1500–1800." *Osiris* 15 (2001): 17–30.

Piquer, Andrés. *Discurso sobre la aplicacion de la Philosophia a los assuntos de religion para la juventud española.* Madrid: Joachin Ibarra, 1757.

———. *Fisica moderna, racional, y experimental.* 2nd ed. Madrid: Joachin Ibarra, 1780.

Pliny the Elder. *The Historie of the World: Commonly Called the Natural History of C. Plinius Secundus.* London: Adam Islip, 1634.

Pollak-Eltz, Angelina. "Woher Stammen die Indianer Amerikas? José Gumilla und das Denken seiner Zeit." *Wiener Ethnohistorische Blätter* 5 (1973): 21–31.

Pomian, Krzysztof. *Collectors and Curiosities: Paris and Venice, 1500–1800.* Cambridge: Polity Press, 1990.

Porter, Roy. Introduction to *Cambridge History of Science, vol. 4: Eighteenth-Century Science,* edited by Roy Porter, 1–20. Cambridge: Cambridge University Press, 2003.

———. Introduction to *The Ferment of Knowledge: Studies in the Historiography of Eighteenth-Century Science,* edited by Roy Porter and George Sebastian Rousseau, 1–10. Cambridge: Cambridge University Press, 1980.

Pratt, Mary Louise. *Imperial Eyes: Travel Writing and Transculturation.* New York: Routledge, 1992.

Puy Múñoz, Francisco. "La comprensión de la moralidad en el pensamiento español del siglo XVIII (1700–1760)." *Anales de la Cátedra Francisco Suárez* 2, no. 1 (1962): 87–118.

———. "El problema del conocimiento en el pensamiento del siglo XVIII (1700–1760)." *Anales de la Cátedra Francisco Suárez* 1, no. 2 (1961): 191–226.

Quijano, Anibal. "Colonialidad y modernidad-racionalidad." In *Los conquistados: 1492 y la población indígena de las Américas,* edited by Robin Blackburn and Heraclio Bonilla, 437–47. Bogotá: Tercer Mundo Editores, 1992.

Quintilian. *Institutes of Oratory.* Translated by John Selby Watson. 2 vols. London: Henry G. Gohn, 1856.

Quiroz-Martínez, Olga Victoria. *La introducción de la filosofía moderna en España (El eclecticismo español de los siglos XVII y XVIII).* Mexico City: Fondo de Cultura Económica, 1949.

Ramos, Demetrio. "El etnógrafo Gumilla y su grupo de historiadores: Nuevos datos sobre las obras misionales de éstos al mediar el siglo XVIII." *Miscellanea P. Rivet, Octogenario dicata* 2 (1958): 857–69.

———. "La geografía de los modos de vida del valle venezolano y el jesuita valenciano P. Gumilla." *SAITABI* 6, nos. 29–30 (July–December 1948): 242–51.

———. "Las ideas geográficas del Padre Gumilla." *Estudios geográficos* 5, no. 14 (1944): 179–99.

———. Introduction to *El Orinoco ilustrado y defendido,* by Joseph Gumilla, xxix–cxxvi. Caracas: Fuentes para la Historia Colonial de Venezuela, 1963.

———. "Un mapa inédito del Río Orinoco es el precedente del de Gumilla y el más antiguo de los conocidos." *Revista de Indias* (January-March 1944): 89–104.

———. "Un plan de inmigración y libre comercio defendido por Gumilla para Guayana en 1739." *Anuario de Estudios Americanos* 15 (1958): 201–24.

Raynal, Abbé [Guillaume-Thomas-François]. *Histoire Philosophique et politique, des etablissemens & du commerce des Europeéns dans les deux Indes.* Amsterdam, 1770.

Real Academia de la Lengua. *Diccionario de la lengua castellana (1726–39).* Madrid: Espasa Calpe, 1992.

Regnault, Noël. *Philosophical Conversations, or a New System of Physics by way of Dialogue.* Translated by Dr. Thomas Dale. London: W. Innys, C. Davis, and N. Prevost, 1731.

Restrepo, Daniel. *Compendio historial y galería de ilustres varones.* Bogotá: Imprenta del Corazón de Jesús, 1940.

Riskin, Jessica. *Science in the Age of Sensibility: The Sentimental Empiricists of the French Enlightenment.* Chicago: University of Chicago Press, 2002.

Rivera, José Eustasio. *La vorágine.* Edited by Montserrat Ordóñez. Madrid: Cátedra, 1998.

Robbins, Jeremy. "From Baroque to Pre-Enlightenment: Resolving the Epistemological Crisis." *Bulletin of Spanish Studies* 87, no. 8 (2005): 225–52.

Röd, Wolfgang. *Die Philosophie der Neuzeit 2: Von Newton bis Rousseau.* Munich: Beck, 1978.

Rodríguez, Manuel. *El Marañon y Amazonas. Historia de los descubrimientos, entradas, y reducción de naciones. Trabajos malogrados de algunos conquistadores, y dichosos de otros, así temporales, como espirituales, en las dilatadas montañas, y mayores rios de la America.* Madrid: Antonio Gonzalez de Reyes, 1684.

Rodríguez Casado, Vicente. "El intento español de una Ilustración cristiana." *Estudios Americanos* 9, no. 42 (1955): 141–69.

Rovira, María del Carmen. *Eclécticos portugueses del siglo XVIII y algunas de sus influencias en América.* Mexico City: El Colegio de México, 1958.

Salmón, Thomas. *Lo stato presente di tutti I paesi e popoli del mondo naturale, político, e morale, con nuove osservazioni e correzioni degli antichi, e moderni viaggiatori* [*Historia universal*]. Venice: Presso giambatista Albrizzi Q. Gir, 1736.

Sánchez-Blanco, Francisco. *El absolutismo y las Luces en el reinado de Carlos III.* Madrid: M. Pons, 2002.

———. *Europa y el pensamiento español del siglo XVIII.* Madrid: Alianza, 1991.

———. *La Ilustración en España.* Madrid: Akal, 1997.

———. *La mentalidad ilustrada.* Madrid: Taurus, 1999.

Santos Hernández, Ángel. *Los Jesuitas en América.* Madrid: Editorial MAPFRE, 1992.

Sarrailh, Jean. *L'Espagne éclairée de la seconde moitié du XVIIIe siècle.* Paris: Impr. nacionale, 1954.

Schaffer, Simon. "Natural Philosophy." In *The Ferment of Knowledge: Studies in the Historiography of Eighteenth-Century Science,* edited by Roy Porter and George Sebastian Rousseau, 55–91. Cambridge: Cambridge University Press, 1980.

———. "Natural Philosophy and Public Spectacle in the Eighteenth Century." *History of Science* 21 (1983): 1–43.

Schlosser, Julius von. *Die Kunst- und Wunderkammern der Spätrenaissance.* Leipzig: Klinkhardt & Biermann, 1908.

Sebold, Russell P. "Colón, Bacon y la metáfora heroica de Feijoo." In *Homenaje a don Agapito Rey,* edited by Joseph Roca-Pons, 333–54. Bloomington: Indiana University Press, 1980.

Shapin, Steven. *The Scientific Revolution.* Chicago: University of Chicago Press, 1996.

———. "Social Uses of Science." In *The Ferment of Knowledge: Studies in the Historiography of Eighteenth-Century Science,* edited by Roy Porter and George Sebastian Rousseau, 93–139. Cambridge: Cambridge University Press, 1980.

Shugar, Debora. "Sacred Rhetoric in the Renaissance." In *Renaissance-Rhetorik,* edited by Heinrich F. Plett, 121–42. Berlin: Walter de Gruyter, 1993.

Silva Cáceres, Raúl. "Una novela de Carpentier." *Mundo nuevo* 17 (1967): 33–37.

Sloan, Hans. "Conjectures on the Charming or Fascinating Power Attributed to the Rattle-Snake: Grounded on Credible Accounts, Experiments and Observations." *Philosophical Transactions* 38 (1733–34): 321–31.

Smith, Adam. *An Inquiry into the Nature and Causes of the Wealth of the Nations.* 1761. London: W. Strahan and T. Cadell, 1776.

Soto Arango, Diana, Miguel Ángel Puig Samper, and Luis Carlos Arboleda, eds. *La Ilustración en América Colonial.* Madrid: Doce Calles, 1995.

Stafford, Barbara Maria. *Artful Science: Enlightenment Entertainment and the Eclipse of Visual Education.* Cambridge, MA: MIT Press, 1994.

———. *Voyage into Substance: Art, Science, Nature, and the Illustrated Travel Account, 1760–1840.* Cambridge, MA: MIT Press, 1984.

———. "Voyeur or Observer? Enlightenment Thoughts on the Dilemmas of Display." *Configurations* 1, no. 1 (1993): 95–128.

Stepan, Nancy Leys. "Going to the Tropics." In *Picturing Tropical Nature.* Ithaca, NY: Cornell University Press, 2001.

Stothers, Richard B. "Ancient Scientific Basis of the 'Giant Serpent' from Historical Evidence." *ISIS* 95, no. 2 (2004): 220–38.

Suárez (Soarez), Ciprian. *De arte rhetorica.* 1586. Translated by Lawrence J. Flynn. Gainesville: University Presses of Florida, 1955.

Suárez, María Matilde. "El contenido etnográfico del *Orinoco Ilustrado.*" *Montalban* 3 (1974): 309–35.

Subirats, Eduardo. *La ilustración insuficiente.* Madrid: Taurus, 1981.

———. *Modernidad truncada de América Latina.* Caracas: CIPOST, Centro de Investigaciones Postdoctorales, 2001.

Sweet, John Wood. *Bodies Politic: Negotiating Race in the American North, 1730–1830.* Baltimore: Johns Hopkins University Press, 2003.

Thorbjarnarson, John. "Trailing the Mythical Anaconda." *Americas* 47, no. 4 (1995): 38(8).

Tietz, Manfred. *Los Jesuitas españoles expulsos: Su imagen y su contribución al saber sobre el mundo hispánico en la Europa del siglo XVIII.* Madrid: Iberoamericana, 2001.

Todorov, Tzvetan. *The Conquest of America: The Question of the Other.* New York: Harper-Perennial, 1984.

Toquita Clavijo, María Constanza. *Desde Roma por Sevilla al Nuevo Reino de Granada: La Compañía de Jesús en tiempos coloniales.* Bogotá: Museo de Arte Colonial, 2004.

Vanpaemal, G. H. W. "Jesuit Science in the Spanish Netherlands." In *Jesuit Science and the Republic of Letters,* edited by Mordechai Feingold, 389–432. Cambridge, MA: MIT Press, 2003.

Vernes, Jules. *Le superbe Orénoque.* Paris: J. Hetzel, 1898.

Vilches, Elvira. "Columbus's Gift: Representations of Grace and Wealth and the Enterprise of the Indies." *MLN* 119, no. 2 (2004): 201–25.

Walker Bynam, Caroline. "Wonder." *American Historical Review* 102 (1997): 1–26.

Whitaker, Arthur P. "Changing and Unchanging Interpretations of the Enlightenment in Spanish America." In *The Ibero-American Enlightenment,* edited by A. Owen Aldridge, 21–57. Urbana: University of Illinois Press, 1971.

———. *Latin America and the Enlightenment.* New York: D. Appleton-Century Company, 1942.

Wilhite, John F. "The Enlightenment in Latin America: Tradition versus Change." *Dieciocho* 3, no. 1 (1980): 18–26.

Wylie, Lesley. "Hearts of Darkness: The Celebration of Otherness in the Latin American *Novela de la selva.*" *Romance Studies* 23, no. 2 (2005): 105–16.

Index